Average Current-Mode Control of DC-DC Power Converters

Average Current-Mode Control of DC-DC Power Converters

Marian K. Kazimierczuk, Dalvir K. Saini, and Agasthya Ayachit

*Department of Electrical Engineering, Wright State University,
Dayton, Ohio, USA*

WILEY

This edition first published 2022

Registered Offices
John Wiley & Sons, Inc., 111 River Street, Hoboken, NJ 07030, USA
John Wiley & Sons Ltd, The Atrium, Southern Gate, Chichester, West Sussex, PO19 8SQ, UK

Editorial Office
The Atrium, Southern Gate, Chichester, West Sussex, PO19 8SQ, UK

For details of our global editorial offices, customer services, and more information about Wiley products visit us at www.wiley.com.

Wiley also publishes its books in a variety of electronic formats and by print-on-demand. Some content that appears in standard print versions of this book may not be available in other formats.

Library of Congress Cataloging-in-Publication Data

Names: Kazimierczuk, Marian K., author. | Saini, Dalvir K., author. | Ayachit, Agasthya, author.
Title: Average current-mode control of DC-DC power converters / Marian K. Kazimierczuk, Dalvir K. Saini, and Agasthya Ayachit.
Description: Hoboken, NJ : Wiley, 2022. | Includes bibliographical references and index.
Identifiers: LCCN 2021039426 (print) | LCCN 2021039427 (ebook) | ISBN 9781119525653 (cloth) | ISBN 9781119525561 (adobe pdf) | ISBN 9781119525684 (epub)
Subjects: LCSH: Electric controllers. | DC-to-DC converters.
Classification: LCC TK2851 .K39 2022 (print) | LCC TK2851 (ebook) | DDC 621.31/32–dc23
LC record available at https://lccn.loc.gov/2021039426
LC ebook record available at https://lccn.loc.gov/2021039427

Cover Design: Wiley
Cover Image: © Image by Dalvir Saini

Set in 9.5/12.5pt STIXTwoText by Straive, Chennai, India

10 9 8 7 6 5 4 3 2 1

Contents

List of Symbols

A_i	load current-to-inductor current transfer function
A_{io}	DC gain of load current-to-inductor current transfer function
BW	bandwidth
C	filter capacitance
C_o	transistor output capacitance
D	DC component of on-switch duty cycle
D_{cr}	critical duty cycle of switch
d	AC component of on-switch duty cycle
d_T	large-signal on-switch duty cycle
f_c	gain crossover frequency
f_z	frequency of zero of transfer function
f_o	corner frequency or natural frequency
f_p	frequency of pole of transfer function
f_s	switching frequency
I_D	average or DC component of diode current
i_d	small-signal AC component of diode current
i_D	large-signal diode current
I_I	average or DC component of input current
i_i	small-signal AC component of input current
i_I	large-signal component of input current
I_L	average or DC component of inductor current
i_l	small-signal AC component of inductor current
i_L	large-signal inductor current
I_O	DC component of load current
i_o	small-signal AC component of load current
i_O	large-signal load current
I_S	DC component of switch current
i_s	small-signal AC component of switch current
i_S	large-signal switch current
i_C	large-signal filter capacitor current
L	inductor or inductance
M_{di}	input voltage to duty-cycle transfer function relevant to inner-current loop

M_{dv} input voltage to duty-cycle transfer function relevant to outer-voltage loop
M_{IDC} DC current transfer function of converter
M_{icl} input voltage-to-inductor current transfer function
M_v open-loop input-to-output voltage transfer function
M_{VDC} DC voltage transfer function of a converter
M_{vcl} input-to-output voltage transfer function relevant to outer-voltage loop
M_{vi} open-loop input voltage-to-inductor current transfer function
M_{vicl} input-to-output voltage transfer function relevant to inner-current loop
M_{vio} open-loop input voltage-to-inductor current transfer function at DC
M_{vo} open-loop input-to-output voltage function at DC
PM phase margin
P_I converter DC input power
P_{LS} converter power loss
P_O converter DC output power
P_{RF} conduction loss in diode forward resistance R_F
P_{rC} conduction loss in equivalent series resistance of the filter capacitor
P_{VF} conduction loss in diode forward voltage V_F
r equivalent averaged resistance
R_{iicl} real component of input impedance relevant to inner-current loop
R_{oicl} real component of output impedance relevant to inner-current loop
R_{ivcl} real component of input impedance relevant to outer-voltage loop
R_{ovcl} real component of output impedance relevant to outer-voltage loop
R_F diode forward resistance
R_L load resistance
r_C equivalent series resistance (ESR) of filter capacitor
r_L equivalent series resistance (ESR) of inductor
r_{DS} ON resistance of switch
R_s sense resistance
T switching period, Loop gain
T_{ci} voltage transfer function of controller relevant to inner-current loop
T_{cv} voltage transfer function of controller relevant to outer-voltage loop
T_f low-pass filter transfer function
T_i loop gain of inner-current loop
T_v loop gain of outer-voltage loop in the presence of inner-current loop
T_{ki} uncompensated loop gain of inner-current loop
T_{kv} uncompensated loop gain of outer-voltage loop in the presence of inner-current loop
T_m transfer function of pulse-width modulator
T_p open-loop duty cycle-to-output transfer function
T_{pi} open-loop duty cycle-to-inductor current transfer function
T_{po} open-loop control-to-output transfer function at DC
T_{pio} open-loop control-to-inductor current transfer function at DC
T_{icl} reference voltage-to-inductor current transfer function

T_{pcl} reference voltage-to-output voltage transfer function relevant to outer-voltage loop
T_{picl} reference voltage-to-output voltage transfer function relevant to inner-current loop
T_v loop gain of outer-voltage loop
V_C DC component of control voltage
v_C total control voltage
v_c AC component of control voltage
V_{CI} DC component of control voltage relevant to inner-current loop
V_{Cpp} peak-to-peak ripple voltage of the filter capacitor
V_D DC component of diode voltage
v_D large-signal component of diode voltage
V_{EI} DC component of error voltage relevant to inner-current loop
v_{EI} total error voltage relevant to inner-current loop
v_{ei} AC component of error voltage relevant to inner-current loop
V_{EV} DC component of error voltage relevant to outer-voltage loop
v_{ev} AC component of error voltage relevant to outer-voltage loop
v_{EV} total error voltage relevant to outer-voltage loop
V_F diode forward voltage
V_{FI} DC component of feedback voltage relevant to inner-current loop
v_{FI} total feedback voltage relevant to inner-current loop
v_{fi} AC component of feedback voltage relevant to inner-current loop
V_{FV} DC component of feedback voltage relevant to outer-voltage loop
v_{FV} total feedback voltage relevant to outer-voltage loop
v_{fv} AC component of feedback voltage relevant to outer-voltage loop
V_I DC component of input voltage
v_i small-signal AC component of input voltage
v_I large-signal component of input voltage
V_O DC component of output voltage
v_o AC component of output voltage
v_O large-signal component of output voltage
V_{RI} reference voltage relevant to inner-current loop
v_{ri} AC component of reference voltage relevant to inner-current loop
V_{RS} DC voltage across sense resistor
V_{RV} reference voltage relevant to outer-voltage loop
v_{rv} AC component of reference voltage relevant to outer-voltage loop
v_{rc} voltage across ESR of filter capacitor
V_r peak-to-peak value of ripple component of the output voltage
V_{Tm} peak ramp voltage of pulse-width modulator
v_L voltage across inductor
X_{iicl} imaginary component of input impedance relevant to inner-current loop
X_{oicl} imaginary component of output impedance relevant to inner-current loop
X_{ivcl} imaginary component of input impedance relevant to outer-voltage loop

X_{ovcl}	imaginary component of output impedance relevant to outer-voltage loop
Z_i	open-loop input impedance
Z_{io}	open-loop input impedance at DC (R_i)
Z_{iicl}	input impedance relevant to inner-current loop
Z_{ivcl}	input impedance relevant to outer-voltage loop
Z_o	open-loop output impedance
Z_{oo}	open-loop output impedance at DC (R_o)
Z_{oicl}	closed-loop output impedance relevant to inner-current loop
Z_{ovcl}	closed-loop output impedance relevant to outer-voltage current
β	transfer function of feedback network
η	efficiency of converter
ξ	damping ratio
σ	real component of complex-conjugate poles
ω	angular frequency
ω_c	unity-gain angular crossover frequency
ω_o	corner frequency or angular frequency
ω_p	angular frequency of simple pole
ω_z	angular frequency of simple zero

About the Authors

Marian K. Kazimierczuk, PhD, Professor of Electrical Engineering, Wright State University, Dayton, Ohio, USA. He has taught undergraduate electronics courses and graduate courses in the area of high-frequency power electronics for more than 35 years. Professor Kazimierczuk has done an extensive research on power converters and has published over 500 hundred papers in *IEEE Transactions* and IEEE international conferences, has written eight textbooks, and holds 7 patents. He is the Life Fellow of the IEEE.

Dalvir K. Saini, PhD, Research Engineer, Failure Analysis Lab, University of Dayton Research Institute, Wright Patterson Air Force Base, Dayton, Ohio, USA. She has been pursuing the area of failure analysis of electrical systems and components related to aircraft safety. She has published several journal and conference publications in the field of analysis, design, and modeling of switched-mode power converters, radio-frequency power amplifiers, and magnetic component design.

Agasthya Ayachit, PhD, Senior Systems Engineer, Mercedes-Benz Research & Development North America, Redford, Michigan, USA. He has been actively contributing to the design and development of power conversion stages in electric vehicle charging and e-drive systems. He has published several journal papers in IEEE Transactions and IET and in various IEEE conferences and has co-authored a laboratory manual on pulse-width-modulated DC–DC power converters. His research interests are in the field of circuit topologies, modeling and design of power converters, wireless charging, and wide-bandgap semiconductor devices (GaN/SiC).

About the Authors

Marian K. Kazimierczuk [illegible] Professor of Electrical Engineering, Wright State University, Dayton, Ohio, USA. He has taught undergraduate and graduate courses [illegible] power electronics [illegible]

[illegible]

Preface

The demand for power converters is growing by the day and is applicable in almost every field of engineering – automotive, consumer applications, datacenters and Internet of things, healthcare, aerospace, and telecommunications. Characterization of the properties of power converters is essential to achieve high efficiency, high power density, and high reliability of power-electronic systems. While circuit topologies and their design, analysis, and performance optimization form one aspect, circuit modeling, closed-loop control design, and converter stability form the other important part of power circuit development.

This textbook *Average Current-Mode Control of DC–DC Power Converters* aims to make the link between the classical circuit theory, circuit modeling, and control of pulse-width-modulated (PWM) DC–DC power converters. The book is envisioned to aid undergraduate and graduate students of electrical engineering, practicing engineers, scientists, and circuit designers to gain an in-depth understanding of modeling and control of PWM DC–DC power converters. One critical development in the field of power electronics has been current-mode control. This book exploits the average current-mode control "for" the classical buck, boost, buck–boost, and flyback PWM DC–DC converters. By developing a small-signal circuit model using the circuit-averaging technique and deriving the related converter power-stage transfer functions, design of higher-order control circuits, and current-mode and voltage-mode schemes, the book characterizes a complete closed-loop converter system. The book also supplies its readers with a comprehensive appendix to get accustomed to different simulation tools such as MATLAB®, Synopsys SABER®, LTSpice®, or any other Spice-based circuit simulation platforms, provides component datasheet parameters for a quick reference, and hosts a set of basic design equations for common DC–DC converters.

This book features the following attributes:

i. Unique in the market of power electronic textbooks.
ii. Ability to adopt as a supplementary material for any commercially available textbooks on power electronics and class notes.
iii. Possible usage for distance-learning power electronic course or e-learning.
iv. Simple and easy-to-understand theoretical framework procedure-sets.
v. Providing quick overview of various power converters and components.

The topics presented in this book have been thoughtfully considered, keeping in mind the benefits it offers to the readers. The primary author of this book has been teaching specialized graduate-level courses in power electronics for more than 35 years and has made significant contributions in the field of modeling of power converters, including circuit-averaging technique, principle of energy conservation, modeling of converters with disjoint switches, and many more. Since then, consistent efforts have been put into developing this material for students and engineers practicing power electronics. The second and third authors have been actively contributing to the areas of aircraft safety systems and electric vehicle (EV) power electronics, respectively, at reputable organizations and add a flavor of industry into the contents of this textbook. The authors of this book recommend the below instructions for instructors and students in making the best use of this book.

The key aspects of this book include and are not limited to:

(1) Derive averaged equations for buck, boost, buck–boost, and flyback power stages operating in continuous-conduction mode (CCM).
(2) Develop DC, averaged, large-signal, and small-signal models for the four converters by the circuit-averaging technique.
(3) Show a methodology of modeling power converters with disjoint switching devices.
(4) Derive small-signal transfer functions and transient responses of the power stage, which are relevant for the design of inner-current and outer-voltage control loops.
(5) Identify the appropriate control circuit (Type I, Type II, or Type III) based on the outcomes of power-stage characteristics for the inner current and outer voltage loops.
(6) Characterize the closed-inner current loop and closed-outer voltage loop.
(7) Identify the key differences between peak current-mode control, classical average current-mode control, and modified average current-mode control.
(8) Give specific examples of transfer functions, input and output impedances, and step responses.
(9) Quick reference to key design equations, MATLAB codes, instructions to operate simulation tools, and component parameters.

Dayton, Ohio
16 October 2021

Marian K. Kazimierczuk
Dalvir K. Saini
Agasthya Ayachit

Acknowledgments

The support offered by John Wiley & Sons was exceptional, and we sincerely appreciate their patience and understanding that has led to this outcome. We wish to express our sincere thanks to Juliet Booker, Managing Editor for authored books, Electrical & Computer Engineering; Becky Cowan, Editorial Assistant; and Sandra Grayson at Wiley for their cooperation. We would also like to extend our thanks and great appreciation to our families for their support.

We have sincerely attempted at making this edition of the book error-free so that readers gain a better understanding of the material. The authors would welcome and greatly appreciate readers' suggestions and corrections for improvements of the technical content and the presentation style and ideas for newer topics, which can be implemented in possible future editions.

Acknowledgments

[illegible]

[illegible]

1
Introduction

Current-mode control of pulse-width modulated (PWM) DC–DC converters has been one of the most challenging and interesting topics of study in the vast field of power electronics for almost two decades. Even to this day, rigorous research to establish a solid and systematic characterization of current-mode controlled DC–DC power converters is being performed. Continuing research in current-mode control is focused toward analysis, design, modeling, and efficient implementation in a wide variety of important applications such as power-factor correctors, battery chargers, and LED drivers. Chapters 2-5 seek to develop a complete characterization of the average current-mode control of basic DC–DC power converter topologies.

Current-mode control, particularly peak current-mode control, was first introduced by T. A. Froeschle in 1967 [1] and Cecil of Bell Labs/Western Electric in 1978 [2]. The invention mainly helped to balance the AC flux in a transformer of a push–pull converter to keep the core from walking off into saturation caused by a minor difference in the volt-sec balance. Core saturation generally occurred in converters with slightly asymmetric drive waveforms or was due to load transients. In view of the growing demand for DC–DC converters, especially in military, telecommunication, and aerospace industries, it became extremely necessary to develop static and dynamic models of power electronic systems. A significant advancement came when Dr. Slobodan Ćuk introduced the state-space averaging technique, where the circuit behavior in switching sub-intervals was mathematically represented in the form of state variables with the state variables being the inductor current and capacitor voltage. Another advancement that came via pioneering research, which realized the important aspects of closed-loop stability of peak current-mode controlled DC–DC converters, was conducted by Dr. David Middlebrook and Dr. F. D. Tan of California Institute of Technology in 1985. To this day, these works serve as a starting point for establishing stability criteria, especially in cascaded or multilevel power converters [3–6].

The concept of average current-mode control was reinstated by Lloyd Dixon from Unitrode Electronics in 1990 [7, 8]. In the following years, closed-loop modeling of the average current-mode controlled AC–DC and DC–C converters was performed by various researchers including Raymond B. Ridley *et al.* [9–11] and Jian Sun and

Average Current-Mode Control of DC-DC Power Converters, First Edition.
Marian K. Kazimierczuk, Dalvir K. Saini, and Agasthya Ayachit.

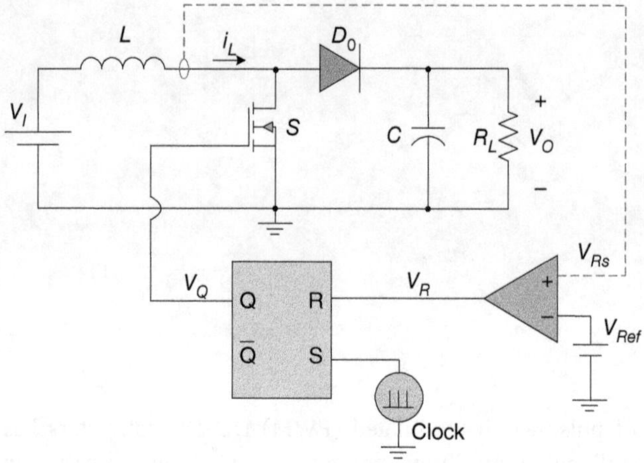

Figure 1.1 Boost DC–DC converter with a peak current-mode control, showing only the inner current loop.

R. M. Bass [12]. The peak current-mode control and the average current-mode control were also studied in [7–11, 13–35].

Before describing the principle of operation of the average current-mode control, it is imperative to discuss properties of peak current-mode control. Figure 1.1 depicts boost DC–DC converter with a peak current-mode control, showing only the inner current loop. In the peak current-mode control, the peak of the inductor current waveform is tracked and controlled [53]. The peak inductor current is compared with the reference current set by the outer-voltage loop of the DC–DC converter (not shown in Figure 1.1). As the peak value of the sensed inductor current becomes equal to the reference current, the comparator outputs a logic low at the instant of comparison and a logic high for all other values. The comparator output is connected to the reset pin R of the RS latch, whereas the set pin S is connected to a clock tuned to operate at the switching frequency. Initially, the set pin S enables the latch output Q to the switch ON state. Eventually, as the reset pin receives a logic high, the output of latch resets the switch to the OFF position and remains in this state until the clock sets it back to high position. The clock determines the beginning of the cycle, and the crossing of the inductor current waveform with the reference current determines the end of the high level of the gate-source voltage. The following operational disadvantages are observed as a consequence of peak current-mode control:

- Switching is highly susceptible to noise. At the instant of comparison of the peak inductor current and the current reference, a noise spike is created. The noise signal can create false triggering of the logic gates, resulting in the undesirable switching events.
- Slope compensation is required. The peak current-mode control needs slope compensation circuitry for duty cycles higher than 0.5 to avoid instability. The fixed compensation ramp provides adequate compensation; however, it will overcompensate much of the time, resulting in the performance degradation and increased distortion [7].

- Peak-to-average current ratio is high. The inductor current peak-to-average ratio is high. Therefore, the outer-voltage loop cannot completely eliminate the error between the peak and average values. This is a problem, especially in power-factor corrector converters, and causes distortion in the input current waveform.
- Uncompensated current loop gain is low. In the classical peak current-mode control, the uncompensated loop gain is very low. Therefore, higher-order compensation schemes such as Type-II and Type-III voltage error amplifiers are needed to boost the loop gain at DC.

1.1 Principle of Operation of Conventional Average Current-Mode Control Technique

In the conventional average current-mode control, the time-varying average value of the sensed inductor current is regulated. Figure 1.2 shows the circuit of a buck DC–DC converter using the conventional average current-mode control scheme. The reference to the current loop is set by the outer loop. The current reference generated by the outer-voltage loop is set to be equal to the desired inductor current average. A high-gain integrating amplifier in the current loop attempts to track the average value of the inductor current and sends the amplified current error to a PWM. The amplified current error is compared with a large amplitude sawtooth waveform oscillating at the switching frequency at the PWM comparator inputs.

Unlike peak current-mode control, the current loop gain can be modified for optimum performance by the compensation network in the current loop. For example, it is possible to have equal loop-gain crossover frequencies in both peak current-mode and average current-mode control schemes, but the gain will be much greater in average current-mode control at lower frequencies. This is essential to reduce the steady-state error significantly.

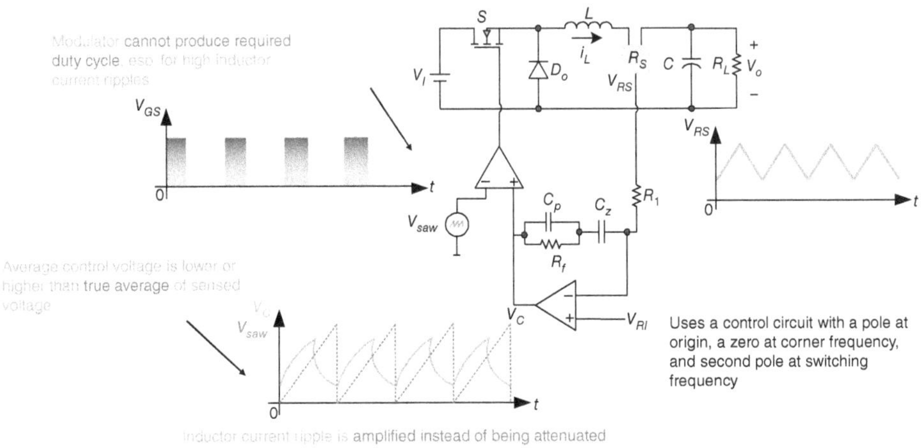

Figure 1.2 Circuit of a buck DC–DC converter with the conventional average current-mode control scheme [7]. Based on L. H. Dixon, 1990.

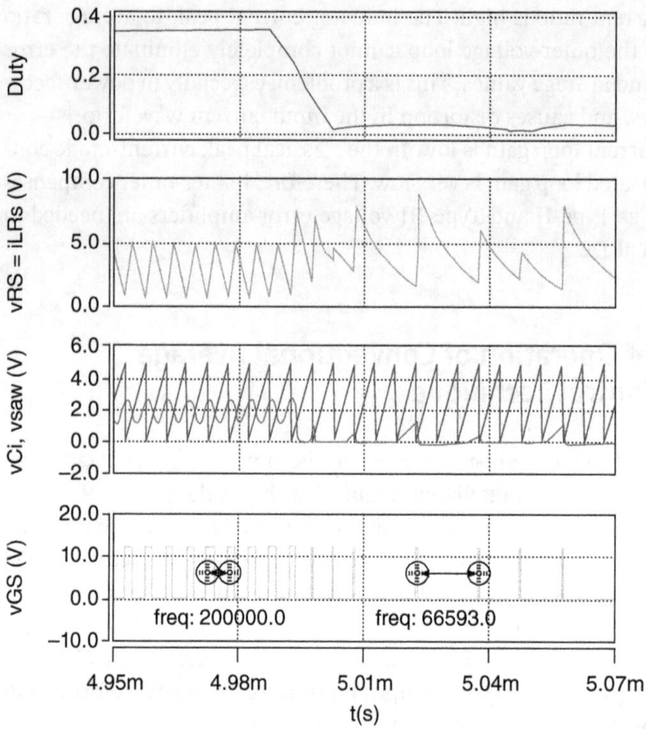

Figure 1.3 Waveforms showing the challenges encountered using conventional average current-mode control.

Although the conventional average current-mode control has unique advantages over peak current-mode, a few challenges are encountered. Figure 1.3 shows the waveforms of duty cycle d, sensed inductor current v_{RS}, output voltage of current error amplifier v_{Ci}, sawtooth voltage v_{saw}, and gate-to-source voltage v_{GS} relevant to a buck DC–DC converter. As the input voltage increases, the duty cycle must reduce. At low duty cycles, the inductor current ripple increases in magnitude. Since the error amplifier basically amplifies the error between the sensed inductor current and the current reference, the large inductor current ripple is also amplified. Therefore, the comparison between the sawtooth waveform and the amplified error output takes place more than once in each switching period resulting in subharmonic oscillations. This drawback cannot be ignored especially in converters used as power-factor correctors. Also, this method limits the designer from implementing switch current sensing due to larger current magnitudes. Moreover, even under normal operating conditions, the current-loop error amplifier must filter the switching frequency component in the sensed inductor current and perform normal compensation functions. In this technique, the inductor current is sensed and converted into sensed voltage by a current-to-voltage transducer. The sensed voltage comprises both DC and switching frequency ripple and is supplied to the control circuit. An integral single-lead (or Type-II) control circuit is used in this technique and performs two tasks:

1. Provides the required compensation for the inner loop.
2. Acts as a low-pass filter to reduce the switching frequency ripple in the sensed voltage. The control circuit comprises a pole at origin and a single pole-zero pair located at high frequencies. The pole at the origin provides a high DC gain to reduce the steady-state error, the high-frequency zero extends the current loop crossover frequency, and the high-frequency pole must filter the switching frequency ripple in the sensed signal [12].

A few evident shortcomings of this technique are as follows:

- The control circuit has a limited degree of freedom. For example, tuning its components for a desired phase margin will compromise the low-pass filter requirements and *vice versa*.
- At a high input voltage, the duty cycle is low and the inductor current ripple is high. This yields a large ripple at the control voltage supplied to the PWM. The slope of the control voltage exceeds the slope of the carrier voltage, causing switching instability [7, 12, 14].
- Due to the above two drawbacks, pulsating or discontinuous currents such as the switch current or diode current in the converter cannot be effectively sensed.
- Since the inductor current is sampled per switching period, the sampling gain needs to be taken into account.
- The location of the high-frequency pole is close to the switching frequency. Therefore, its contribution in reducing the switching frequency ripple and providing any compensation for the inner loop at high frequencies is limited.

Therefore, a modified average current-mode control scheme is developed and is shown in Figure 1.4. Here, the compensation circuit and the low-pass filter are decoupled, and they function independently. The sensed voltage that is dependent on the inductor current is provided to a low-pass filter dedicated only to cancel its switching frequency component. The output voltage of the low-pass filter with

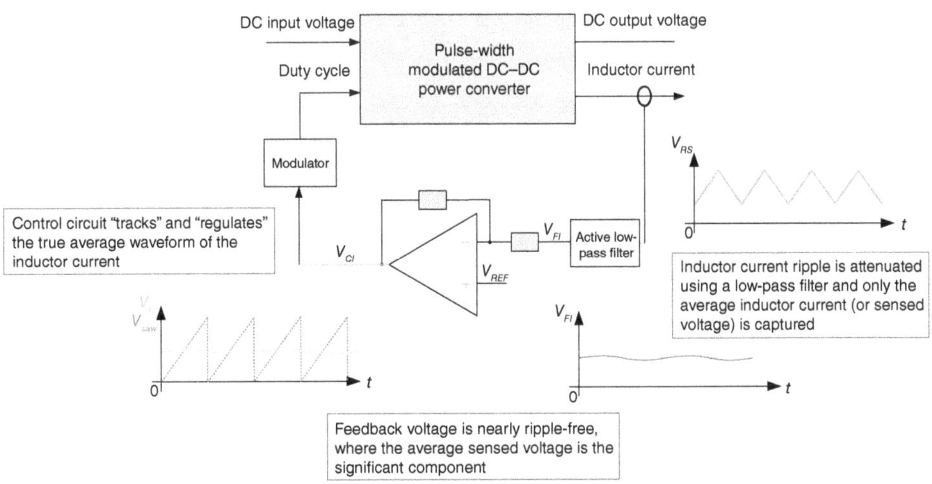

Figure 1.4 Block diagram of the average current-mode controlled DC–DC power converter.

a reduced switching frequency ripple component is then provided to the control circuit. Thus, the control voltage also exhibits a reduced ripple [52]. A detailed operation of the modified average current-mode control scheme is discussed in Section 1.2.

1.2 Principle of Operation of Modified Average Current-Mode Control Technique

Figure 1.4 shows the block diagram of the modified average current-mode controlled DC–DC converter. Figure 1.5 illustrates the key waveforms related to the sensing and feedback path. The sense resistor or current sensor is placed in the inductor branch to sense the inductor current. The sensed voltage is v_{RS}. The sensed voltage is provided to a low-pass filter composed of a resistor and capacitor network. The switching frequency component is eliminated by the low-pass filter. Therefore, the filter output voltage is DC with negligible ripple. The effective feedback voltage to the inner loop is v_{FI}. The reference voltage to the inner-current loop v_{RI} is governed by the outer-voltage loop. Based on the desired parameters such as high bandwidth and high DC gain of the inner-current loop, an appropriate control circuit with forward path impedance Z_{ii} and the feedback path impedance Z_{fi} will be derived and is a topic of discussion in Chapters 2–5. The control voltage v_{CI} at the output of the controller is compared with a sawtooth waveform v_{saw}, which generates the duty cycle d_T. By virtue of circuit operation, the exact average value of the inductor current (or any branch current) is sensed and controlled by the inner-current loop.

The major advantages of using the modified average current-mode control technique are as follows:

1. The compensation and filtering processes are performed independently. A good selection of the filter cutoff frequency assists the dynamic response and will be discussed in Chapters 2–5.

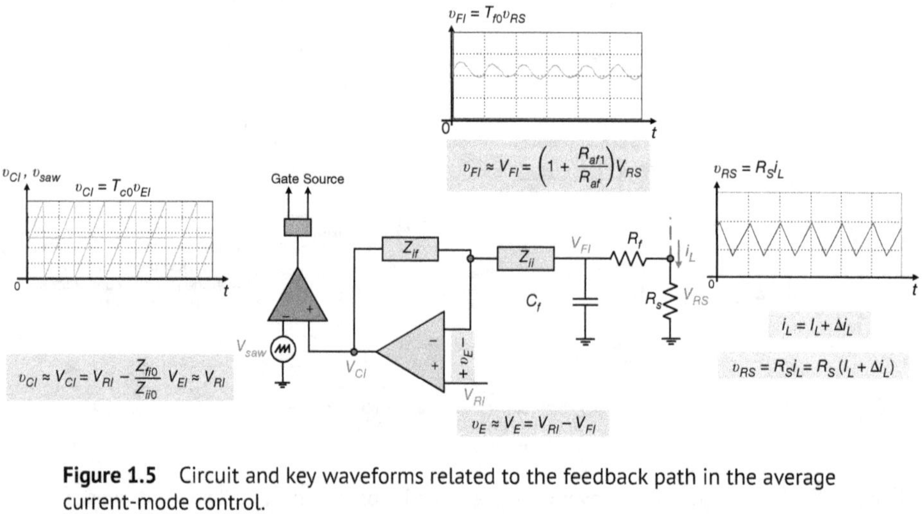

Figure 1.5 Circuit and key waveforms related to the feedback path in the average current-mode control.

2. The feedback voltage to the control circuit primarily contains the average value of the sensed current, and the switching frequency component is negligible. Therefore, a discontinuous switch or diode current with a large amplitude, a continuous or discontinuous inductor current of any amplitude, or the load current can be sensed by placing the current sensor in the corresponding branch.
3. The low-pass filter can be placed in either the feedback path or the forward path to reduce the ripple in the control voltage.
4. Due to the presence of the low-pass filter, the current loop does not behave as a sampling system. Thus, the effect of sampling gain is eliminated.
5. The problems due to the switching instability are completely avoided as the control voltage intersects the carrier voltage only once during each switching period.

This book endeavors to present a complete inner-loop analysis of different power converters whose average branch current such as inductor current, switch current, etc. is regulated by the modified current-mode control scheme. The steady-state large-signal operation, open and closed-loop transfer functions, and transient characteristics are presented.

1.3 Steady-State Operation

The inductor current comprises switching frequency component Δi_L superimposed on DC value I_L. From Figure 1.5, the potential difference across the sensed resistor R_S is given as

$$v_{RS} = R_s i_L = R_s(I_L + \Delta i_L). \tag{1.1}$$

The filter eliminates the switching frequency component of v_{RS}. By selecting appropriate sense resistance R_s, V_{RS} can be made equal to V_{RI}, where V_{RI} is the current reference voltage. The output voltage of the active filter is then compared with the reference voltage V_{RI}. The error signal is given as

$$V_{EI} = V_{RI} - V_{FI}. \tag{1.2}$$

The error signal is the input to the controller. The output of the controller is

$$V_{CI} = V_{RI} + T_{cio} V_{EI}, \tag{1.3}$$

where T_{cio} is the DC gain of the controller. The pulse-width modulator circuit is an inverting op-amp comparator. The control voltage is applied to the non-inverting input, and the sawtooth voltage signal v_{saw} is applied to the inverting input of the comparator. For any change in the control voltage v_C from V_C to $V_C + v_c$, the duty cycle d_T also changes from D to $D + d$. The control voltage-to-duty cycle transfer function of the PWM is expressed as

$$T_m = \frac{D}{V_{CI}} = \frac{1}{V_{Tm}}, \tag{1.4}$$

where V_{Tm} is the amplitude of the sawtooth waveform. In steady state, $V_{EI} \approx 0$, which yields

$$V_{CI} = V_{RI} = DV_{Tm}. \tag{1.5}$$

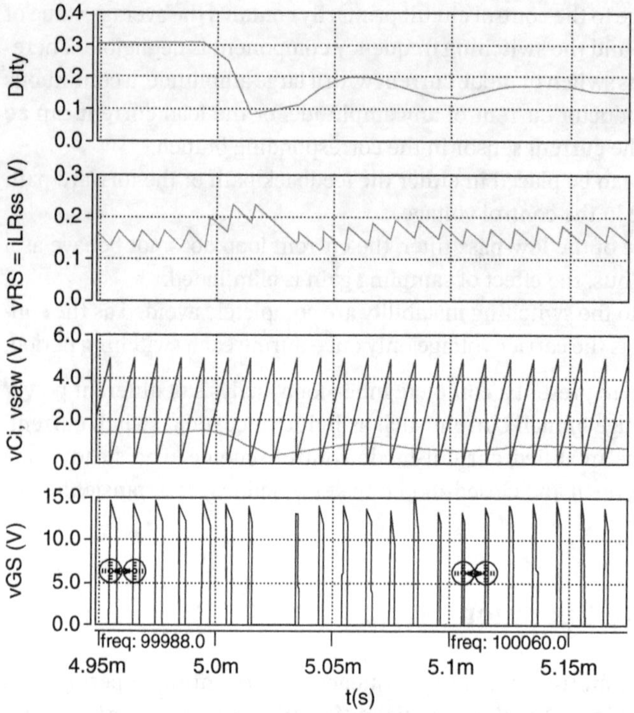

Figure 1.6 Waveforms related to the average current-mode controlled DC–DC buck converter with modified average current-mode control technique.

A buck DC–DC converter circuit with the modified average current-mode control was simulated, and the waveforms of duty cycle d, sensed inductor current v_{RS}, output of current error amplifier v_{CI}, sawtooth waveform v_{saw}, and gate-to-source voltage v_{GS} are shown in Figure 1.6. It can be seen that as the duty cycle decreases, the inductor current ripple stays unchanged. The comparator is able to track the amplified error voltage and produce gate-to-source voltage oscillating at the required switching frequency.

Chapters 2–5 discuss the closed-loop small-signal analysis of closed-loop buck, boost, and buck-boost PWM DC–DC converter with the modified average current-mode control technique. The DC large-signal and AC small-signal models are derived using circuit averaging technique [36–44]. The open-loop and closed-loop small-signal transfer functions are derived for the DC–DC converters with modified average current-mode control. Block diagrams are developed for each converter to characterize the inner-current loop for the power stage and the outer-voltage loop. Techniques to design the inner current-loop and outer voltage-loop are provided, and the responses of open loop, closed inner-current loop, and closed outer-voltage loop are compared. Through the step-by-step approach presented in this book, the readers can gain a deeper insight into the steady-state and small-signal operation of the fundamental DC–DC converters used today in various electronic applications.

2

Average Current-Mode Control of Buck DC–DC Converter

This chapter explains the average current-mode control of pulse-width-modulated (PWM) buck DC–DC converter operated in continuous-conduction mode (CCM). The circuit-averaging technique is presented, which is based on the principle of conservation of energy. The small-signal model of the buck converter is derived taking into account the relevant losses. The open-loop and closed-loop transfer functions and transient responses are derived and analyzed. A design example is utilized throughout the chapter to understand open-loop and closed-loop converter responses. The objectives of this chapter are as follows:

1. To describe the DC characteristics of the buck converter.
2. To develop the small-signal linear model of the buck converter.
3. To derive and analyze the open-loop power stage transfer functions of the buck converter.
4. To derive and analyze the open-loop input and output impedances of the buck converter.
5. To design a control circuit for the buck DC–DC converter.
6. To demonstrate the behavior of average current-mode controlled buck converter.
7. To derive and analyze the closed-inner-current loop transfer functions of the buck converter.
8. To derive and analyze the closed-outer-voltage loop transfer functions of the buck converter.

As the outcome of this chapter, the reader will gain sufficient knowledge in analysis, design, and characterization of buck DC–DC converter employing average current-mode control. The subsequent analysis in this chapter is based on the following assumptions:

1. The diode capacitance and the lead inductance are zero, and thereby the switching losses are neglected.
2. Passive components are linear, time invariant, and frequency dependent.
3. The output impedance of the supply voltage source is zero for both DC and AC components.
4. The converter operates in steady state.

Average Current-Mode Control of DC-DC Power Converters, First Edition.
Marian K. Kazimierczuk, Dalvir K. Saini, and Agasthya Ayachit.

5. The switching period is much shorter than the time constants of the reactive components designed for CCM operation.
6. The DC output voltage is constant; however, the DC supply voltage and the load resistance are variable.

2.1 Circuit Description, DC Characteristics, and Design

2.1.1 Circuit Description

The circuit of PWM buck converter is shown in Figure 2.1. The operation of the buck converter in CCM is given in detail in [45]. The circuit consists of a power metal-oxide semiconductor field-effect transistor (MOSFET) MOSFET S, a diode D_o, an inductor L, a filter capacitor C, and a load resistor R_L. The switch is turned ON and OFF at a switching frequency f_s and at a duty cycle D.

2.1.2 DC Model

The derivation of DC voltage transfer function of the buck DC–DC converter for CCM is given in detail in [45]. The ideal current and voltage waveforms of the switching network of the buck converter are shown in Figure 2.2. The circuit averaging technique is applied to the switching network waveforms. The DC voltage transfer function of the ideal converter is

$$M_{VDC} = \frac{V_O}{V_I} = D, \tag{2.1}$$

and the DC current transfer function of the ideal converter is

$$M_{IDC} = \frac{I_O}{I_I} = \frac{1}{D}. \tag{2.2}$$

As shown in Figure 2.2b, the waveform of the switch current can be approximated by

$$i_S = \begin{cases} I_L & 0 < t \leq DT \\ 0 & DT < t \leq T. \end{cases} \tag{2.3}$$

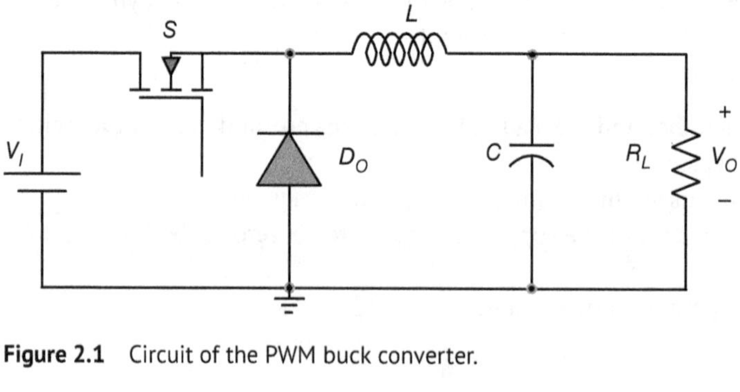

Figure 2.1 Circuit of the PWM buck converter.

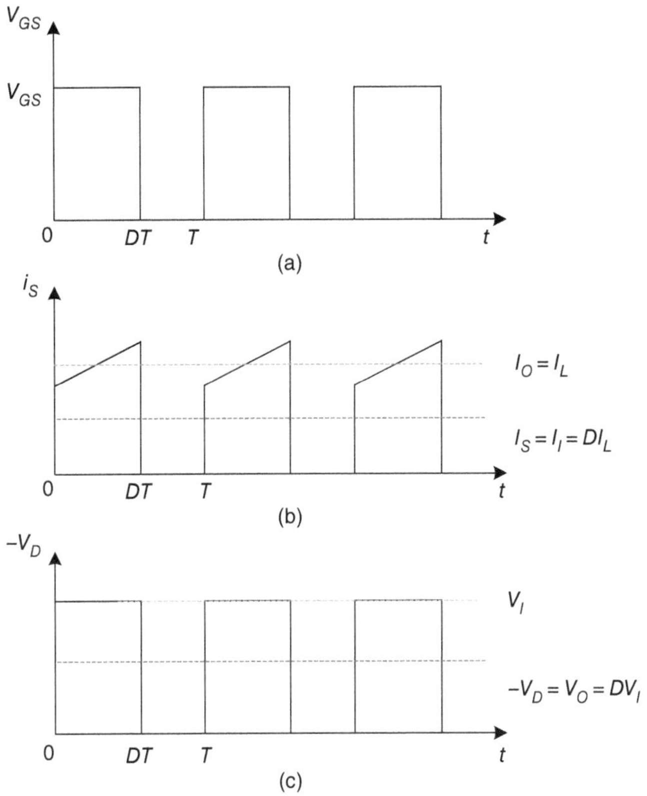

Figure 2.2 Waveforms of the ideal switching network. (a) Gate-to-source voltage of switch. (b) Switch current. (c) Diode voltage.

The DC component of the switch current is

$$I_S = \frac{1}{T}\int_0^T i_S dt = \frac{1}{T}\int_0^{DT} I_L dt = DI_L. \tag{2.4}$$

Hence, the MOSFET can be replaced by an ideal DC current-dependent current source.

As illustrated in Figure 2.2c, the waveform of the diode inverted voltage, equal to the voltage between the inductor and diode terminals, is given by

$$-v_D(t) = v_{KA}(t) = v_{LD} = \begin{cases} V_I & \text{for } 0 < t \leq DT \\ 0 & \text{for } DT < t \leq T. \end{cases} \tag{2.5}$$

The DC component of diode inverted voltage is

$$-V_D = V_{KA} = V_{LD} = \frac{1}{T}\int_0^T v_{KA} dt = \frac{1}{T}\int_0^{DT} V_{SD} dt = DV_{SD} = DV_I, \tag{2.6}$$

where V_{KA} is the DC component of diode branch voltage between the cathode and anode terminals, V_{SD} is the DC component of voltage between the drain of the MOSFET and anode of the diode, and V_{LD} is the DC component of voltage between the inductor (positive terminal) and the anode of the diode. The diode can be

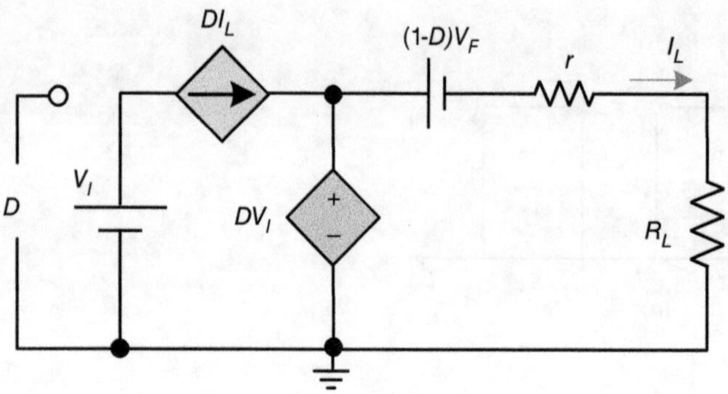

Figure 2.3 DC model of PWM buck DC–DC converter in CCM.

replaced by an ideal DC voltage-dependent voltage source. The DC averaged model of the buck converter in CCM is shown in Figure 2.3. The inductor is a short circuit, and the filter capacitor is an open circuit in the DC model. This model describes the behavior of the converter at steady state. The DC component of the input current is

$$I_I = I_S = DI_L, \tag{2.7}$$

and the DC component of output current is

$$I_O = I_L. \tag{2.8}$$

Applying the principle of conservation of energy [37, 45], the equivalent averaged resistance r placed in series with the inductance L is the sum of the averaged resistance of the switch, diode, and inductor branch given by

$$r = Dr_{DS} + (1-D)R_F + r_L, \tag{2.9}$$

where r_L, r_{DS}, and R_F are the equivalent series resistance (ESR) of the inductor, ON-state resistance of the MOSFET, and diode forward resistance, respectively. The resistance r produces the same loss in the form of heat as that produced by the individual resistances placed in their respective branches.

The efficiency of the converter is

$$\begin{aligned}\eta &= \frac{P_O}{P_I} = \frac{V_O I_O}{V_I I_I} = M_{IDC} M_{VDC} \\ &= \frac{1}{1 + \frac{Dr_{DS} + (1-D)R_F + r_L}{R_L} + \frac{(1-D)V_F}{V_O} + \frac{f_s C_o R_L}{M_{VDC}^2} + \frac{r_C R_L (1-D)^2}{12 f_s^2 L^2}},\end{aligned} \tag{2.10}$$

where r_L, r_{DS}, V_F, R_F, r_C, and C_o are the ESR of the inductor, ON-state resistance of the MOSFET, diode forward voltage drop, diode forward resistance, ESR of filter capacitor, and MOSFET output capacitance, respectively.

2.1.3 Design Example

The design equations for power converters in CCM are given in Appendix A. Parameters of selected power MOSFETs and power diodes are listed in Appendices B and C, respectively. Spice model parameters of selected power MOSFETs and power diodes are given in Appendices D and E, respectively. Introduction to SPICE, introduction to MATLAB (including an example code to obtain DC values, magnitude, phase, and step plots), and introduction to SABER circuit simulator are given in Appendix F.

The subsequent analysis is done for the buck converter designed with the following specifications:

1. Supply voltage $V_I = 28$ V.
2. Output voltage $V_O = 14$ V.
3. Output power $P_O = 20$ W.
4. Load resistance $R_L = 10\ \Omega$.
5. Switching frequency $f_s = 100$ kHz.
6. The inductance to ensure CCM operation is $L = 301\ \mu$H.
7. The capacitance to ensure CCM operation is $C = 68\ \mu$F.

The ESRs of the inductor is $r_L = 0.19\ \Omega$, and the capacitor is $r_C = 0.11\ \Omega$. The ON-state resistance of the MOSFET and the on resistance of the diode are $r_{DS} = 0.11\ \Omega$ and $R_F = 0.022\ \Omega$, and the output capacitance of the MOSFET is $C_o = 100$ pF. Diode forward voltage is $V_F = 0.7$ V. The calculated value of converter efficiency is $\eta = 90\%$. The required duty cycle to achieve the rated output voltage is $D = 0.556$. The calculated value of equivalent averaged resistance using (2.9) is $r \approx 0.26\ \Omega$.

The calculated values of open-loop DC quantities used Section 2.1.2 are input current $I_I = 0.78$ A, output current $I_O = 1.4$ A, dc current transfer function $M_{IDC} = 1.8$, DC voltage transfer function $M_{VDC} = 0.5$, DC inductor current $I_L = 1.4$ A, DC switch current $I_S = 0.78$ A, and DC diode inverted voltage $V_D = -15.56$ V.

2.2 Large-Signal and Small-Signal Models of PWM Buck Converter in CCM

A low-frequency perturbation is applied to the DC model about the steady-state operating point to yield a large-signal nonlinear model. The large-signal quantities can be expressed as the sum of DC and AC components and are given as

$$i_S = I_S + i_s, \tag{2.11}$$

$$i_L = I_L + i_l, \tag{2.12}$$

$$v_D = V_D + v_d, \tag{2.13}$$

$$v_I = V_I + v_i, \tag{2.14}$$

$$v_O = V_O + v_o, \tag{2.15}$$

$$d_T = D + d, and \tag{2.16}$$

where i_S, i_L, v_D, v_I, v_O, and d_T are the large-signal switch current, inductor current, diode voltage, input voltage, output voltage, and duty cycle, respectively. The quantities i_s, i_l, v_d, v_i, v_o, and d are the small-signal switch current, inductor current, diode voltage, input voltage, output voltage, and duty cycle, respectively. Hence, from (2.4) and (2.6),

$$i_S = d_T i_L = (D + d)(I_L + i_l) = DI_L + Di_l + I_L d + i_l d \tag{2.17}$$

and

$$-v_D = d_T v_I = (D + d)(V_I + v_i) = DV_I + Dv_i + V_I d + v_i d. \tag{2.18}$$

The large-signal nonlinear model of a buck DC–DC converter in CCM is shown in Figure 2.4. In (2.17) and (2.18), the nonlinear small-signal products $i_l d$ and $v_i d$ are eliminated using small-signal conditions ($d << D, v_i << V_I, i_l << I_L$), yielding a linearized large-signal model [46]. Hence, the linearized large-signal expressions for switch current and diode inverted voltage are given as

$$i_S = DI_L + Di_l + I_L d \tag{2.19}$$

and

$$-v_D = DV_I + Dv_i + V_I d. \tag{2.20}$$

Figure 2.5 shows the large-signal linear model of the buck DC–DC converter. The AC terms are separated from DC components to obtain the linear, averaged, small-signal model as shown in Figure 2.6.

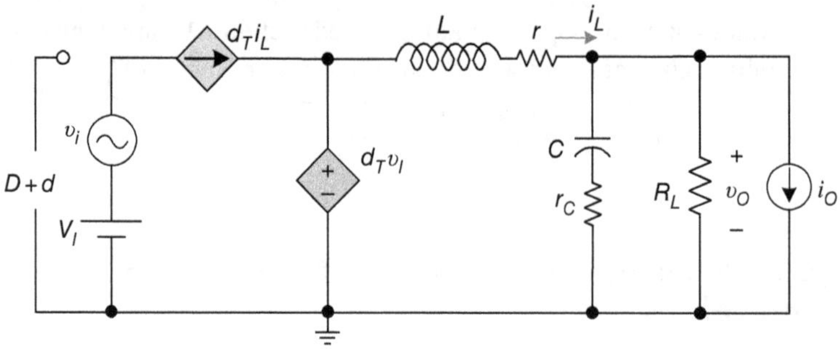

Figure 2.4 Large-signal nonlinear model of the PWM buck DC–DC converter in CCM.

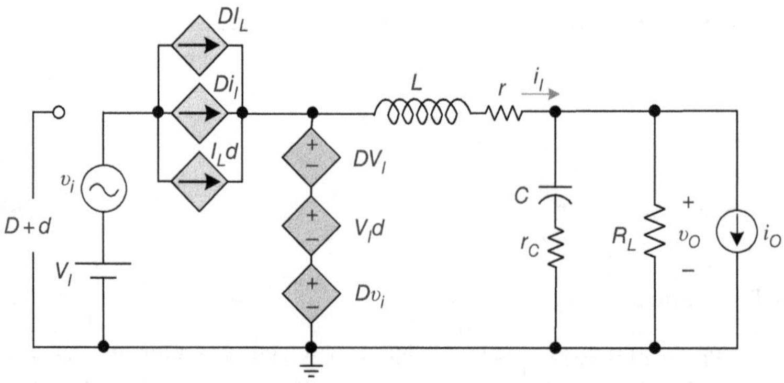

Figure 2.5 Large-signal linear model of the PWM buck DC–DC converter in CCM.

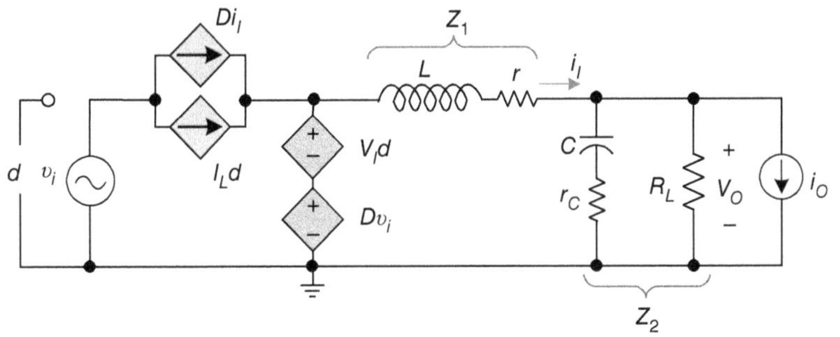

Figure 2.6 Small-signal model of the PWM buck DC–DC converter in CCM.

2.3 Power Stage Transfer Functions

The following power stage transfer functions are discussed in this section:

1. Duty cycle-to-output voltage transfer function T_p (control).
2. Duty cycle-to-inductor current transfer function T_{pi} (control).
3. Input voltage-to-output voltage transfer function M_v (disturbance).
4. Input voltage-to-inductor current transfer function M_{vi} (disturbance).
5. Reverse current gain A_i (disturbance).
6. Input impedance Z_i (disturbance).
7. Output impedance Z_o (disturbance).

The small-signal model of the PWM buck converter is shown in Figure 2.6. Using this model, the governing equations to derive the transfer functions are developed in this section.

The impedances Z_1 and Z_2 are

$$Z_1 = r + sL \tag{2.21}$$

and

$$Z_2 = R_L || \left(r_C + \frac{1}{sC}\right) = \frac{R_L\left(r_C + \frac{1}{sC}\right)}{R_L + r_C + \frac{1}{sC}}. \tag{2.22}$$

The current through the parallel combination of the load resistance and the filter capacitor is

$$i_{Z_2} = \frac{v_o}{Z_2}. \tag{2.23}$$

The current through the inductor is

$$i_l = i_{Z2} + i_o = \frac{v_o}{Z_2} + i_o. \tag{2.24}$$

The input and the switch currents are

$$i_i = i_s = Di_l + I_L d. \tag{2.25}$$

Applying Kirchhoff's voltage law, we obtain

$$v_o = Dv_i + V_I d - i_l Z_1. \tag{2.26}$$

Substituting (2.26) in (2.24), we get

$$i_l = \frac{Dv_i + V_I d - i_l Z_1}{Z_2} + i_o = \frac{Dv_i + V_I d + Z_2 i_o}{Z_1 + Z_2}. \tag{2.27}$$

Applying Kirchhoff's voltage law, the output voltage is

$$v_o = (Dv_i + V_I d)\frac{Z_2}{Z_1 + Z_2}. \tag{2.28}$$

2.3.1 Duty Cycle-to-Output Voltage Transfer Function T_p

The duty cycle-to-output voltage transfer function is obtained by setting $v_i = 0$ and $i_o = 0$ in Figure 2.6. Setting $v_i = 0$ in (2.28) and rearranging, we get

$$v_o = V_I d \frac{Z_2}{Z_1 + Z_2}. \tag{2.29}$$

Hence, the duty cycle-to-output transfer function is

$$T_p(s) = \left.\frac{v_o(s)}{d(s)}\right|_{v_i = i_o = 0} = V_I \frac{Z_2}{Z_1 + Z_2}. \tag{2.30}$$

Substituting (2.21) and (2.22) into (2.30) gives the control-to-output transfer function in the s domain as

$$T_p(s) = \left.\frac{v_o(s)}{d(s)}\right|_{v_i = i_o = 0} = T_{px} \frac{s + \omega_{zn}}{s^2 + 2\xi\omega_o s + \omega_o^2} = T_{po} \frac{1 + \frac{s}{\omega_{zn}}}{\left(\frac{s}{\omega_o}\right)^2 + \frac{2\xi s}{\omega_o} + 1}, \tag{2.31}$$

where

$$T_{px} = \frac{V_I R_L r_C}{L(R_L + r_C)}, \tag{2.32}$$

the DC gain T_{po} is

$$T_{po} = T_p(0) = \frac{V_I R_L}{R_L + r}, \tag{2.33}$$

the angular corner frequency or the angular undamped natural frequency is

$$\omega_o = \sqrt{\frac{R_L + r}{LC(R_L + r_C)}}, \tag{2.34}$$

the damping ratio is

$$\xi = \frac{L + C[R_L(r_C + r) + r_C r]}{2\sqrt{LC(R_L + r_C)(R_L + r)}}, \tag{2.35}$$

and the angular frequency of the left-half plane (LHP) zero is

$$\omega_{zn} = \frac{1}{r_C C}. \tag{2.36}$$

The magnitude of the transfer function is

$$|T_p| = T_{po}\sqrt{\frac{\left[1+\left(\frac{\omega}{\omega_{zn}}\right)^2\right]}{\left[1-\left(\frac{\omega}{\omega_o}\right)^2\right]^2 + 4\xi^2\left(\frac{\omega}{\omega_o}\right)^2}}, \tag{2.37}$$

and the phase of the transfer function is

$$\phi_{Tp} = -180° + \tan^{-1}\left(\frac{\omega}{\omega_{zn}}\right) - \tan^{-1}\left[\frac{2\xi\left(\frac{\omega}{\omega_o}\right)}{1-\left(\frac{\omega}{\omega_o}\right)^2}\right]. \tag{2.38}$$

For converters, whose damping coefficient is closer to unity, the converter bandwidth f_{BW} is nearly equal to the natural frequency f_o. In contrast, for transfer functions with a lower ξ, the bandwidth is higher than the natural frequency. The 3-dB bandwidth of the converter power stage determines the speed of the outer-voltage loop in voltage-mode control. The gain at f_{BW} is

$$\frac{|T_p(f_{BW})|}{T_{po}} = \frac{1}{\sqrt{2}}. \tag{2.39}$$

The following analysis for the transfer function bandwidth holds true for $\omega_{zn} >> \omega_o$. The bandwidth of the T_p transfer function is governed by the dominant poles. Using the condition (2.39) in (2.37) and assuming dominant poles,

$$\frac{T_p(f_{BW})}{T_{po}} = \frac{1}{\sqrt{2}} = \sqrt{\frac{1}{\left[1-\left(\frac{\omega_{BW}}{\omega_o}\right)^2\right]^2 + 4\xi^2\left(\frac{\omega_{BW}}{\omega_o}\right)^2}}, \tag{2.40}$$

yielding

$$\frac{1}{\left[1-\left(\frac{f_{BW}}{f_o}\right)^2\right]^2 + 4\xi^2\left(\frac{f_{BW}}{f_o}\right)^2} = \frac{1}{2} \tag{2.41}$$

and producing the bandwidth as

$$f_{BW} = f_o\sqrt{1 - 2\xi^2 + \sqrt{4\xi^4 - 2\xi^2 + 2}}\,. \tag{2.42}$$

For the buck converter designed in Section 2.1.3, the DC value of duty cycle-to-output voltage transfer function is $T_{po} = 27.2$ V = 28.69 dBV, natural corner frequency is $f_o = \omega_o/2\pi = 1.12$ kHz, frequency of the LHP zero is $f_{zn} = \omega_{zn}/2\pi = 21.28$ kHz, and the damping coefficient is $\xi = 0.19$. The gain decreases by 40 dB/dec beyond f_o. The zero frequency is caused by the series combination of the filter capacitance and the ESR of the filter capacitance. The phase is 0° at DC and low frequencies. In other words, as duty cycle increases or decreases, the output voltage increases or decreases, respectively. At high frequencies, the phase flattens at −90° due to the frequency of zero f_{zn}. The phase dips toward −180° and achieves a minimum value of −160° at 3.2 kHz. The minimum phase depends largely on the load resistance R_L. The lower is the load resistance, and the closer is the phase to −180°. As the load resistance decreases, the damping coefficient becomes small and the phase approaches −180°, causing the ringing in the output voltage waveform at frequency f_o. Figure 2.7 shows the theoretically obtained magnitude and phase plots of duty cycle-to-output voltage transfer function.

2.3.2 Duty Cycle-to-Inductor Current Transfer Function T_{pi}

The duty cycle-to-inductor current transfer function is obtained by setting $v_i = 0$ and $i_o = 0$ in Figure 2.6. Setting $v_i = 0$ in (2.24) and rearranging, we get

$$i_l = \frac{V_I d}{Z_1 + Z_2}. \tag{2.43}$$

Hence, the duty cycle-to-inductor current transfer function is

$$T_{pi}(s) = \left.\frac{i_l(s)}{d(s)}\right|_{v_i=i_o=0} = \frac{V_I}{Z_1 + Z_2}. \tag{2.44}$$

Substituting (2.21) and (2.22) into (2.44) gives the duty cycle-to-inductor current transfer function in the s-domain as

$$T_{pi}(s) = \left.\frac{i_l(s)}{d(s)}\right|_{v_i=i_o=0} = T_{pix}\frac{s + \omega_{zi}}{s^2 + 2\xi\omega_o s + \omega_o^2} = T_{pio}\frac{1 + \frac{s}{\omega_{zi}}}{\left(\frac{s}{\omega_o}\right)^2 + \frac{2\xi s}{\omega_o} + 1}, \tag{2.45}$$

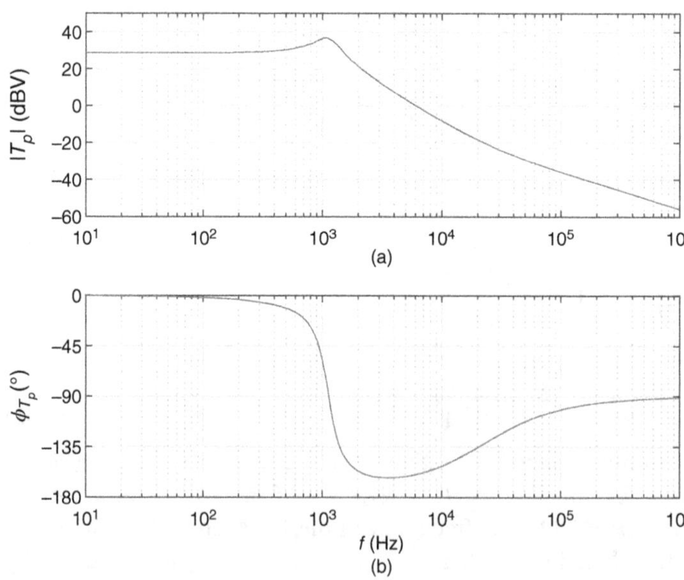

Figure 2.7 Theoretically obtained plots of duty cycle-to-output voltage transfer function T_p. (a) Magnitude. (b) Phase.

where

$$T_{pix} = \frac{V_I}{L}, \tag{2.46}$$

the DC gain T_{pio} is

$$T_{pio} = T_{pi}(0) = \frac{V_I}{R_L + r}, \tag{2.47}$$

and the angular frequency of the LHP zero is

$$\omega_{zi} = \frac{1}{C(R_L + r_C)}. \tag{2.48}$$

The magnitude of the duty cycle-to-inductor current transfer function is

$$|T_{pi}| = T_{pio}\sqrt{\frac{1 + \left(\frac{\omega}{\omega_{zi}}\right)^2}{\left[1 - \left(\frac{\omega}{\omega_o}\right)^2\right]^2 + 4\xi^2\left(\frac{\omega}{\omega_o}\right)^2}}, \tag{2.49}$$

and the phase of the transfer function is

$$\phi_{Tpi} = -180^\circ + \tan^{-1}\left(\frac{\omega}{\omega_{zi}}\right) - \tan^{-1}\left[\frac{2\xi\left(\frac{\omega}{\omega_o}\right)}{1 - \left(\frac{\omega}{\omega_o}\right)^2}\right]. \tag{2.50}$$

Figure 2.8 shows the theoretically obtained magnitude and phase plots of duty cycle-to-inductor current transfer function. The gain at DC and low frequencies is

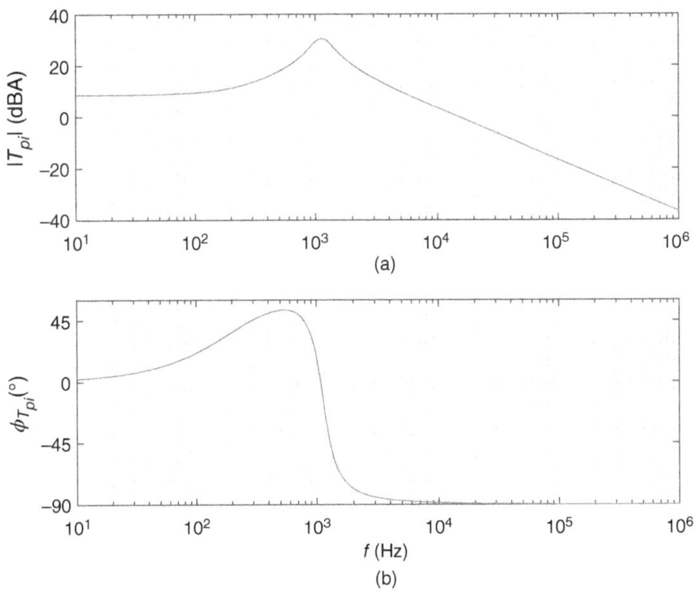

Figure 2.8 Theoretically obtained plots of duty cycle-to-inductor current transfer function T_{pi}. (a) Magnitude. (b) Phase.

nearly $T_{pio} = 2.8$ A = 8.94 dBA. The natural corner frequency of the second-order low-pass filter formed by the inductor, capacitor, and load resistor is $f_o = 1.12$ kHz. The frequency of the LHP zero is $f_{zi} = 232$ Hz. The gain decreases at the rate of 20 dB/dec beyond f_o. The duty cycle-to-inductor current transfer function for the buck converter exhibits a low-frequency zero f_{zi} and a corner frequency f_o due to the pair of complex-conjugate poles. The phase starts at zero and rises to 45° due to f_{zi}. This means that at low frequencies, the duty cycle signal leads the inductor current by a given phase angle. The bandwidth of T_{pi} is wider than T_p. Therefore, the inner-current loop presents a faster response in inductor current for change in duty cycle. The low-frequency zero f_{zi} depends on the load resistance R_L and the parameters of the filter capacitance, namely, C and r_C. As the value of R_L is decreased at a fixed C, f_{zi} moves toward f_o. The nature of the roots of the characteristic equation (denominator of T_{pi}) or the poles of T_{pi} converts from complex-conjugate to real. At a specific minimum load resistance, $f_{zi} \approx f_o$ and f_{zi} cancel the effect of one of the complex-conjugate poles at f_o producing a first-order dominant pole system. The dominant pole is caused by the inductor and load resistance.

2.3.3 Input Voltage-to-Output Voltage Transfer Function M_v

The input voltage-to-output voltage transfer function is obtained by setting $d = 0$ and $i_o = 0$ in Figure 2.6. Setting $d = 0$ in (2.28) and rearranging gives

$$v_o = D\frac{Z_2}{Z_1 + Z_2}v_i. \tag{2.51}$$

Hence, the input voltage-to-output voltage transfer function M_v is

$$M_v(s) = \left.\frac{v_o(s)}{v_i(s)}\right|_{d=i_o=0} = D\frac{Z_2}{Z_1 + Z_2}. \tag{2.52}$$

Substituting (2.21) and (2.22) into (2.52) gives the input voltage-to-output voltage transfer function in the s-domain as

$$M_v(s) = \left.\frac{v_o(s)}{v_i(s)}\right|_{d=i_o=0} = M_{vx}\frac{s + \omega_{zn}}{s^2 + 2\xi\omega_o s + \omega_o^2} = M_{vo}\frac{1 + \dfrac{s}{\omega_{zn}}}{\left(\dfrac{s}{\omega_o}\right)^2 + \dfrac{2\xi s}{\omega_o} + 1}, \tag{2.53}$$

where

$$M_{vx} = \frac{DR_L r_C}{L(R_L + r_C)} \tag{2.54}$$

and the DC gain M_{vo} is

$$M_{vo} = \frac{DR_L}{R_L + r}. \tag{2.55}$$

The angular frequency of the LHP zero is given in (2.36).

Figure 2.9 shows the theoretically obtained magnitude and phase plots of input voltage-to-output voltage transfer function. The gain at DC and low frequencies is $M_{vo} = 0.54$ V/V = −5.35 dB, which is nearly equal to the duty cycle D. The

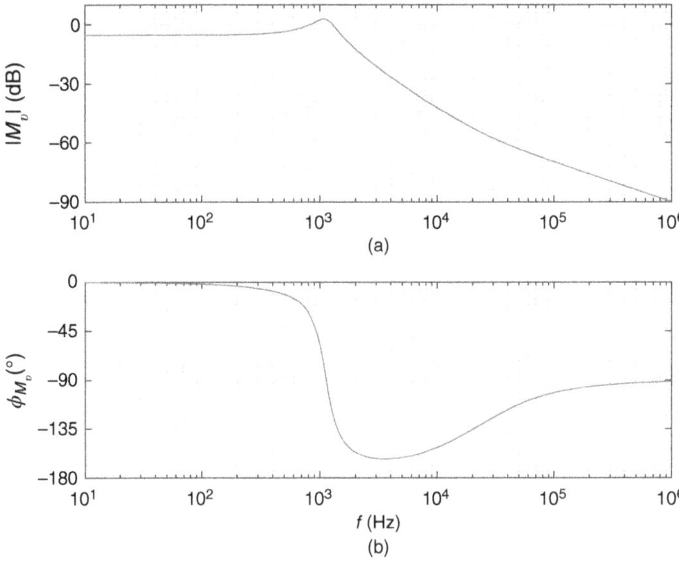

Figure 2.9 Theoretically obtained plots of input-to-output voltage transfer function M_v. (a) Magnitude. (b) Phase.

open-loop buck converter inherently offers high supply voltage rejection especially at low duty cycles. This is because, at low duty cycle, the power supply is connected to the load, however only for a portion of the switching period. The phase starts at zero and decreases toward −180°. However, the high-frequency LHP zero at ω_{zn} increases the phase to −90°.

2.3.4 Input Voltage-to-Inductor Current Transfer Function M_{vi}

The input voltage-to-inductor current transfer function is obtained by setting $d = 0$ and $i_o = 0$ in Figure 2.6. Setting $d = 0$ in (2.24) and rearranging gives

$$i_l = \frac{Dv_i}{Z_1 + Z_2}. \tag{2.56}$$

Hence, the input voltage-to-inductor current transfer function M_{vi} is

$$M_{vi}(s) = \left.\frac{i_l(s)}{v_i(s)}\right|_{d=i_o=0} = \frac{D}{Z_1 + Z_2}. \tag{2.57}$$

Substituting (2.21) and (2.22) into (2.57) gives the input voltage-to-inductor current transfer function in the s-domain as

$$M_{vi}(s) = \left.\frac{i_l(s)}{v_i(s)}\right|_{d=i_o=0} = M_{vix}\frac{s + \omega_{zi}}{s^2 + 2\xi\omega_o s + \omega_o^2} = M_{vio}\frac{1 + \frac{s}{\omega_{zi}}}{\left(\frac{s}{\omega_o}\right)^2 + \frac{2\xi s}{\omega_o} + 1}, \tag{2.58}$$

where

$$M_{vix} = \frac{D}{L}, \tag{2.59}$$

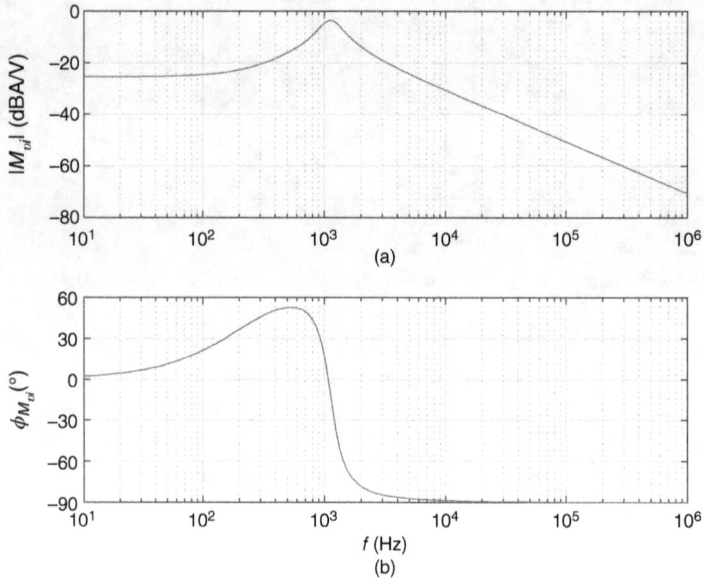

Figure 2.10 Theoretically obtained plots of input voltage-to-inductor current transfer function M_{vi}. (a) Magnitude. (b) Phase.

the DC gain M_{vio} is

$$M_{vio} = \frac{D}{R_L + r}, \tag{2.60}$$

and the angular frequency of the LHP zero is

$$\omega_{zi} = \frac{1}{C(R_L + r_C)}. \tag{2.61}$$

Figure 2.10 shows the theoretically obtained magnitude and phase plots of input voltage-to-inductor current transfer function. The gain at DC is $M_{vio} = 0.054$ A/V = −25.35 dBA/V. Through the M_{vi}, a current-loop transfer function, one can observe that the current loop offers a lower audio-susceptibility at DC than that provided by the voltage-loop transfer function M_v. All other frequency-domain properties remain identical to T_{pi}.

2.3.5 Reverse Current Gain A_i

The small-signal model to derive the output-to-inductor current transfer function is shown in Figure 2.11. This model is obtained by setting $v_i = 0$ and $d = 0$. An independent voltage source v_t is applied at the output, which forces a current i_t. Hence, the output voltage and output current are

$$v_o = -v_t = -i_l Z_1 \tag{2.62}$$

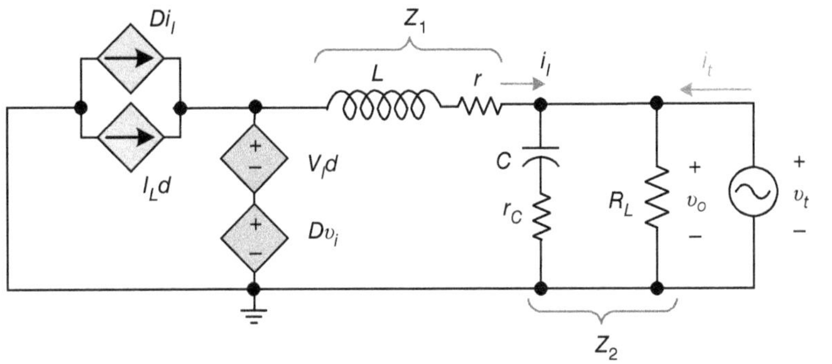

Figure 2.11 Small-signal model of the PWM buck DC–DC converter in CCM to derive the output current-to-inductor current transfer function A_i.

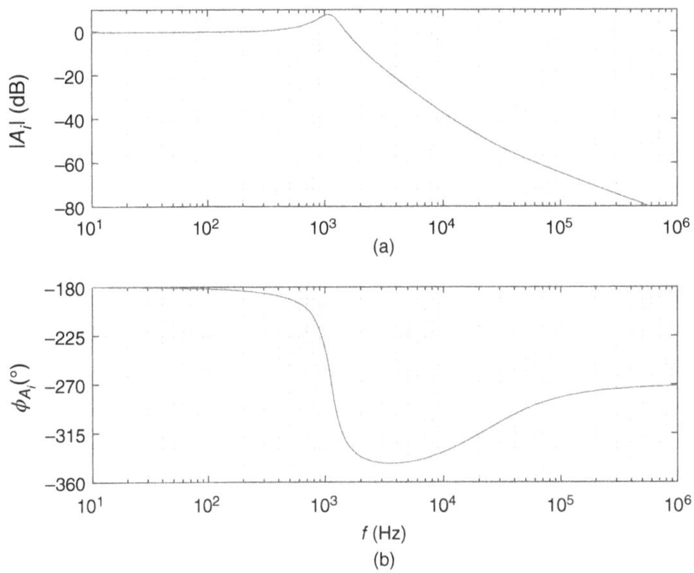

Figure 2.12 Theoretically obtained plots of output current-to-inductor current transfer function A_i. (a) Magnitude. (b) Phase.

and

$$\begin{aligned} i_l &= i_o + \frac{v_o}{Z_2} = i_o + \frac{-i_l Z_1}{Z_2} \\ i_o &= i_l\left(1 + \frac{Z_1}{Z_2}\right) = i_l \frac{Z_1 + Z_2}{Z_2}. \end{aligned} \tag{2.63}$$

The current i_t is

$$i_t = -i_o = -i_l \frac{Z_1 + Z_2}{Z_2}. \tag{2.64}$$

Hence, the output current-to-inductor current transfer function or reverse current transfer function A_i is

$$A_i(s) = \left.\frac{i_l(s)}{i_o(s)}\right|_{d=v_i=0} = \frac{Z_2}{Z_1 + Z_2}. \tag{2.65}$$

Substituting (2.21) and (2.22) into (2.65) gives the output current-to-inductor current transfer function in the s-domain as

$$A_i(s) = \left.\frac{i_l(s)}{i_o(s)}\right|_{d=v_i=0} = A_{ix}\frac{s + \omega_{zn}}{s^2 + 2\xi\omega_o s + \omega_o^2} = A_{io}\frac{1 + \frac{s}{\omega_{zn}}}{\left(\frac{s}{\omega_o}\right)^2 + \frac{2\xi s}{\omega_o} + 1}, \tag{2.66}$$

where

$$A_{ix} = \frac{R_L r_C}{L(R_L + r_C)} \tag{2.67}$$

and the DC gain A_{io} is

$$A_{io} = \frac{R_L}{R_L + r}. \tag{2.68}$$

The angular frequency of the LHP zero ω_{zn} is given in (2.36).

Figure 2.12 shows the theoretically obtained magnitude and phase plots of output-to-inductor current transfer function. The gain at DC is nearly $A_{io} = 0$ dB $= 1$ A/A, i.e. the reverse current gain is unity at DC and low frequencies up to nearly f_o. Beyond f_o, the impedance offered by the capacitor branch is lower than the inductor, and the small-signal output current begins to flow into the capacitor. Therefore, the gain reduces by −40 dB per decade.

2.3.6 Open-Loop Input Impedance Z_i

The open-loop input impedance is obtained by setting $d = 0$ and $i_o = 0$ in (2.25) and (2.28) and rearranging as

$$i_i = D i_l \tag{2.69}$$

and

$$i_l = \frac{D v_i}{Z_1 + Z_2}, \tag{2.70}$$

yielding

$$i_i = D\frac{D v_i}{Z_1 + Z_2}. \tag{2.71}$$

Hence, the open-loop input impedance is

$$Z_i(s) = \left.\frac{v_i(s)}{i_i(s)}\right|_{d=i_o=0} = \frac{Z_1 + Z_2}{D^2}. \tag{2.72}$$

Substituting (2.21) and (2.22) into (2.72) gives the input impedance in the s-domain as

$$Z_i(s) = \left.\frac{v_i(s)}{i_i(s)}\right|_{d=i_o=0} = L_i\frac{s^2 + 2\xi\omega_o s + \omega_o^2}{s + \omega_{zi}} = R_{io}\frac{\left(\frac{s}{\omega_o}\right)^2 + \frac{2\xi s}{\omega_o} + 1}{1 + \frac{s}{\omega_{zi}}}, \tag{2.73}$$

where L_i is the high-frequency inductance and R_{io} is the DC input resistance. At high frequencies,

$$Z_i(\infty) = sL_i = \frac{L}{D^2}. \tag{2.74}$$

At DC,

$$R_{io} = Z_i(0) = \frac{R_L + r}{D^2}. \tag{2.75}$$

The angular frequency of the LHP zero ω_{zi} is given in (2.61).

Figure 2.13 shows the theoretically obtained magnitude and phase plots of input impedance. The gain at DC is nearly $R_{io} = 33.4\ \Omega$, and the high-frequency inductance is $L_i = 975\ \mu$H. The phase starts at zero and decreases toward $-90°$. However, the LHP zero at ω_{zi} increases the phase to 90°. The input resistance R_{io} at DC is proportional to the duty cycle and the load resistance. With an increase in frequency, the input impedance shows capacitive nature. Beyond f_o, the inductive reactance takes over, and the input impedance is mainly dominated by the inductive reactance. In open-loop circuit, the input current is in phase with the input voltage, i.e. a step change in the input voltage in positive direction, causing a corresponding change in the input current in the same direction.

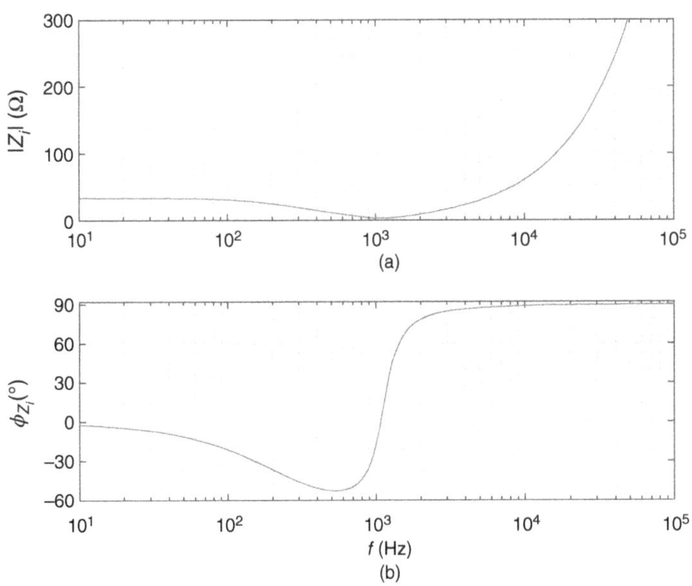

Figure 2.13 Theoretically obtained plots of input impedance Z_i. (a) Magnitude. (b) Phase.

2.3.7 Open-Loop Output Impedance Z_o

The small-signal model to derive the output impedance is shown in Figure 2.11. This model is obtained by setting $v_i = 0$ and $d = 0$. The open-loop output impedance is

$$Z_o(s) = \left.\frac{v_o(s)}{i_t(s)}\right|_{d=i_o=0} = Z_1||Z_2. \tag{2.76}$$

Substituting (2.21) and (2.22) into (2.76) gives the open-loop output impedance in the s domain as

$$Z_o(s) = \left.\frac{v_o(s)}{i_o(s)}\right|_{d=i_o=0} = R_{ox}\frac{(s+\omega_{zn})(s+\omega_{rl})}{s^2+2\xi\omega_o s+\omega_o^2} = R_{oo}\frac{\left(1+\frac{s}{\omega_{zn}}\right)\left(1+\frac{s}{\omega_{rl}}\right)}{\left(\frac{s}{\omega_o}\right)^2+\frac{2\xi s}{\omega_o}+1}, \tag{2.77}$$

where the high-frequency output resistance R_{ox} is

$$R_{ox} = Z_o(\infty) = \frac{R_L r_C}{R_L + r_C}, \tag{2.78}$$

the DC gain R_{oo} is

$$R_{oo} = Z_o(0) = \frac{rR_L}{r + R_L}, \tag{2.79}$$

and the angular frequency of the LHP zero ω_{rl} is

$$\omega_{rl} = \frac{r}{L}. \tag{2.80}$$

The angular frequency of the LHP zero ω_{zn} is given in (2.36). Figure 2.14 shows the theoretically obtained magnitude and phase plots of output impedance. The gain

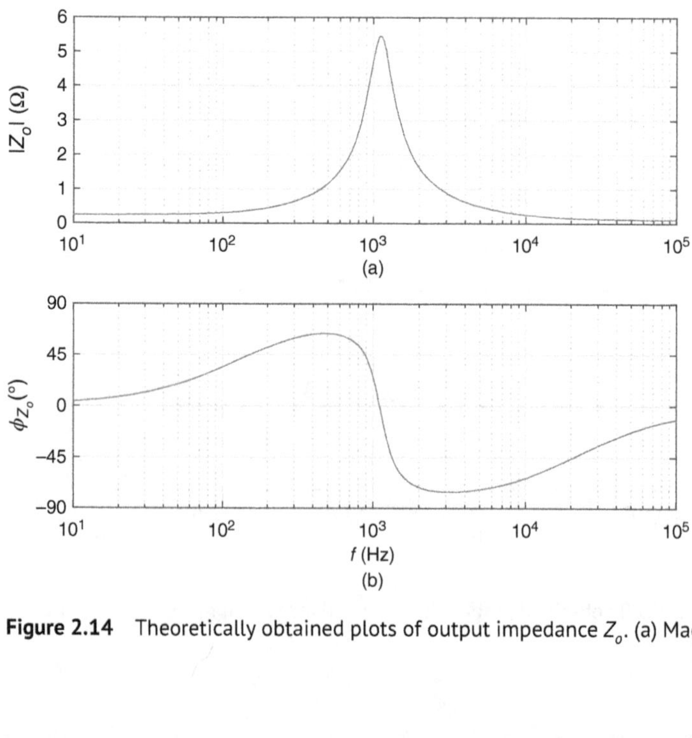

Figure 2.14 Theoretically obtained plots of output impedance Z_o. (a) Magnitude. (b) Phase.

Table 2.1 Summary of calculated values for open-loop transfer functions.

Variable	Value
$\|T_{po}\|$	27.2 V = 28.69 dBV
$\|T_{pio}\|$	2.8 A = 8.94 dBA
$\|M_{vo}\|$	0.54 V/V = −5.35 dB
$\|M_{vio}\|$	0.054 A/V = −25.35 dBA/V
$\|Z_{io}\|$	33.4 Ω
$\|Z_{oo}\|$	0.25 Ω
ω_o	7.04 krad/s
f_o	1.12 kHz
ξ	0.19
ω_{zi}	1.45 krad/s
f_{zi}	232 Hz
ω_{zn}	133.7 krad/s
f_{zn}	21.28 kHz
ω_{rl}	867 rad/s
f_{rl}	138 Hz

at DC is nearly $R_{oo} = 0.25\ \Omega$. The high-frequency output resistance is $R_{ox} = 0.11\ \Omega$. At DC, the inductor is a short circuit, and the filter capacitor is an open circuit. The DC resistance is due to the parallel combination of the load resistance R_L and the equivalent averaged resistance r. With the increase in frequency, the capacitive reactance is high, and the reactance is mostly offered by the inductive reactance. At a frequency f_o, the inductive and capacitive reactances are equal, and the converter filter network undergoes resonance. Beyond f_o, the capacitive reactance dominates and reduces the output impedance to nearly zero. The phase starts at zero and increases toward 90° due to the zero at frequency w_{rl}. The phase decreases toward −90° around the frequency of poles. However, the LHP zero at ω_{zn} increases the phase back to 0°. Table 2.1 provides a summary of the characteristics values for open-loop transfer functions.

2.4 Inner-Current Loop

The architecture of the inner-current loop of the buck converter with the low-pass filter in the current loop is shown in Figure 2.15. Figure 2.16 shows the circuit of the buck converter with the low-pass filter in the current loop. The current through inductor branch is sensed using resistor R_s. The analysis in this chapter assumes $R_s = 1\ \Omega$. The low-pass filter attenuates the switching frequency components and

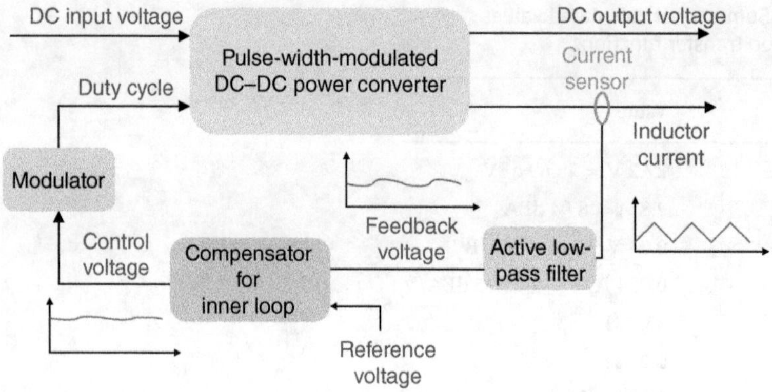

Figure 2.15 Architecture of the inner-current loop with low-pass filter in the feedback path.

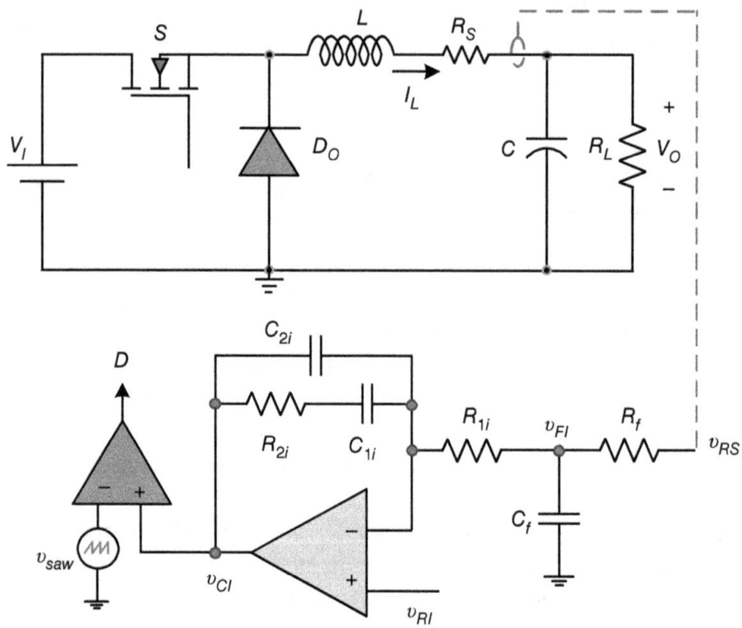

Figure 2.16 Circuit of the buck DC–DC converter with inner-current loop.

its harmonics, which are present in the sensed voltage $v_{RS} = R_s i_L$. Therefore, the theoretically obtained output voltage of the low-pass filter is $V_{FI} = R_s I_L$ and is governed by the DC component of the sensed inductor current. Thus, in this method, the *average* component of the inductor current is sensed, thereby reducing the problems due to the subharmonic instability and transistor misgating. The transfer functions related to the inner-current loop, which include filter circuit in the feedback path, are derived and discussed in this section.

2.4.1 Transfer Function of Filter and Non-inverting Amplifier T_f

The low-pass filter is composed of a resistor R_f and a capacitor C_f. The transfer function of two stages is

$$T_f(s) = \frac{v_{fi}(s)}{v_{rs}(s)} = \frac{Z_{Cf}}{R_f + Z_{Cf}} = \frac{\frac{1}{sC_f}}{R_f + \frac{1}{sC_f}} = \frac{1}{sR_fC_f + 1}$$
$$= \frac{1}{R_fC_f}\frac{1}{\frac{1}{R_fC_f} + s} = \frac{\omega_{pf}}{\omega_{pf} + s} = \frac{1}{1 + \frac{s}{\omega_{pf}}}. \tag{2.81}$$

Setting $s = j\omega$,

$$T_f(j\omega) = \frac{1}{1 + j\frac{\omega}{\omega_{pf}}}, \tag{2.82}$$

yielding

$$T_f(f) = \frac{1}{1 + j\frac{f}{f_{pf}}} = |T_f(f)|e^{j\phi(f)}, \tag{2.83}$$

where

$$|T_f(f)| = \frac{1}{\sqrt{1 + \left(\frac{f}{f_{pf}}\right)^2}} \tag{2.84}$$

and

$$\phi(f) = -\tan^{-1}\left(\frac{f}{f_{pf}}\right). \tag{2.85}$$

Using (2.84),

$$|T_f(f)|^2 = \frac{1}{1 + \left(\frac{f}{f_{pf}}\right)^2} \tag{2.86}$$

yields

$$\left(\frac{f}{f_{pf}}\right)^2 = \frac{1}{|T_f(f)|^2} - 1 = \frac{1 - |T_f(f)|^2}{|T_f(f)|^2} \tag{2.87}$$

and

$$\frac{f_{pf}}{f} = \sqrt{\frac{|T_f(f)|^2}{1 - |T_f(f)|^2}}. \tag{2.88}$$

Hence,

$$f_{pf} = f\sqrt{\frac{|T_f(f)|^2}{1 - |T_f(f)|^2}}. \tag{2.89}$$

The magnitude of the transfer function T_f in terms of the filter input and output voltages is also given as

$$|T_f(f)| = \frac{\Delta V_{FI}}{\Delta V_{RS}}. \tag{2.90}$$

The filter frequency can be set by the amount of attenuation desired. As an example, to achieve an attenuation of 50% for $f = f_s = 100$ kHz, the filter upper cutoff frequency can be calculated using (2.89) as

$$f_{pf} = f_s\sqrt{\frac{|T_f(f)|^2}{1-|T_f(f)|^2}} = 100\times 10^3 \times \sqrt{\frac{0.50^2}{1-0.50^2}} = 57 \text{ kHz}. \tag{2.91}$$

Assuming $C_f = 1$ nF, the value of the filter capacitor can be calculated as $R_f = 1/(2\pi f_{pf} C_f) \approx 2.8$ kΩ. A 3.2 kΩ/0.25 W/5% resistor is selected.

2.4.2 Transfer Function of Pulse-Width Modulator T_m

The control voltage-to-duty cycle transfer function of the pulse-width modulator is

$$T_m = \frac{D}{V_{CI}} = \frac{d}{v_{ci}} = \frac{1}{V_{Tm}}, \tag{2.92}$$

where V_{Tm} is the amplitude of the sawtooth waveform.

The steady-state average inductor current is $I_L = V_O/R_L = 1.4$ A. At a duty cycle $D = 0.556$ and assuming the peak sawtooth voltage of the pulse-width modulator is $V_{Tm} = 2.5$ V, the required steady-state control voltage is $V_{CI} = DV_{Tm} = 0.556 \times 2.5 = 1.4$ V. The reference voltage to the current loop is $V_{RI} = V_{CI}$ at steady-state since the error voltage is zero, i.e. $V_{RI} = 1.4$ V. The analysis in this chapter assumes $R_s = 1$ Ω. From Section 2.4.1, $R_f = 3.2$ kΩ and $C_f = 1$ nF.

2.4.3 Uncompensated Loop Gain T_{ki}

The natural behavior of the inner-current loop can be determined using the uncompensated loop gain as

$$T_{ki} = \frac{v_{fi}}{v_{ei}} = T_m T_{pi} R_s T_f = T_{kio}\frac{\left(1+\frac{s}{\omega_{zi}}\right)}{\left(1+\frac{s}{\omega_{pf}}\right)\left(1+\frac{2\xi s}{\omega_o}+\frac{s^2}{\omega_o^2}\right)}, \tag{2.93}$$

where T_m, T_{pi}, and T_f are given in (2.92), (2.45), and (2.81), respectively. The value of sense resistor is $R_s = 1$ Ω. The DC gain T_{kio} is given by

$$T_{kio} = \frac{R_s}{V_{Tm}}\frac{V_I}{R_L + r}. \tag{2.94}$$

Figure 2.17 shows the theoretically obtained magnitude and phase plots of the uncompensated loop gain of the inner-current loop T_{ki}. The gain at DC is $T_{kio} = 1.08 = 0.67$ dB. A high value of DC gain is required to reduce the steady-state error in the output voltage due to the change in the input voltage or load current. The control circuit must be designed to provide high gain at DC and to boost the phase at crossover frequency of the loop gain. An integral single-lead control circuit is designed for the inner-current loop in Section 2.4.4.

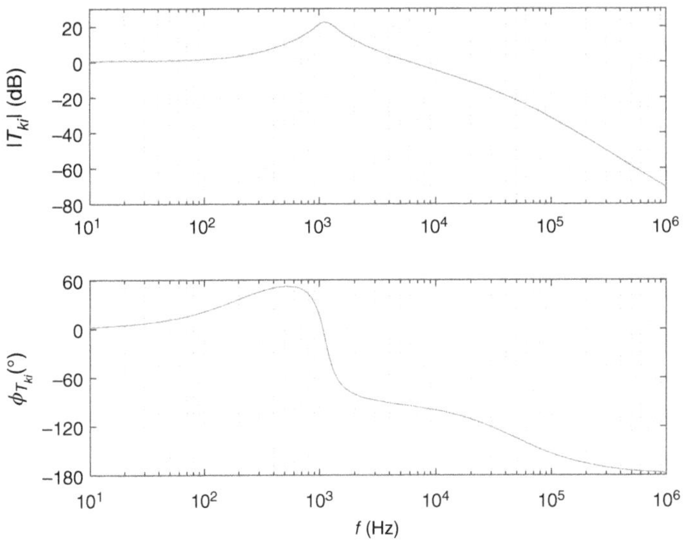

Figure 2.17 Theoretically obtained plots of uncompensated loop gain T_{ki}. (a) Magnitude. (b) Phase.

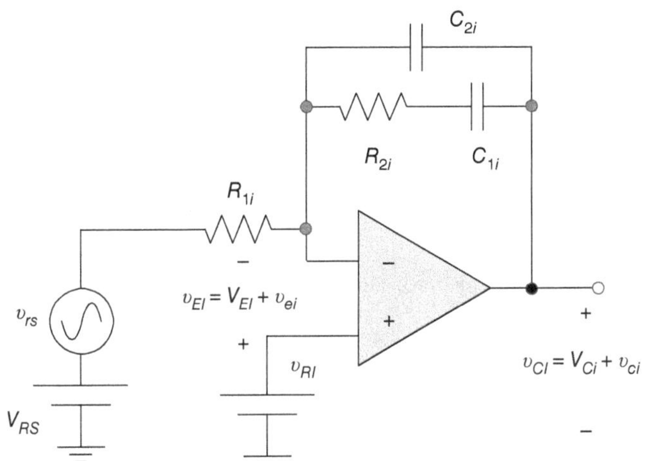

Figure 2.18 Circuit of Type-II controller.

2.4.4 Transfer Function of Control Circuit for Inner-Current Loop T_{ci}

Figure 2.18 shows circuit of a single-lead integral controller, which is also called a Type-II controller. A single-lead integral controller introduces a pole at the origin required to boost the DC gain and a pole–zero pair to adjust crossover frequency f_c to maintain required phase margin (PM). A detailed procedure to determine the transfer function of the controller to improve the DC gain (> 20 dB) and to achieve an inner-current loop phase margin $PM = 60°$ is also explained in [45, 51]. The remaining section details the Type-II controller design for the inner-current loop.

The magnitude of the uncompensated loop gain transfer function can be determined using

$$|T_{ki}(f)| = T_{kio}\frac{\sqrt{1+\left(\frac{f}{f_{zi}}\right)^2}}{\sqrt{1+\left(\frac{f}{f_{pf}}\right)^2}\sqrt{1-\left(\frac{f}{f_o}\right)^2+(4\xi^2\left(\frac{f}{f_o}\right)^2}}, \tag{2.95}$$

and the phase is

$$\phi_{Tk}(f) = -180° + \tan^{-1}\left(\frac{f}{f_{zi}}\right) - \tan^{-1}\left(\frac{f}{f_{pf}}\right) - \tan^{-1}\frac{2\xi\frac{f}{f_o}}{1-\left(\frac{f}{f_o}\right)^2}. \tag{2.96}$$

The desired crossover frequency for the loop gain is chosen as $f_c = 10$ kHz. At this frequency, the magnitude of the uncompensated loop gain transfer function is $|T_{ki}(f_c)| = 0.583 = -4.68$ dB, and its phase is $\phi_{Tk}(f_c) = -100°$. The required phase margin is $PM = 60°$. Therefore, the phase boost required at f_c is

$$\phi_m = PM - \phi_{Tk}(f_c) - 90° = 70°, \tag{2.97}$$

yielding the geometric mean between the controller corner frequencies as [47]

$$K = \tan\left(\frac{\phi_m}{2} + 45°\right) = 5.7. \tag{2.98}$$

Assuming $R_{1i} = 1$ kΩ/0.25 W/2%, the value of C_{2i} is

$$C_{2i} = \frac{|T_k(f_c)|}{2\pi f_c K R_{1i}} = 1.6 \text{ nF}. \tag{2.99}$$

A 1.6 nF/25 V standard capacitor is selected. The expression for C_{1i} is

$$C_{1i} = C_{2i}(K^2 - 1) = 51 \text{ nF}. \tag{2.100}$$

A 51 nF/25 V standard capacitor is selected. The expression for R_{2i} is

$$R_{2i} = \frac{K}{2\pi f_c C_{1i}} = 1.8 \text{ k}\Omega. \tag{2.101}$$

A 1.8 kΩ/0.25 W/2% standard resistor is selected. The transfer function T_{ci} of the current-loop control circuit is

$$T_{ci}(s) = \frac{v_{ci}(s)}{v_{ei}(s)} = T_{cio}\frac{1+\frac{s}{\omega_{zci}}}{s\left(1+\frac{s}{\omega_{pci}}\right)}, \tag{2.102}$$

where

$$T_{cio} = \frac{1}{R_{1i}(C_{1i}+C_{2i})} = 19 \text{ kV/V}, \tag{2.103}$$

$$\omega_{zci} = \frac{1}{R_{2i}C_{1i}} = 11 \text{ krad/s}, \tag{2.104}$$

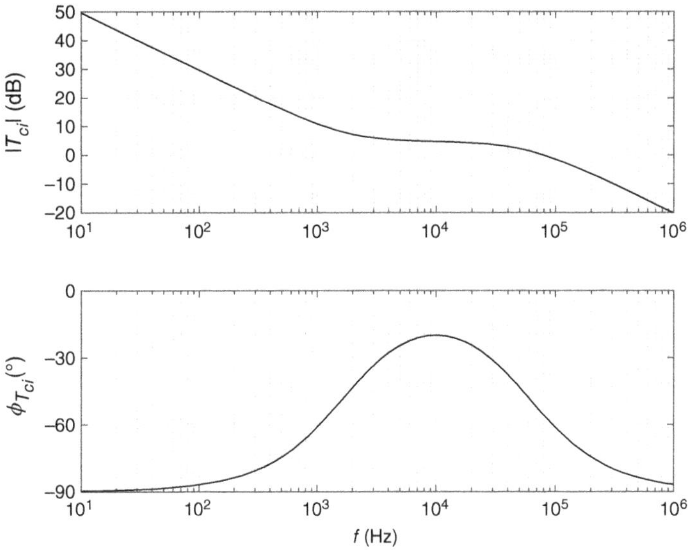

Figure 2.19 Theoretically obtained plots of controller transfer function T_{ci}. (a) Magnitude. (b) Phase.

and

$$\omega_{pci} = \frac{C_{1i} + C_{2i}}{R_{2i}C_{1i}C_{2i}} = 358 \text{ krad/s}. \tag{2.105}$$

Figure 2.19 shows the theoretically obtained magnitude and phase plots of the designed Type-II controller transfer function used in the inner-current loop. The control transfer function T_{ci} is incorporated into the critical path to determine the inner-loop gain and is discussed in Section 2.4.5.

2.4.5 Compensated Loop Gain of Inner-Current Loop T_i

The loop gain of the compensated inner-current loop is

$$T_i = \frac{v_{ei}}{v_{fi}} = T_{ki}T_{ci} = T_m T_{pi} R_s T_f T_{ci}, \tag{2.106}$$

to yield

$$T_i = T_{io}\frac{\left(1+\frac{s}{\omega_{zi}}\right)\left(1+\frac{s}{\omega_{zci}}\right)}{s\left(1+\frac{s}{\omega_{pf}}\right)\left(1+\frac{s}{\omega_{pci}}\right)\left(1+\frac{2\xi s}{\omega_o}+\frac{s^2}{\omega_o^2}\right)}, \tag{2.107}$$

where T_{io} is the gain at $s = 0$ given as

$$T_{io} = T_{kio}T_{cio} = \frac{V_I R_s}{V_{Tm}(R_L + r)}\frac{1}{R_{1i}(C_{1i} + C_{2i})}. \tag{2.108}$$

Figure 2.21 shows the theoretically obtained magnitude and phase plots of compensated loop gain of the inner-current loop. Using the controller designed

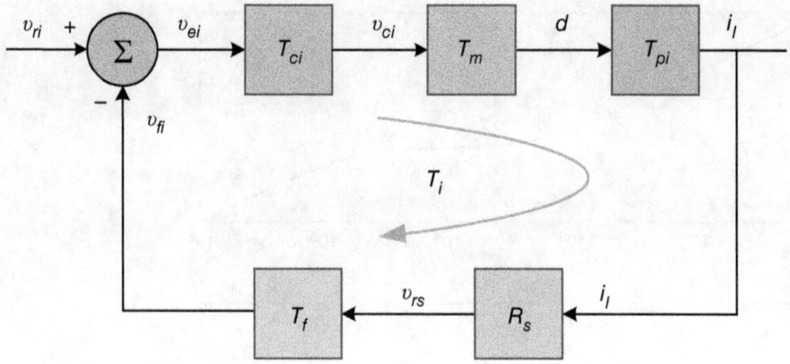

Figure 2.20 Block diagram of inner-current loop.

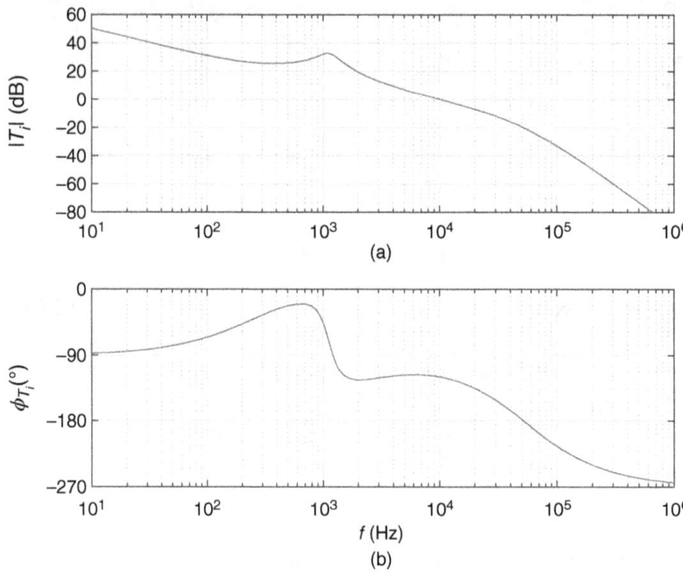

Figure 2.21 Theoretically obtained plots of compensated loop gain T_i of the inner current loop. (a) Magnitude. (b) Phase.

in Section 2.4.4, the calculated value of compensated loop gain at DC is $T_{io} > 50$ dB. The phase margin is $PM = 60°$ and the gain margin is 20 dB. The closed-loop transfer functions for the inner-current loop are derived and discussed in Section 2.5.

2.5 Closed-Loop Transfer Functions for Inner-Current Loop

The following closed-loop control transfer functions are derived:

- Reference voltage-to-inductor current transfer function T_{icl} (control).
- Reference voltage-to-output voltage transfer function T_{picl} (control).

- Input voltage-to-inductor current transfer function M_{icl} (disturbance).
- Input voltage-to-output voltage transfer function M_{vicl} (disturbance).
- Input Impedance Z_{iicl} (disturbance).
- Output Impedance Z_{oicl} (disturbance).

The small-signal current reference v_{ri} for the inner loop is set by the outer loop. This section discusses only the inner-current loop without the presence of outer-voltage loop. The inner-current loop in the presence of outer-voltage loop will be discussed in Section 2.6. Figure 2.20 shows the block diagram required to characterize the inner-current loop and derive the relevant transfer functions.

2.5.1 Reference Voltage-to-Inductor Current Transfer Function T_{icl}

Using the block diagram shown in Figure 2.22, the closed-loop reference voltage-to-inductor current transfer function is

$$T_{icl}(s) = \left.\frac{i_l(s)}{v_{ri}(s)}\right|_{v_i=i_o=0} = \frac{T_{ci}T_mT_{pi}}{1+T_i}. \tag{2.109}$$

Figure 2.23 shows the theoretically obtained magnitude and phase plots of the inner-current loop reference voltage-to-inductor current transfer function.

2.5.2 Reference Voltage-to-Output Voltage Transfer Function T_{picl}

Using the block diagram shown in Figure 2.24, the closed-loop reference voltage-to-output voltage transfer function is

$$T_{picl}(s) = \left.\frac{v_o(s)}{v_{ri}(s)}\right|_{v_i=i_o=0} = \frac{T_{ci}T_mT_{pi}T_z}{1+T_i}, \tag{2.110}$$

where T_z is the inductor current-to-output voltage transfer function given as

$$T_z(s) = \left.\frac{v_o(s)}{i_l(s)}\right|_{v_i=d=0} = Z_2 = R_L||\left(r_C+\frac{1}{sC}\right) = \frac{R_L\left(r_C+\frac{1}{sC}\right)}{R_L+r_C+\frac{1}{sC}}. \tag{2.111}$$

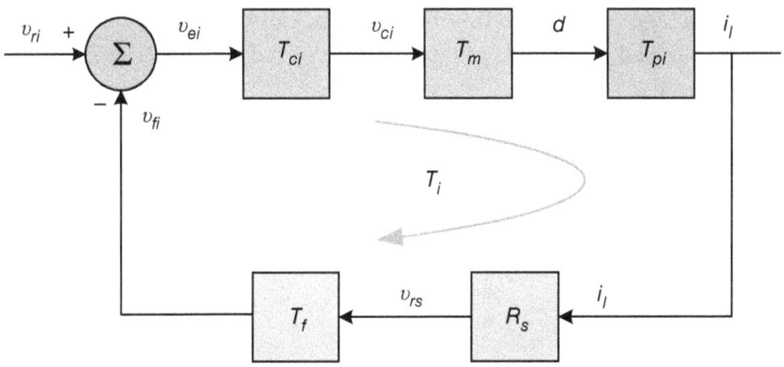

Figure 2.22 Block diagram used to derive the inner-loop control-to-inductor current transfer function T_{icl}.

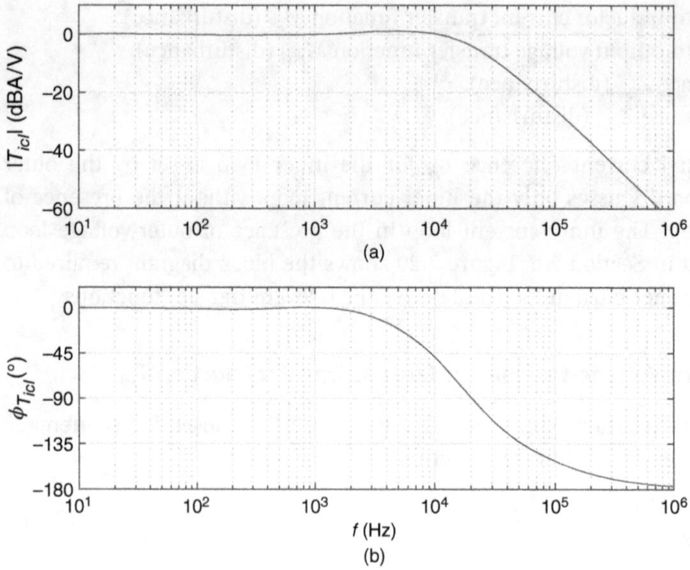

Figure 2.23 Theoretically obtained plots of inner-current loop reference voltage-to-inductor current transfer function T_{icl}. (a) Magnitude. (b) Phase.

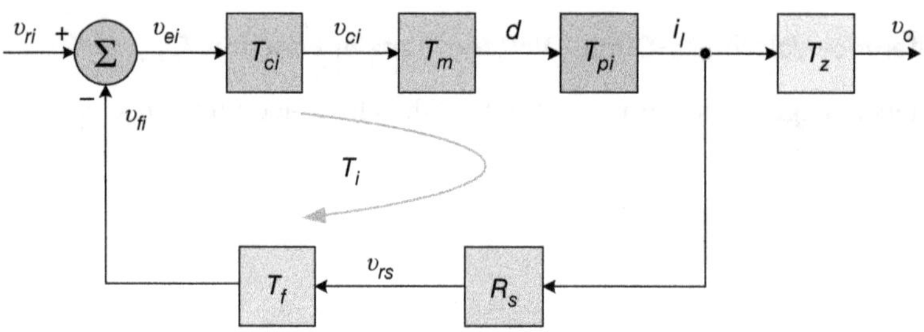

Figure 2.24 Block diagram required to derive the inner-loop control-to-output voltage transfer function T_{picl}.

Figure 2.24 shows the critical path between the reference voltage and output voltage. The output voltage is controlled by the inductor current. Figure 2.25 shows the theoretically obtained magnitude and phase plots of the inner-current-loop reference voltage-to-output voltage transfer function.

2.5.3 Input Voltage-to-Inductor Current Transfer Function M_{icl}

Using the block diagram shown in Figure 2.26, the closed-loop input voltage to inductor current transfer function is

$$M_{icl}(s) = \left.\frac{i_l(s)}{v_i(s)}\right|_{v_{ri}=i_o=0}. \tag{2.112}$$

From the block diagram,

$$i_l = i_l' + i_l'' = -T_{ci}T_mT_{pi}T_fR_si_l + M_{vi}v_i = -T_ii_l + M_{vi}v_i. \tag{2.113}$$

$$i_l(1+T_i) = M_{vi}v_i. \tag{2.114}$$

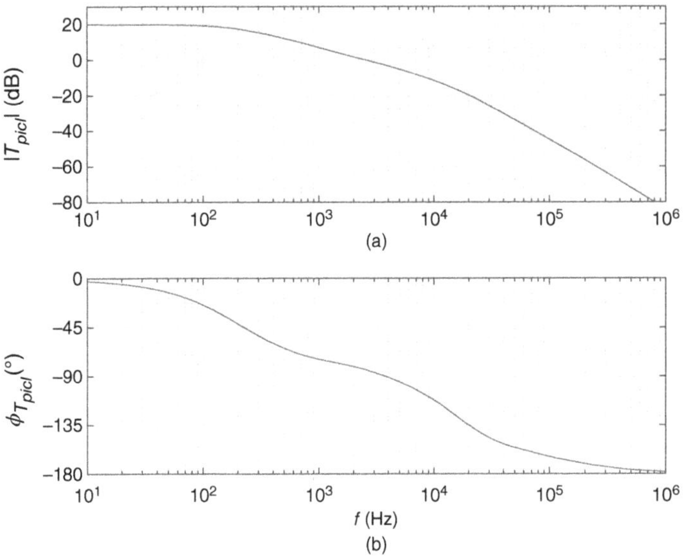

Figure 2.25 Theoretically obtained plots of reference voltage-to-output voltage transfer function T_{picl}. (a) Magnitude. (b) Phase.

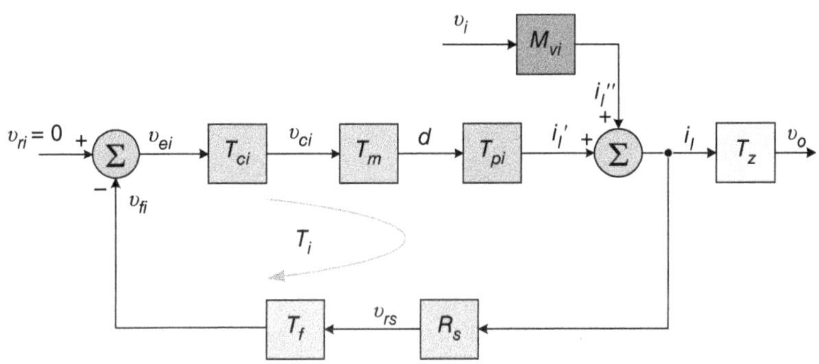

Figure 2.26 Block diagram required to derive the inner-loop input voltage-to-inductor current transfer function M_{icl} and inner-loop input voltage-to-output voltage transfer function M_{vicl}.

Hence,

$$M_{icl}(s) = \frac{i_l(s)}{v_i(s)} = \frac{M_{vi}}{1 + T_i}. \tag{2.115}$$

Figure 2.27 shows the theoretically obtained magnitude and phase plots of the inner-current-loop input voltage-to-inductor current transfer function.

2.5.4 Input Voltage-to-Output Voltage Transfer Function M_{vicl}

Using the block diagram shown in Figure 2.26, the closed-loop input voltage-to-output voltage transfer function is

$$M_{vicl}(s) = \left.\frac{v_o(s)}{v_i(s)}\right|_{v_{ri}=i_o=0}. \tag{2.116}$$

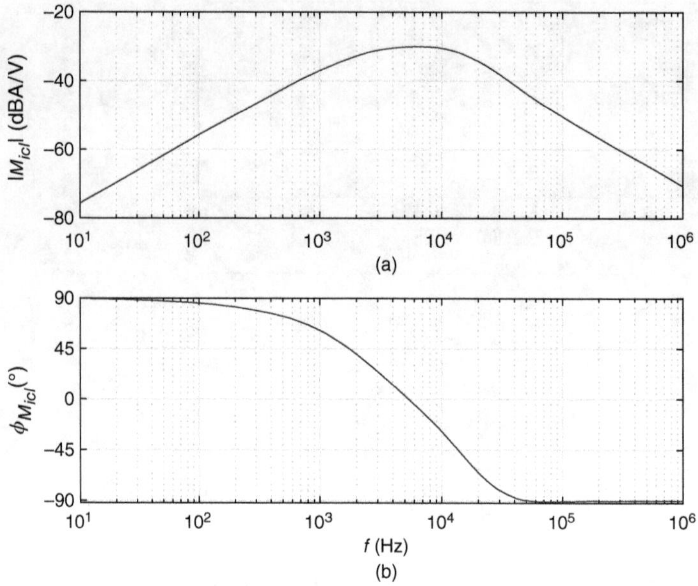

Figure 2.27 Theoretically obtained plots of input voltage-to-inductor current transfer function M_{icl}. (a) Magnitude. (b) Phase.

From (2.115), the closed-loop input-to-output transfer function is

$$M_{vicl}(s) = T_z M_{icl} = T_z \frac{M_{vi}}{1 + T_i}. \tag{2.117}$$

Figure 2.28 shows the theoretically obtained magnitude and phase plots of the inner-current loop input voltage-to-output voltage transfer function.

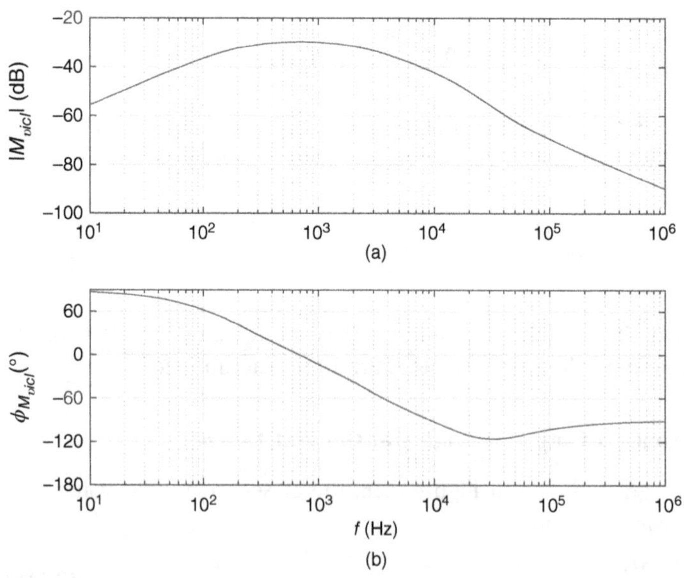

Figure 2.28 Theoretically obtained plots of input voltage-to-output voltage transfer function M_{vicl}. (a) Magnitude. (b) Phase.

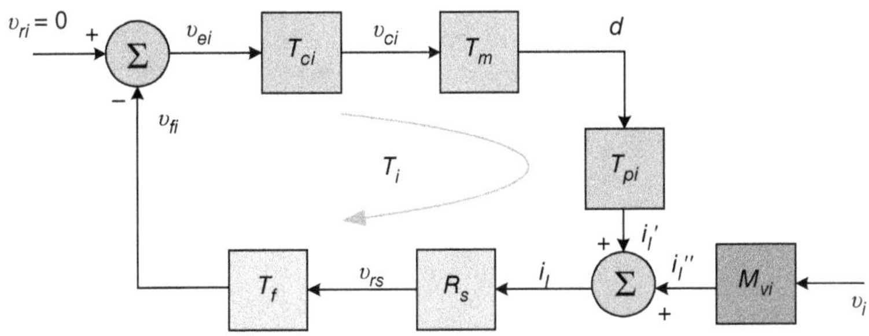

Figure 2.29 Block diagram used to derive the closed-inner-loop input impedance Z_{iicl}.

2.5.5 Input Impedance Z_{iicl}

Figure 2.29 shows the block diagram used to derive the closed-inner-loop input impedance Z_{iicl}, which is

$$Z_{iicl}(s) = \left.\frac{v_i(s)}{i_i(s)}\right|_{i_o=d=0}. \tag{2.118}$$

From Figure 2.29, the inductor current is

$$i_l = i_l' + i_l'' = T_{pi}d + M_{vi}v_i, \tag{2.119}$$

and the duty cycle is

$$d = T_{ci}T_m v_{ei} = T_{ci}T_m(-T_f R_s i_l) = -T_{ci}T_m T_f R_s i_l. \tag{2.120}$$

Substituting (2.120) into (2.119),

$$i_l = -T_{pi}T_{ci}T_m T_f R_s i_l + M_{vi}v_i = -T_i i_l + M_{vi}v_i. \tag{2.121}$$

Rearranging (2.121),

$$i_l = \frac{M_{vi}}{1+T_i}v_i. \tag{2.122}$$

From (2.119),

$$d = \frac{i_l}{T_{pi}} - \frac{M_{vi}v_i}{T_{pi}}. \tag{2.123}$$

The input current is given as

$$\begin{aligned} i_i &= Di_l + I_L d = Di_l + I_L\left(\frac{i_l}{T_{pi}} - \frac{M_{vi}v_i}{T_{pi}}\right) \\ &= \left(D + \frac{I_L}{T_{pi}}\right)i_l - \frac{M_{vi}I_L}{T_{pi}}v_i \\ &= \left[\left(D + \frac{I_L}{T_{pi}}\right)\frac{M_{vi}}{1+T_i} - \frac{M_{vi}I_L}{T_{pi}}\right]v_i. \end{aligned} \tag{2.124}$$

Hence, the closed-inner-loop input impedance is

$$Z_{iicl}(s) = \left.\frac{v_i(s)}{i_i(s)}\right|_{i_o=d=0} = \frac{T_{pi}(1+T_i)}{M_{vi}(DT_{pi} - I_L T_i)}. \tag{2.125}$$

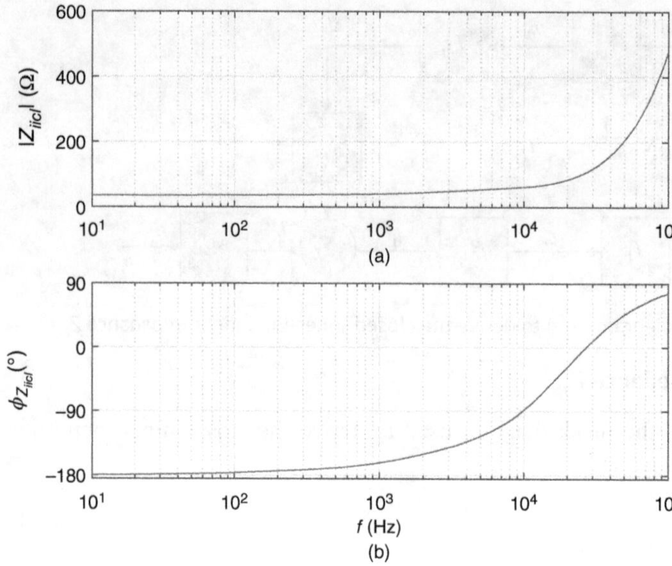

Figure 2.30 Theoretically obtained plots of closed-inner-loop input impedance Z_{iicl}. (a) Magnitude. (b) Phase.

The closed-loop input admittance is

$$Y_{iicl}(s) = \frac{i_i(s)}{v_i(s)} = \frac{1}{Z_{iicl}(s)}. \tag{2.126}$$

Figure 2.30 shows the theoretically obtained magnitude and phase plots of the closed-inner-current-loop input impedance.

2.5.6 Output Impedance Z_{oicl}

Using the block diagram shown in Figure 2.31, the closed-loop output impedance is

$$Z_{oicl}(s) = \left.\frac{v_o(s)}{i_o(s)}\right|_{v_{ri}=v_i=0}. \tag{2.127}$$

From the block diagram,

$$i_l = i_l' + i_l'' = T_{pi}d + A_i i_o \tag{2.128}$$

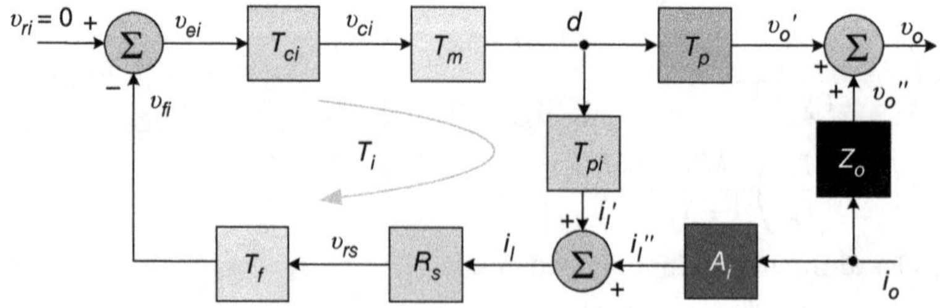

Figure 2.31 Block diagram required to derive the closed-inner-loop output impedance Z_{oicl}.

and

$$\frac{-d}{T_{ix}} = T_{pi}d + A_i i_o. \tag{2.129}$$

Rearranging (2.129),

$$-d\left(\frac{1}{T_{ix}} + T_{pi}\right) = A_i i_o \tag{2.130}$$

and

$$d = -\frac{A_i T_{ix}}{1 + T_i} i_o. \tag{2.131}$$

From the block diagram in Figure (2.31),

$$v_o = v_o' + v_o'' = T_p d + Z_o i_o. \tag{2.132}$$

Substituting d from (2.131) into (2.132), we get

$$v_o = T_p\left(-\frac{A_i T_{ix}}{1 + T_i} i_o\right) + Z_o i_o = \left(Z_o - \frac{A_i T_p T_{ix}}{1 + T_i}\right) i_o. \tag{2.133}$$

The closed-inner-current-loop output impedance is

$$Z_{oicl}(s) = \frac{v_o(s)}{i_o(s)} = Z_o - \frac{A_i T_p T_{ix}}{1 + T_i}. \tag{2.134}$$

Figure 2.32 shows the theoretically obtained magnitude and phase plots of the closed-inner-current-loop output impedance.

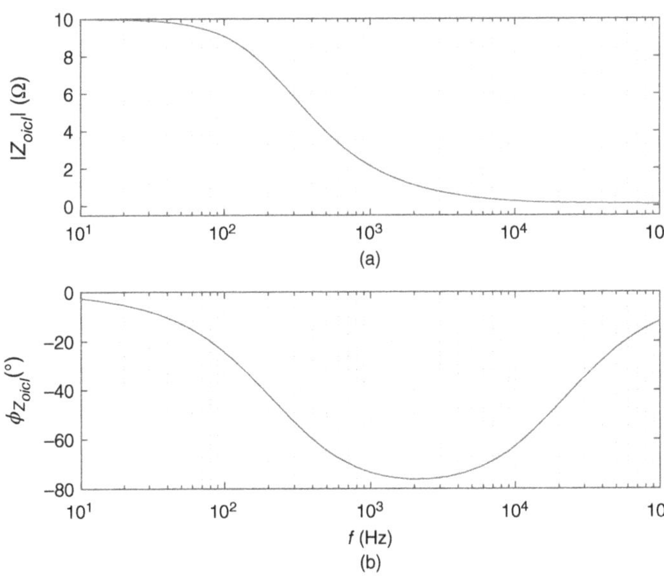

Figure 2.32 Theoretically obtained plots of closed-inner-loop output impedance Z_{oicl}. (a) Magnitude. (b) Phase.

2.6 Outer-Voltage Loop

To maintain the output voltage constant, it is essential to introduce the outer-voltage loop set with a reference voltage V_{RV} that regulates the output voltage. In a typical two-loop system, the reference voltage to the outer loop is a fraction of the rated output voltage. The fraction is determined by the feedback network, whose voltage gain β is determined by a network of resistors. Figure 2.33 shows the architecture of average current-mode controlled buck converter with outer-voltage loop. Figure 2.34 shows the circuit of the average current-mode controlled buck converter with outer-voltage loop.

2.6.1 Transfer Function of Feedback Network β

The transfer function of the feedback network is given as

$$\beta = \frac{V_{RV}}{V_O}, \tag{2.135}$$

where V_{RV} is the reference voltage to the outer-voltage loop. Assuming the reference voltage V_{RV} = 2.5 V, the transfer function of the feedback network is

$$\beta = \frac{V_{RV}}{V_O} = \frac{2.5}{1.4} = 0.179. \tag{2.136}$$

2.6.2 Uncompensated Loop Gain for Outer-Voltage Loop T_{kv}

The natural behavior of the outer-voltage loop can be determined using the uncompensated loop gain as

$$T_{kv} = \frac{v_{fv}}{v_{ev}} = \frac{\beta v_o}{v_{ri}} = \beta T_{picl}. \tag{2.137}$$

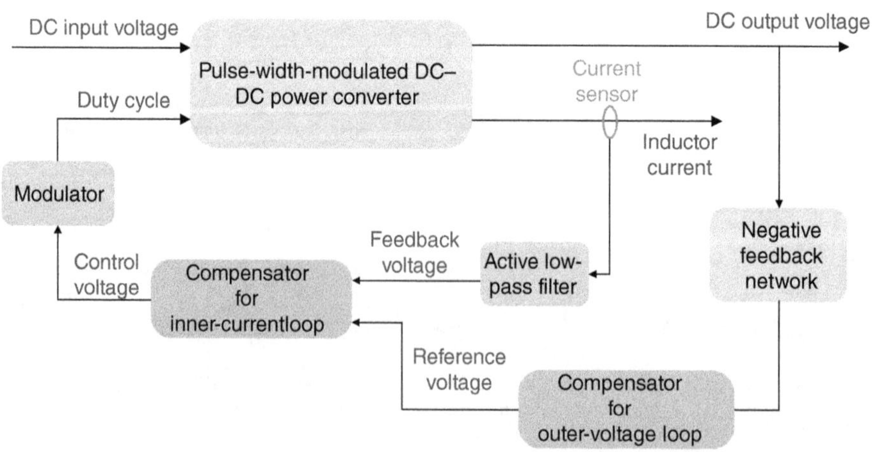

Figure 2.33 Architecture of average current-mode controlled buck converter with outer-voltage loop.

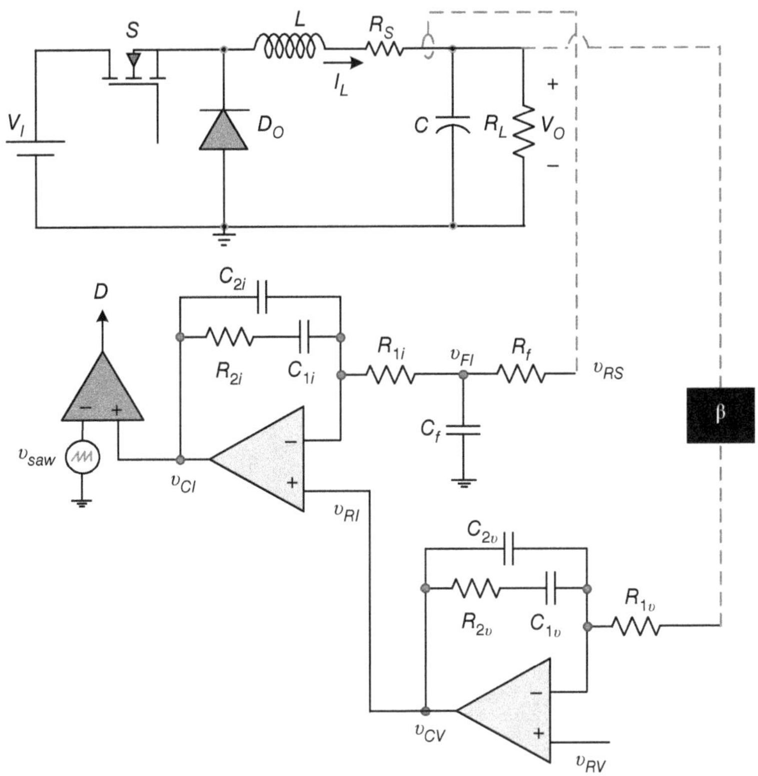

Figure 2.34 Circuit of the average current-mode controlled buck converter with outer-voltage loop.

Figure 2.35 shows the theoretically obtained magnitude and phase plots of the uncompensated loop gain of the outer-voltage loop.

2.6.3 Transfer Function of Control Circuit for Outer-Voltage Loop T_{cv}

Figure 2.36 shows the circuit of a Type-II controller used in outer-voltage loop. The Type-II controller introduces a pole at the origin, which is required to boost the DC gain and a pole–zero pair to adjust the crossover frequency f_c, thus maintaining the required PM. A detailed procedure to determine the transfer function of the controller to improve the DC gain (> 20 dB) and achieve an inner-current loop phase margin $PM = 60°$ is explained in [45].

The desired crossover frequency for the loop gain is chosen as $f_c = 2$ kHz. At this frequency, the magnitude of the uncompensated loop gain transfer function is $|T_{kv}(f_c)| = 0.22$ V/V $= -13.1$ dB and its phase is $\phi_{Tkv}(f_c) = -81°$. The required phase margin is $PM = 60°$. Therefore, the phase boost required at f_c is

$$\phi_m = PM - \phi_{Tkv}(f_c) - 90° = 51°, \tag{2.138}$$

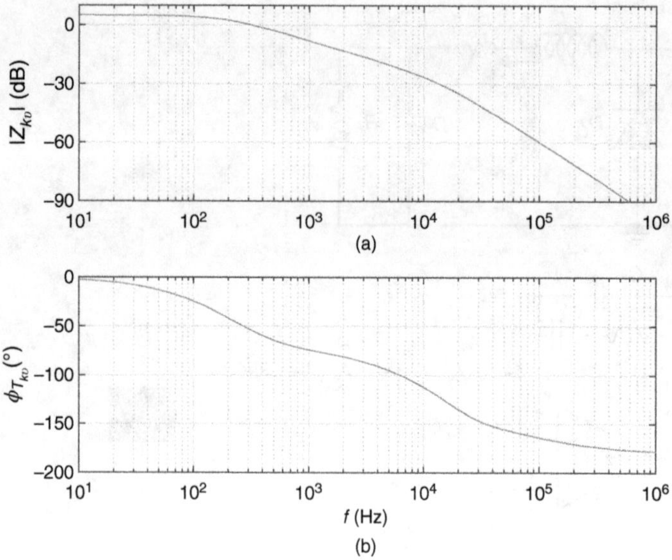

Figure 2.35 Theoretically obtained plots of uncompensated loop gain T_{kv}. (a) Magnitude. (b) Phase.

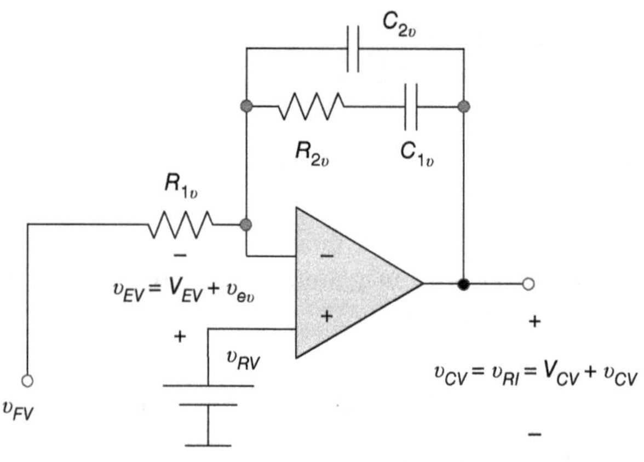

Figure 2.36 Circuit of Type-II controller used in outer-voltage loop.

giving the geometric mean between the controller corner frequencies as

$$K = \tan\left(\frac{\phi_m}{2} + 45^\circ\right) = 2.83. \tag{2.139}$$

Assuming $R_{1v} = 1$ kΩ/0.25 W/2%, the value of C_{2v} is

$$C_{2v} = \frac{|T_{kv}(f_c)|}{2\pi f_c K R_{1v}} = 6.2 \text{ nF}. \tag{2.140}$$

A 6.2 nF/25 V standard capacitor is selected. The expression for C_{1v} is

$$C_{1v} = C_{2v}(K^2 - 1) = 42 \text{ nF}. \tag{2.141}$$

A 43 nF/50 V standard capacitor is selected. The expression for R_{2v} is

$$R_{2v} = \frac{K}{2\pi f_c C_{1v}} = 5.2\ \text{k}\Omega. \tag{2.142}$$

A 5.2 kΩ/0.5 W/5% standard resistor is selected. The transfer function T_{cv} of the voltage-loop control circuit is

$$T_{cv}(s) = \frac{v_{cv}(s)}{v_{ev}(s)} = T_{cvo}\frac{1 + \dfrac{s}{\omega_{zcv}}}{s\left(1 + \dfrac{s}{\omega_{pcv}}\right)}, \tag{2.143}$$

where

$$T_{cvo} = \frac{1}{R_{1v}(C_{1v} + C_{2v})} = 20.7\ \text{kV/V}, \tag{2.144}$$

$$\omega_{zcv} = \frac{1}{R_2 C_1} = 4.4\ \text{krad/s}, \tag{2.145}$$

and

$$\omega_{pcv} = \frac{C_{1v} + C_{2v}}{R_{2v} C_{1v} C_{2v}} = 36\ \text{krad/s}. \tag{2.146}$$

Figure 2.37 shows the theoretically obtained magnitude and phase plots of the designed Type-II controller transfer function used in the outer-voltage loop. The control transfer function T_{cv} is incorporated into the critical path to determine the outer-voltage loop gain and is discussed in Section 2.6.4.

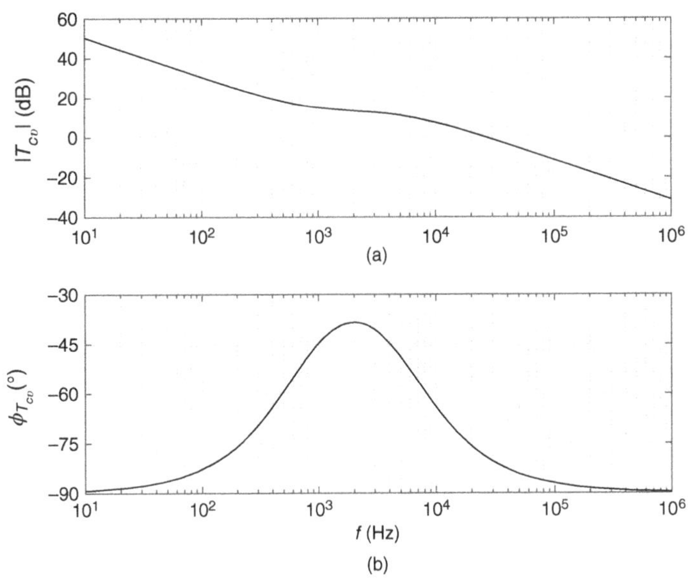

Figure 2.37 Theoretically obtained plots of controller transfer function T_{cv} of the outer-voltage loop. (a) Magnitude. (b) Phase.

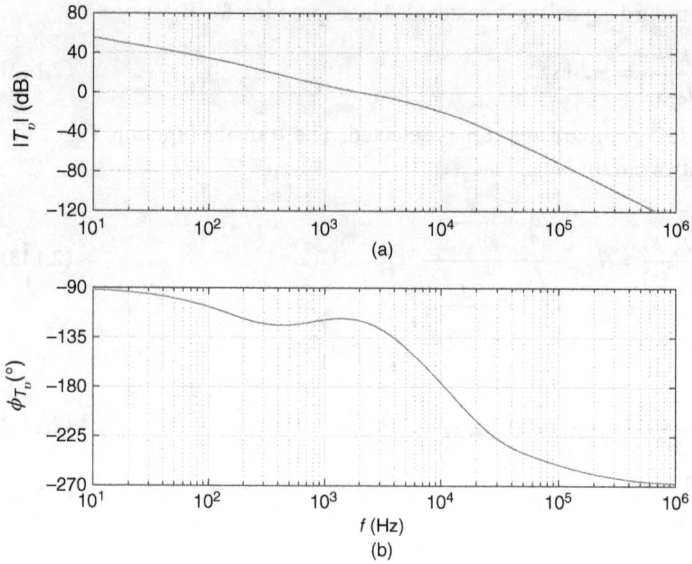

Figure 2.38 Theoretically obtained plots of loop gain T_v of the compensated outer-voltage loop. (a) Magnitude. (b) Phase.

2.6.4 Compensated Loop Gain of Outer-Voltage Loop T_v

The loop gain of the compensated inner-current loop is

$$T_v = \frac{v_{ev}}{v_{fv}} = T_{kv}T_{cv} = \beta T_{picl}T_{cv}. \tag{2.147}$$

Figure 2.38 shows the theoretically obtained magnitude and phase plots of the loop gain of the compensated outer-voltage loop.

2.7 Closed-Loop Transfer Functions for Outer-Voltage Loop

The following closed-loop transfer functions for the outer-voltage loop in the presence of the inner-current loop are derived:

- Reference voltage-to-output voltage transfer function T_{pcl} (control).
- Input voltage-to-duty-cycle transfer function M_{dv} (disturbance).
- Input voltage-to-output voltage transfer function M_{vcl} (disturbance).
- Input impedance Z_{ivcl} (disturbance).
- Output impedance Z_{ovcl} (disturbance).

2.7.1 Reference Voltage-to-Output Voltage Transfer Function T_{pcl}

Using the block diagram shown in Figure 2.39, the closed-loop reference voltage-to-output voltage transfer function is

$$T_{pcl}(s) = \left.\frac{v_o(s)}{v_{rv}(s)}\right|_{v_i=i_o=0} = \frac{T_{cv}T_{picl}}{1+T_v}. \tag{2.148}$$

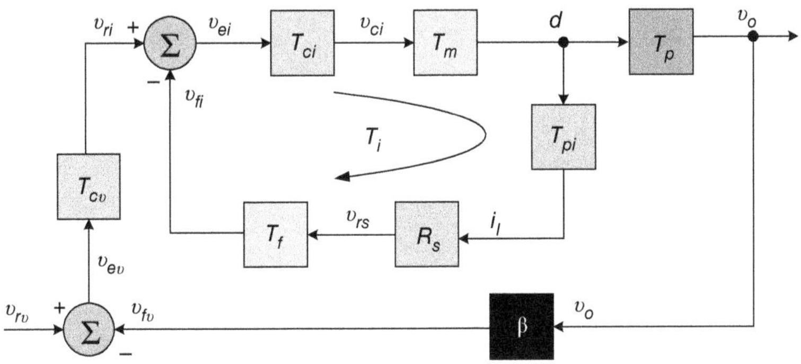

Figure 2.39 Block diagram used to derive the control-to-output voltage transfer function T_{pcl}.

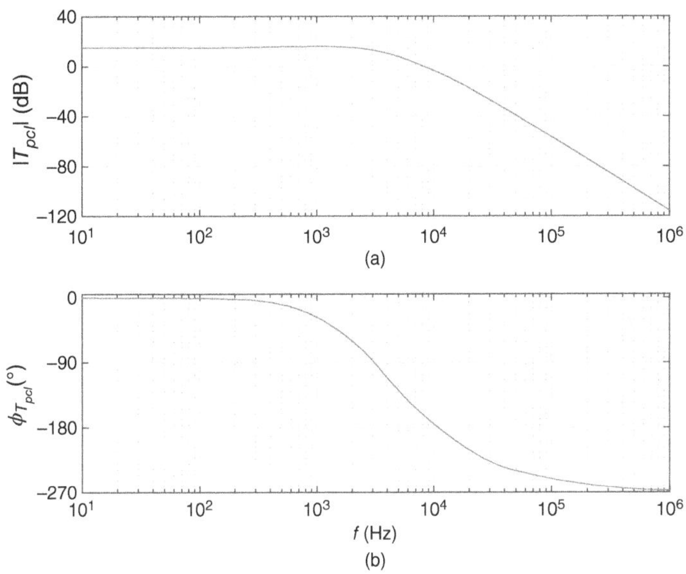

Figure 2.40 Theoretically obtained plots of the reference voltage-to-output voltage T_{pcl}. (a) Magnitude. (b) Phase.

Figure 2.40 shows the theoretically obtained magnitude and phase plots of the reference voltage-to-output voltage transfer function.

2.7.2 Input Voltage to Duty-Cycle Transfer Function M_{dv}

Using the block diagram shown in Figure 2.41, the closed-loop input voltage to duty-cycle transfer function is

$$M_{dv}(s) = \left. \frac{d(s)}{v_i(s)} \right|_{v_{rv}=i_o=0}. \tag{2.149}$$

From the block diagram,

$$v_o = v_o' + v_o'' = T_p d + M_v v_i. \tag{2.150}$$

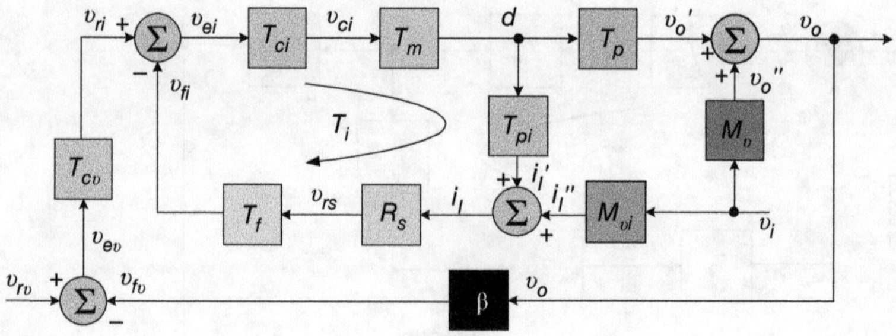

Figure 2.41 Block diagram used to derive the input voltage to duty-cycle transfer function M_{dv}.

Also,

$$v_{fv} = -\beta v_o \tag{2.151}$$

and

$$v_{cv} = v_{ri} = -\beta T_{cv} v_o. \tag{2.152}$$

The error voltage v_{ei} is

$$v_{ei} = v_{ri} - v_{fi} = -T_{cv}\beta v_o - R_s T_f i_l \tag{2.153}$$

and also given in (2.120) as

$$v_{ei} = \frac{d}{T_m T_{ci}}. \tag{2.154}$$

The inductor current is

$$i_l = T_{pi} d + M_{vi} v_i. \tag{2.155}$$

Substituting (2.155) and (2.154) into (2.153) yields

$$\frac{d}{T_m T_{ci}} = -T_{cv}\beta v_o - R_s T_f (T_{pi} d + M_{vi} v_i). \tag{2.156}$$

Rearranging (2.156), we get

$$v_o = -\frac{d}{T_{cv}\beta}\left(\frac{1}{T_m T_{ci}} + R_s T_f T_{pi}\right) - \frac{v_i}{T_{cv}\beta}\left(R_s T_f M_{vi}\right). \tag{2.157}$$

Equating (2.150) and (2.157) and rearranging,

$$v_i\left(-\frac{M_{vi} R_s T_f}{T_{cv}\beta} - M_v\right) = d\left(T_p + \frac{1}{T_m T_{ci} T_{cv}\beta} + \frac{R_s T_f T_{pi}}{T_{cv}\beta}\right), \tag{2.158}$$

$$-\frac{v_i}{T_{cv}\beta}\left(M_{vi} R_s T_f + M_v T_{cv}\beta\right) = \frac{d}{T_{cv}\beta}\left(T_p T_{cv}\beta + \frac{1 + R_s T_f T_{pi} T_m T_{ci}}{T_m T_{ci}}\right). \tag{2.159}$$

Rearranging further,

$$-v_i\left(M_{vi} R_s T_f + M_v T_{cv}\beta\right) = d\left(T_p T_{cv}\beta + \frac{1 + R_s T_f T_{pi} T_m T_{ci}}{T_m T_{ci}}\right), \tag{2.160}$$

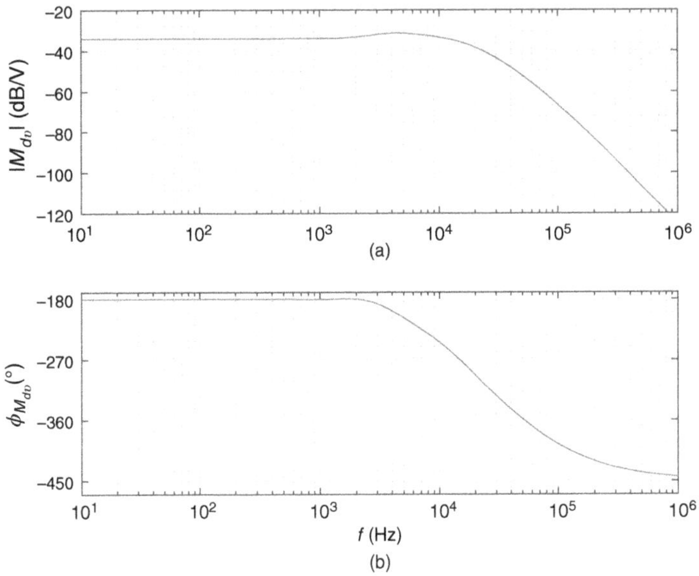

Figure 2.42 Theoretically obtained plots of the input voltage-to-duty cycle transfer function M_{dv}. (a) Magnitude. (b) Phase.

and

$$M_{dv}(s) = \frac{d(s)}{v_i(s)} = -\frac{M_{vi}R_sT_f + \beta M_v T_{cv}}{\beta T_p T_{cv} + \dfrac{1+T_i}{T_m T_{ci}}}. \tag{2.161}$$

Figure 2.42 shows the theoretically obtained magnitude and phase plots of the input voltage-to-duty cycle transfer function.

2.7.3 Input Voltage-to-Output Voltage Transfer Function M_{vcl}

Using the block diagram shown in Figure 2.43, the closed-loop input voltage-to-output voltage transfer function with outer-voltage loop is

$$M_{vcl}(s) = \left.\frac{v_o(s)}{v_i(s)}\right|_{v_{rv}=i_o=0}. \tag{2.162}$$

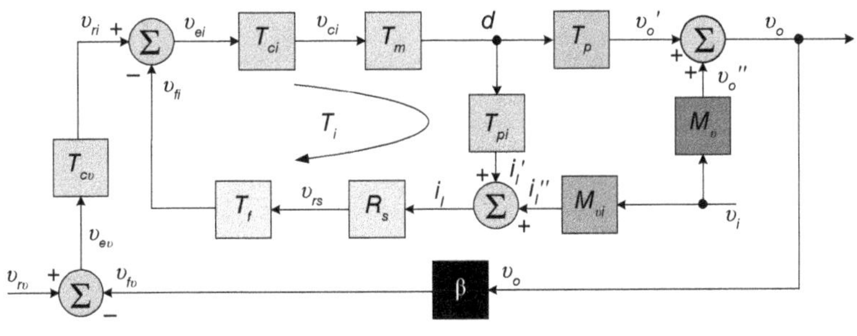

Figure 2.43 Block diagram used to derive the input voltage-to-output voltage transfer function M_{vcl}.

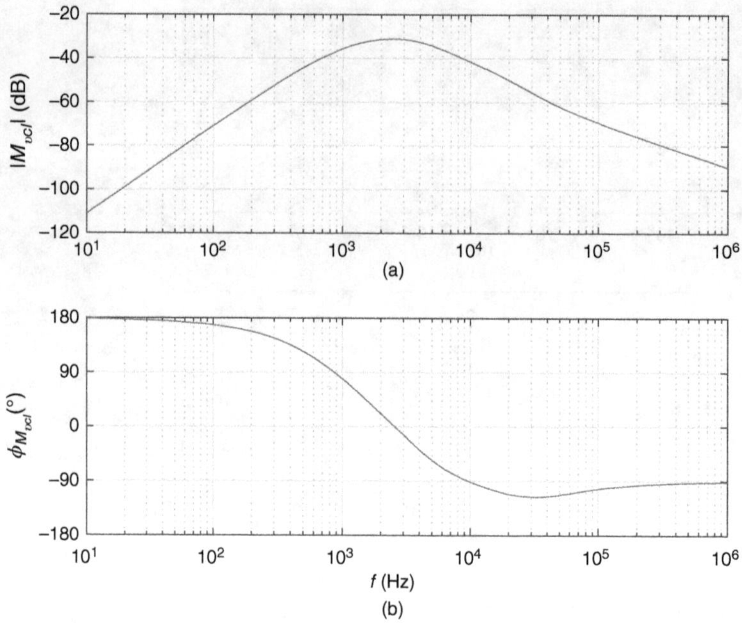

Figure 2.44 Theoretically obtained plots of the input voltage-to-output voltage transfer function M_{vcl}. (a) Magnitude. (b) Phase.

From (2.149),

$$d(s) = M_{dv}v_i(s). \tag{2.163}$$

Substituting (2.163) into (2.150),

$$v_o = v_o' + v_o'' = T_p d + M_v v_i = T_p M_{dv} v_i + M_v v_i. \tag{2.164}$$

Rearrangement of (2.164) yields

$$M_{vcl}(s) = \frac{v_o(s)}{v_i(s)} = T_p M_{dv} + M_v. \tag{2.165}$$

Figure 2.44 shows the theoretically obtained magnitude and phase plots of the input voltage-to-output voltage transfer function M_{vcl}.

2.7.4 Input Impedance Z_{ivcl}

Using the block diagram shown in Figure 2.45, the closed-loop input impedance with outer-voltage loop is

$$Z_{ivcl}(s) = \left.\frac{v_i(s)}{i_i(s)}\right|_{v_{rv}=i_o=0}. \tag{2.166}$$

From the Figure 2.45,

$$v_o = v_o' + v_o'' = T_p d + M_v v_i, \tag{2.167}$$

and

$$i_l = i_l' + i_l'' = T_{pi} d + M_{vi} v_i. \tag{2.168}$$

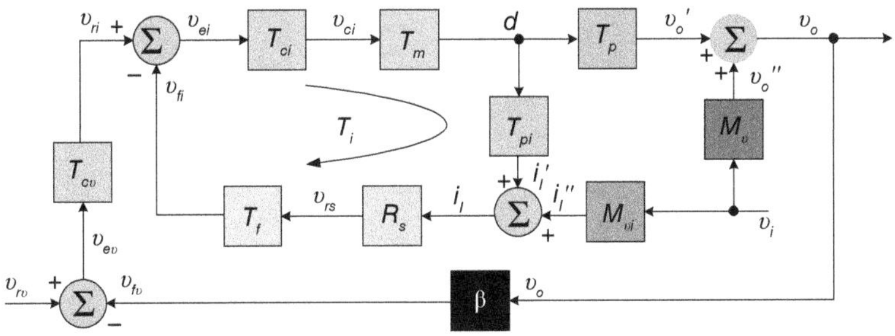

Figure 2.45 Block diagram used to derive the closed-outer-voltage-loop input impedance Z_{ivcl}.

Also,

$$v_{fv} = \beta v_o, \tag{2.169}$$

$$v_{ev} = v_{rv} - v_{fv} = 0 - \beta v_o = -\beta v_o, \tag{2.170}$$

and

$$v_{cv} = v_{ri} = T_{cv} v_{ev} = -\beta T_{cv} v_o. \tag{2.171}$$

The error voltage v_{ei} is

$$v_{ei} = \frac{d}{T_{ci} T_m}. \tag{2.172}$$

The feedback voltage v_{fi} is

$$v_{fi} = T_f R_s i_l. \tag{2.173}$$

The reference voltage to inner-current loop is

$$v_{ri} = v_{ei} + v_{fi}. \tag{2.174}$$

Substituting (2.171), (2.172), and (2.173) into (2.174),

$$-\beta T_{cv} v_o = \frac{d}{T_{ci} T_m} + T_f R_s i_l. \tag{2.175}$$

Rearranging (2.175) gives

$$v_o = -\frac{d}{\beta T_{cv} T_{ci} T_m} - \frac{T_f R_s i_l}{\beta T_{cv}}. \tag{2.176}$$

Substituting (2.167) into (2.176),

$$T_p d + M_v v_i = -\frac{d}{\beta T_{cv} T_{ci} T_m} - \frac{T_f R_s i_l}{\beta T_{cv}}, \tag{2.177}$$

and rearranging,

$$d\left(T_p + \frac{1}{\beta T_{cv} T_{ci} T_m}\right) = -\frac{T_f R_s i_l}{\beta T_{cv}} - M_v v_i. \tag{2.178}$$

Substituting (2.168) in (2.178),

$$d\left(T_p + \frac{1}{\beta T_{cv}T_{ci}T_m}\right) = -\frac{T_f R_s(T_{pi}d + M_{vi}v_i)}{\beta T_{cv}} - M_v v_i, \tag{2.179}$$

$$d\left(T_p + \frac{1}{\beta T_{cv}T_{ci}T_m} + \frac{T_f R_s T_{pi}}{\beta T_{cv}}\right) = \left(-\frac{T_f R_s M_{vi}}{\beta T_{cv}} - M_v\right), \tag{2.180}$$

$$d\left(\frac{\beta T_p T_{ci}T_m T_{cv} + 1 + T_f R_s T_{pi} T_{ci} T_m}{\beta T_{ci}T_m T_{cv}}\right) = \left(\frac{-T_f R_s M_{vi} - \beta T_{cv}M_v}{\beta T_{cv}}\right), \text{ and} \tag{2.181}$$

$$d = -\frac{T_{ci}T_m(T_f R_s M_{vi} + \beta M_v T_{cv})}{\beta T_p T_{ci}T_m T_{cv} + 1 + T_f R_s T_{pi} T_{ci} T_m} v_i. \tag{2.182}$$

The input current i_i is

$$i_i = I_L d + D i_l = I_L d + D(T_{pi}d + M_{vi}v_i) = (I_L + DT_{pi})d + DM_{vi}v_i. \tag{2.183}$$

Substituting (2.182) into (2.183),

$$i_i = (I_L + DT_{pi})\left[-\frac{T_{ci}T_m(T_f R_s M_{vi} + \beta M_v T_{cv})}{\beta T_p T_{ci}T_m T_{cv} + 1 + T_f R_s T_{pi} T_{ci} T_m} v_i\right] + DM_{vi}v_i, \text{ and} \tag{2.184}$$

$$i_i = \left[-(I_L + DT_{pi})\frac{T_{ci}T_m(T_f R_s M_{vi} + \beta M_v T_{cv})}{\beta T_p T_{ci}T_m T_{cv} + 1 + T_f R_s T_{pi} T_{ci} T_m} + DM_{vi}\right] v_i. \tag{2.185}$$

The closed-loop input admittance with outer-voltage loop is

$$Y_{ivcl}(s) = \frac{i_i(s)}{v_i(s)} = -(I_L + DT_{pi})\frac{T_{ci}T_m(T_f R_s M_{vi} + \beta M_v T_{cv})}{\beta T_p T_{ci}T_m T_{cv} + 1 + T_f R_s T_{pi} T_{ci} T_m} + DM_{vi}. \tag{2.186}$$

The closed-loop input impedance with outer-voltage loop is

$$Z_{ivcl}(s) = \frac{v_i(s)}{i_i(s)} = \frac{1}{Y_{ivcl}(s)}. \tag{2.187}$$

Figure 2.46 shows the theoretically obtained magnitude and phase plots of the closed-loop input impedance Z_{ivcl}.

2.7.5 Output Impedance Z_{ovcl}

Using the block diagram shown in Figure 2.47, the closed-loop output impedance with outer-voltage loop is

$$Z_{ovcl}(s) = \left.\frac{v_o(s)}{i_o(s)}\right|_{v_{rv}=i_o=0}. \tag{2.188}$$

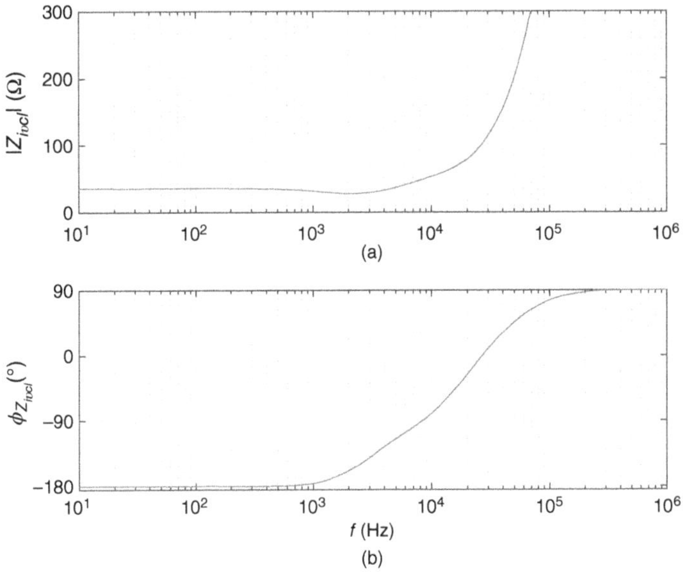

Figure 2.46 Theoretically obtained plots of the closed-outer-voltage loop input impedance Z_{ivcl}. (a) Magnitude. (b) Phase.

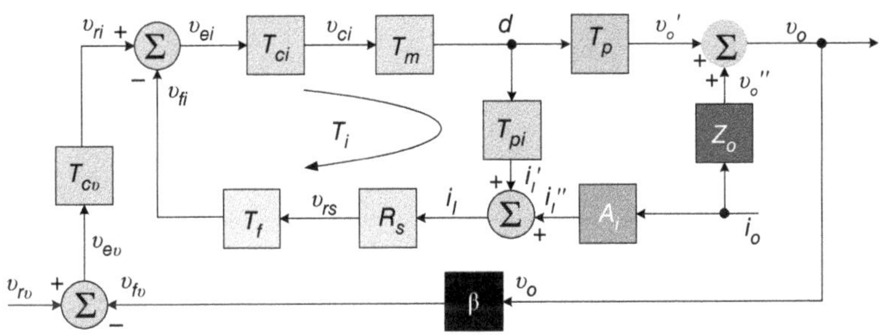

Figure 2.47 Block diagram used to derive the closed-outer-voltage-loop output impedance Z_{ovcl}.

From the Figure 2.47,

$$v_o = v_o' + v_o'' = T_p d + Z_o i_o, \tag{2.189}$$

and

$$i_l = i_l' + i_l'' = A_i i_o + T_{pi} d. \tag{2.190}$$

Also,

$$v_{fv} = \beta v_o, \tag{2.191}$$

and

$$v_{cv} = v_{ri} = -\beta T_{cv} v_o. \tag{2.192}$$

The error voltage v_{ei} is

$$v_{ei} = v_{ri} - v_{fi} = -T_{cv}\beta v_o - R_s T_f i_l \tag{2.193}$$

and

$$v_{ei} = \frac{d}{T_m T_{ci}}. \tag{2.194}$$

Substituting (2.190) and (2.194) into (2.193) yields

$$\frac{d}{T_m T_{ci}} = -T_{cv}\beta v_o - R_s T_f (A_i i_o + T_{pi} d). \tag{2.195}$$

Rearranging (2.195),

$$d = -\frac{(T_{cv}\beta v_o + R_s T_f A_i i_o) T_{ci} T_m}{1 + T_i}. \tag{2.196}$$

Substituting (2.196) into (2.189) yields

$$v_o = T_p \left[-\frac{(T_{cv}\beta v_o + R_s T_f A_i i_o) T_{ci} T_m}{1 + T_i} \right] + Z_o i_o. \tag{2.197}$$

Rearranging further gives

$$v_o \left(1 + \frac{\beta T_p T_{cv} T_{ci} T_m}{1 + T_i} \right) = i_o \left(Z_o - \frac{T_p R_s T_f A_i T_{ci} T_m}{1 + T_i} \right). \tag{2.198}$$

Hence, the closed-loop output impedance with outer-voltage loop is

$$Z_{ovcl}(s) = \frac{v_o(s)}{i_o(s)} = \frac{Z_o(1 + T_i) - A_i T_p T_{ix}}{1 + T_i + T_p T_{cv} \beta T_{ci} T_m}. \tag{2.199}$$

Figure 2.48 shows the theoretically obtained magnitude and phase plots of the closed-loop output impedance Z_{ovcl}.

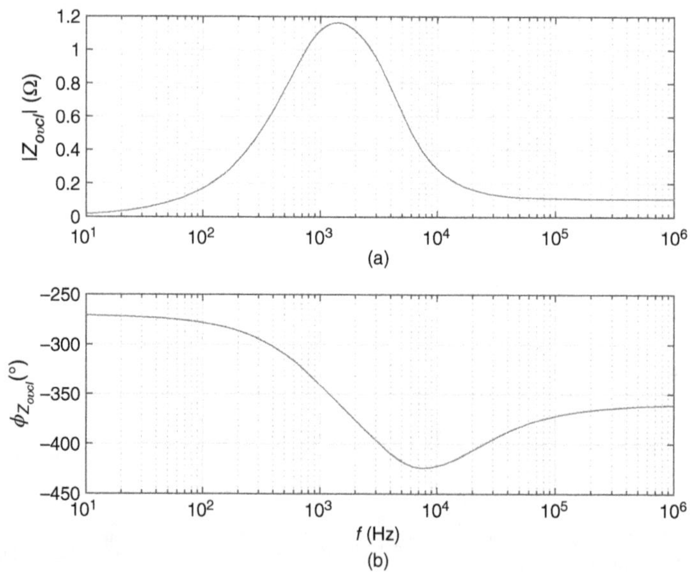

Figure 2.48 Theoretically obtained plots of the closed-outer-voltage loop output impedance Z_{ovcl}. (a) Magnitude. (b) Phase.

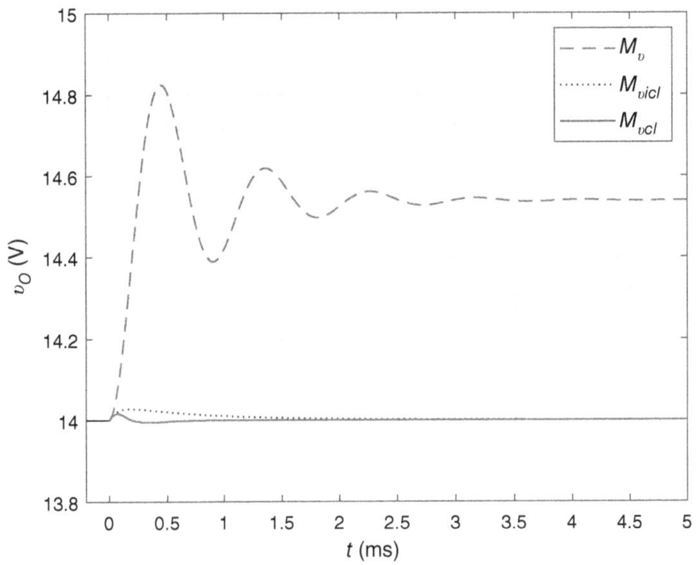

Figure 2.49 Comparison of responses in output voltage for step changes in input voltage by 1 V obtained using M_v, M_{vicl}, and M_{vcl} transfer functions, respectively.

2.8 Comparison of Closed-Loop and Open-Loop Step Responses

2.8.1 Response of Output Voltage to Step Change in Input Voltage

Figure 2.49 compares the responses of the output voltage for unit step changes in input voltage by 1 V for (a) open loop (M_v), (b) closed-inner-current loop (M_{vicl}), and (c) closed-outer-voltage loop in the presence of the inner-current loop (M_{vcl}). In the open-loop configuration, a step change $\Delta V_I = 1$ V in the input voltage causes the output voltage to achieve a new steady-state value $V_O(\infty) = 14.54$ V. Therefore, the step change in the output voltage is $\Delta V_O = V_O(\infty) - V_O(0) = 14.54 - 14 = 0.54$ V, yielding $M_{vo} = (\Delta V_O)/(\Delta V_I) = 0.54/1 = 0.54$ V/V, which was also observed in the magnitude response of M_v. However, for closed-loop responses determined by current-loop transfer function M_{vicl} and voltage-loop transfer function M_{vcl}, step changes in input voltage cause the output voltage to return to its steady-state value $V_O(\infty) = 14$ V within 2 ms and 0.15 ms, respectively.

2.8.2 Response of Output Voltage to Step Change in Duty Cycle, Current-Loop reference Voltage, and Voltage-Loop Reference Voltage

Figure 2.50 compares the responses of the output voltage for step changes in (a) duty cycle for open-loop by 0.1 (T_p), (b) current-loop reference voltage for the closed-inner-current loop by 1 V (T_{picl}), and (c) voltage-loop reference voltage for the closed-outer-voltage loop by 1 V (T_{pcl}). In the open-loop configuration, a step

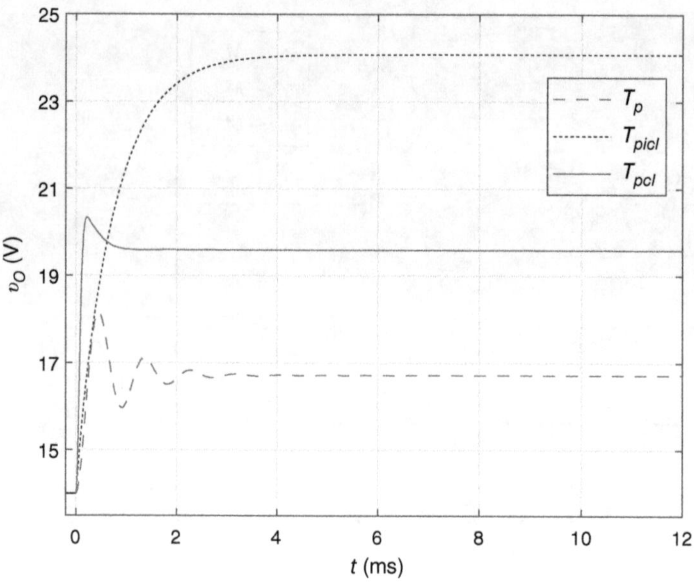

Figure 2.50 Comparison of responses in output voltage for step changes in duty cycle by 0.1, current-loop reference voltage by 1 V, and voltage-loop reference voltage by 1 V, obtained using T_p, T_{picl}, and T_{pcl} transfer functions, respectively.

change in duty cycle $\Delta D = 0.1$ causes the output voltage to achieve a new steady-state value $V_O(\infty) = 16.72$ V. Therefore, the step change in the output voltage is $\Delta V_O = V_O(\infty) - V_O(0) = 16.72 - 14 = 2.72$ V, yielding $T_{po} = (\Delta V_O)/\Delta D = 2.72/0.1 = 27.2$ V, which was also observed in the magnitude plot of T_p. The step response shows a settling time of $\approx$ 4 ms. The output voltage response for the closed-inner-current loop achieves a steady-state value $V_O(\infty) = 24$ V in a settling time of 4 ms. The output voltage response for the closed-outer-voltage loop achieves a steady-state value $V_O(\infty) = 19.6$ V in a rapid settling time of nearly 1 ms.

2.8.3 Response of Input Current to Step Change in Input Voltage

Figure 2.51 compares the responses of the input current for unit step change in input voltage by 1 V for (a) open-loop (Y_i), (b) closed-inner-current loop (Y_{iicl}), and (c) closed-outer-voltage loop (Y_{ivcl}). In the open-loop configuration, a step change in input voltage $\Delta V_I = 1$ V causes the input current to achieve a new steady-state value $I_I(\infty) = 0.81$ A. Therefore, the step change in the input current is $\Delta I_I = I_I(\infty) - I_I(0) = 0.81 - 0.78 = 0.03$ A, yielding $Z_{io} = R_i = (\Delta V_I)/(\Delta I_I) = 1/0.03 = 33.3\ \Omega$, which was also observed in the magnitude plot of Z_i. The step response shows a settling time of 4 ms. The input current response for the closed-inner-current loop achieves a steady-state value

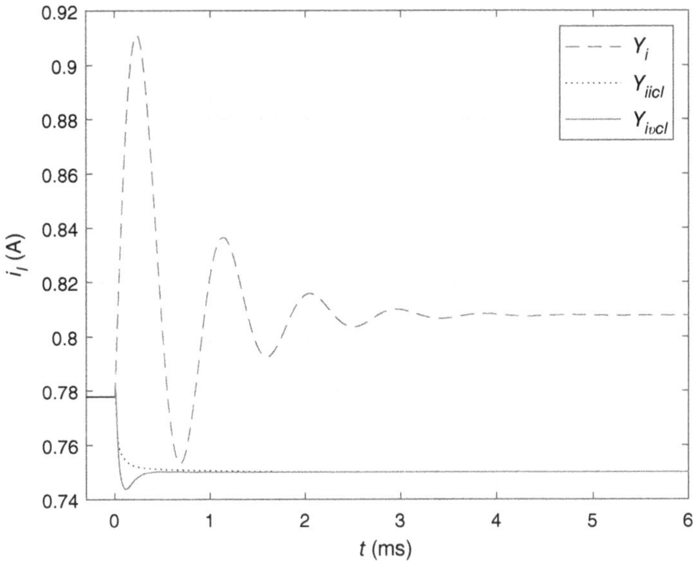

Figure 2.51 Comparison of responses in input current for step changes in the input voltage by 1 V obtained using Y_i, Y_{iicl}, and Y_{ivcl} transfer functions.

$I_I(\infty) = 0.75$ A in a settling time of 0.5 ms. The input current response for the closed-outer-voltage loop achieves a steady-state value $I_I(\infty) = 0.75$ A in a rapid settling time of nearly 0.25 ms. The reduction in the input current in negative direction for positive step change in output voltage is due to the negative input resistance caused at DC by the closed-inner-current and closed-outer-voltage loops.

2.8.4 Response of Output Voltage to Step Change in Load Current

Figure 2.52 compares the responses of the output voltage for unit step changes in output current by 1 A for (a) open-loop (Z_o), (b) closed-inner-current loop (Z_{oicl}), and (c) closed-outer-voltage loop (Z_{ovcl}). In the open-loop configuration, a step change in output current $\Delta I_O = 1$ A causes the output voltage to achieve a new steady-state value $V_O(\infty) = 14.25$ V. Therefore, the step change in the output voltage is $\Delta V_O = V_O(\infty) - V_O(0) = 14.25 - 14 = 0.25$ V, yielding $Z_{oo} = R_o = (\Delta V_O)/(\Delta I_O) = 0.25/1 = 0.25$ V/A, which was also observed in the magnitude plot of Z_o. The step response shows a settling time of 3 ms. The output voltage response for the closed-inner-current loop achieves a steady-state value $V_O(\infty) = 24$ V in a settling time of 4 ms. The output voltage for the closed-outer-voltage loop returns to the steady-state value $V_O(\infty) = V_O(0) = 14$ V in a settling time of 1 ms.

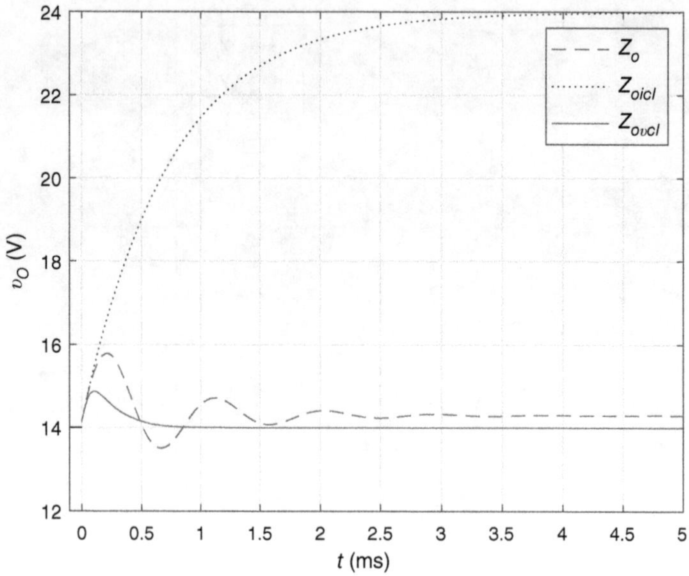

Figure 2.52 Comparison of responses in the output voltage for step changes in the output current by 1 A obtained using Z_o, Z_{oicl}, and Z_{ovcl} transfer functions.

2.9 Summary

1. The small-signal, linear model of the buck converter in CCM consists of a current-controlled current source representing the transistor and a voltage-controlled voltage source representing the diode, in addition to the input AC voltage source, inductor, and filter capacitor.
2. The small-signal duty cycle-to-output voltage transfer function of a lossy buck converter is a second-order transfer function exhibiting an LHP real zero and a pair of complex-conjugate poles.
3. The small-signal duty cycle-to-inductor transfer function of a lossy buck converter is a second-order transfer with a real LHP zero and a pair of complex-conjugate LHP poles.
4. The small-signal input-to-output voltage transfer function of a lossy buck converter is a second-order transfer exhibiting an LHP zero and a pair of complex-conjugate LHP poles. The location of the LHP zero and complex poles is identical to that of the duty cycle-to-output voltage transfer function.
5. The small-signal input voltage-to-inductor transfer function of a lossy buck converter is a second-order transfer exhibiting an LHP zero and a pair of complex-conjugate LHP poles.
6. The input impedance of a lossy buck converter exhibits a pair of LHP complex-conjugate zeros and an LHP real zero. The input resistance is a function of the load resistance and the duty cycle. The location of the complex zero is identical to the complex pole produced by the duty cycle-to-output voltage transfer function.

7. The output impedance of a lossy buck converter has two LHP real zeros and an LHP complex-conjugate poles. The output resistance is a function of the load resistance. The location of the complex pole is identical to that produced by the duty cycle-to-output voltage transfer function.
8. An integral single-lead controller for the inner-current loop of the buck converter is an effective compensation scheme to achieve a significant reduction in the DC steady-state error and improve the closed-loop bandwidth.

3

Average Current-Mode Control of Boost DC–DC Converter

The boost converter is a fundamental topology belonging to a class of single-stage, voltage step-up, high-efficiency power supplies. The boost converter is widely accepted because of its unique advantages such as continuous input current and low-side controllable switch, making it an ideal choice as power-factor correctors, pre-regulators for onboard automobile power supplies, renewable energy applications for voltage step-up from solar photovoltaic (PV) output of 3.3 V to nearly hundreds of volts, and light-emitting diode (LED) drivers. The average current-mode controlled boost converter is an attractive choice due to the combined advantages of the power-stage and control method. The objective of this chapter is to provide a comprehensive discussion on the dynamic properties of the boost converter and provide a procedure to design a well-regulated power supply. Many modeling methods have been employed in the literature to accomplish the task of modeling the nonlinear switching network in the boost converter. In this chapter, the circuit-averaging technique is presented, which is based on the principle of conservation of energy. The main objectives of this chapter are as follows:

1. To develop the circuit-averaged model describing the DC characteristics of the converter power stage in continuous-conduction mode (CCM).
2. To discuss the principle of voltage or current perturbation leading to the development of a large-signal model.
3. To show the development of the nonlinear large-signal and linearized small-signal model and discuss the principle of *linearization*.
4. To derive, analyze, and characterize the open-loop power-stage transfer functions related to the current and voltage loops, including the small-signal input and output impedances.
5. To identify the optimal controller topology that satisfies the needs of the inner-current and outer-voltage loops.
6. To derive and analyze the closed-inner-current loop and closed-outer-voltage loop transfer functions of the boost converter.

As the outcome of this chapter, the reader will gain sufficient knowledge in analysis, design, and characterization of boost DC–DC converter employing average current-mode control. The subsequent analysis in this chapter is based on the following assumptions:

Average Current-Mode Control of DC-DC Power Converters, First Edition.
Marian K. Kazimierczuk, Dalvir K. Saini, and Agasthya Ayachit.

1. The transistor output capacitance, diode capacitance, and lead inductance are zero, and thereby the switching losses are neglected.
2. Passive components are linear, time invariant, and frequency dependent.
3. The output impedance of the supply voltage source is zero for both DC and AC components.
4. The converter operates in steady state.
5. The switching period is much shorter than the time constants of the reactive components designed for CCM operation.
6. The DC output voltage is constant; however, the DC supply voltage and the load resistance are variable.

3.1 Circuit Description, DC Characteristics, and Design

3.1.1 Circuit Description

The pulse-width-modulated (PWM) boost converter circuit is shown in Figure 3.1. The circuit consists of a power MOSFET S, a diode D_o, an inductor L, a filter capacitor C, and a load resistor R_L. The switch is turned ON and OFF at a switching frequency f_s and a duty cycle D. The parasitic resistances of the passive and semiconductor components are considered in the analysis. The equivalent series resistance of the inductor and capacitor are r_L and r_C, respectively. The ON-state resistance of the MOSFET and the ON resistance of the diode are r_{DS} and R_F, respectively, and the diode forward voltage is V_F.

3.1.2 DC Model

Figure 3.2 shows the ideal current and voltage waveforms of the switching network of the boost converter. The small-ripple approximation is made, where the amplitude of the inductor current ripple due to the switching action is considered to be much lower than the DC component of the inductor current. As a result, the peak value of the switch current is equal to the average value of the inductor current I_L. The DC voltage transfer function of the ideal boost converter in CCM is

$$M_{V\,DC(lossless)} = \frac{V_O}{V_I} = \frac{1}{1-D}, \tag{3.1}$$

which produces the duty cycle for the lossless boost converter, given as

$$D = 1 - \frac{V_I}{V_O}. \tag{3.2}$$

As shown in Figure 3.2b, the waveform of the switch current can be approximated by

$$i_S = \begin{cases} I_L & 0 < t \leq DT \\ 0 & DT < t \leq T. \end{cases} \tag{3.3}$$

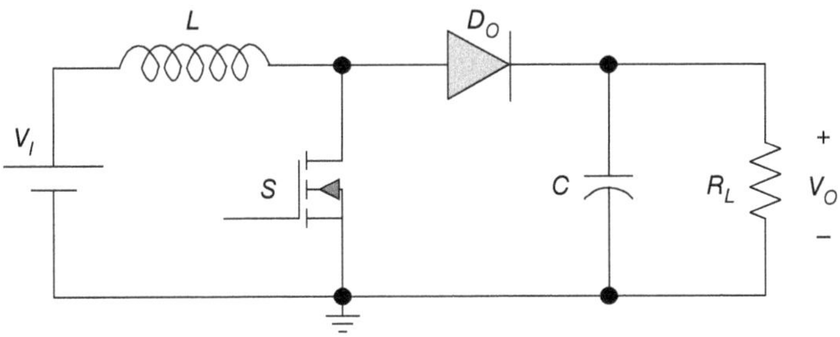

Figure 3.1 Circuit of the PWM boost converter.

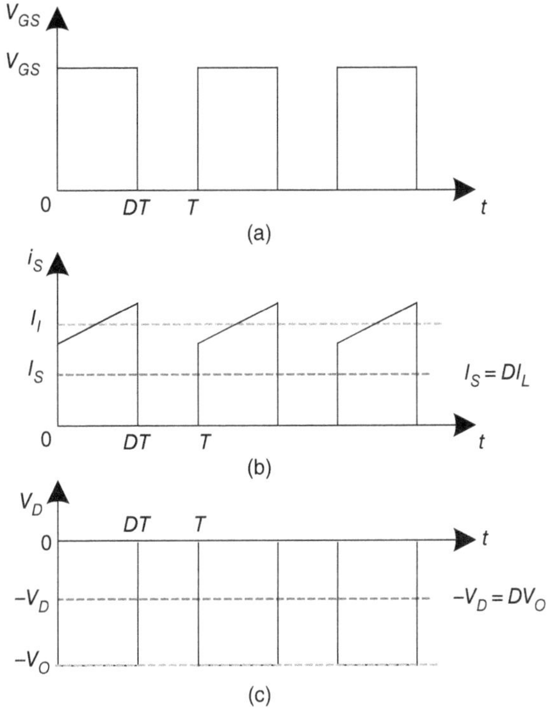

Figure 3.2 Key waveforms of the switching network. (a) Gate-to-source voltage of switch. (b) Switch current. (c) Diode voltage.

The average DC component of the switch current is

$$I_S = \frac{1}{T}\int_0^T i_S \, dt = \frac{1}{T}\int_0^{DT} I_L \, dt = DI_L. \tag{3.4}$$

Notice that the average switch current is a function of the average inductor current and the duty cycle. By the principle of circuit averaging, the switch can be replaced by a current source dependent on the inductor current and the duty cycle. A similar approach can be adopted to model the diode.

As illustrated in Figure 3.2c, the waveform of the cathode-to-anode diode voltage, equal to the diode inverted voltage, is given by

$$-v_D(t) = v_{KA}(t) = \begin{cases} V_O & \text{for } 0 < t \leq DT \\ 0 & \text{for } DT < t \leq T. \end{cases} \tag{3.5}$$

The DC component of diode inverted voltage is

$$-V_D = V_{KA} = \frac{1}{T}\int_0^{DT} v_{KA}\, dt = \frac{1}{T}\int_0^{DT} V_{SD}\, dt = DV_{SD} = DV_O, \tag{3.6}$$

where V_{KA} is the DC component of diode branch voltage between the cathode and anode terminals and $V_{SD} = V_O$ is the DC component of node voltage between the input voltage terminal V_I and anode terminal.

The diode can be replaced by a voltage source dependent on the duty cycle and the output voltage. Since the voltage across the diode is negative, when the switch is ON, the negative sign can be incorporated into the model by reversing the polarities of the controlled voltage source. The switch and diode are first replaced by their equivalent current-controlled current source and voltage-controlled voltage source, respectively. At DC, the capacitor is an open circuit, and the inductor is a short circuit. Such substitutions result in the DC, averaged model of the boost converter as shown in Figure 3.3. This form of circuit modeling presents another advantage where the component parasitic resistances can be lumped and placed in the inductor branch. The equivalent averaged resistance placed in the inductor branch is

$$r = Dr_{DS} + (1 - D)R_F + r_L. \tag{3.7}$$

The steady-state performance of the converter can be analyzed using the DC model. Applying Kirchhoff's current law (KCL) at the node connecting the inductor, switch, and diode results in

$$I_L = DI_L + I_O = \frac{I_O}{1 - D} = \frac{V_O}{(1 - D)R_L}. \tag{3.8}$$

The DC input current I_I and the average inductor current I_L are equal. Therefore, The dc current transfer function is

$$M_{IDC} = \frac{I_O}{I_L} = 1 - D. \tag{3.9}$$

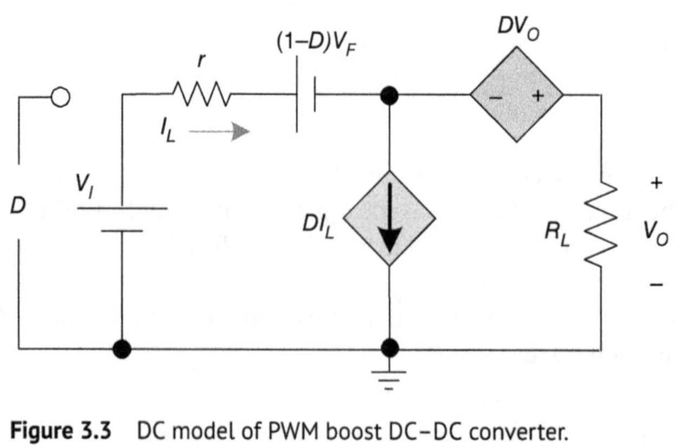

Figure 3.3 DC model of PWM boost DC–DC converter.

Applying Kirchhoff's voltage law (KVL) to the loop between the supply voltage and the load,

$$V_I = I_L r - DV_O + V_O = \frac{I_O}{1-D} r + V_O(1-D) = \frac{V_O}{R_L(1-D)} r + V_O(1-D) \tag{3.10}$$

to yield the DC voltage transfer function of the lossy boost DC–DC converter as

$$M_{V\,DC(lossy)} = \frac{V_O}{V_I} = \frac{1}{\frac{r}{R_L(1-D)} + (1-D)} = \frac{1}{(1-D)\left[1 + \frac{r}{R_L(1-D)^2}\right]} = \eta M_{V\,DC(lossless)}, \tag{3.11}$$

where the converter efficiency, excluding the switching loss and the loss in the filter capacitor, is [45]

$$\eta = \frac{P_O}{P_I} = \frac{1}{1 + \frac{r}{R_L(1-D)^2}} = \frac{1}{1 + \frac{Dr_{DS} + (1-D)R_F + r_L}{R_L(1-D)^2}}. \tag{3.12}$$

Therefore, the duty cycle of a lossy boost DC–DC converter is

$$D = 1 - \frac{\eta}{M_{V\,DC(lossless)}}. \tag{3.13}$$

3.1.3 Design Example

This chapter analyzes a boost converter in CCM designed for the following specifications:

1. Supply voltage $V_I = 12$ V.
2. Output voltage $V_O = 20$ V.
3. Output power $P_O = 10$ W.
4. Load resistance $R_L = 40\ \Omega$.
5. Switching frequency $f_s = 100$ kHz.
6. The inductance to ensure CCM operation is $L = 50\ \mu$H.
7. The capacitance to ensure CCM operation and to obtain an output voltage ripple $V_r < V_O/100$ [45] is $C = 100\ \mu$F.

It is assumed that the equivalent series resistances of the inductor and the capacitor are $r_L = 30$ mΩ and $r_C = 25$ mΩ, respectively. The ON-state resistance of the MOSFET and the ON-resistance of the diode are $r_{DS} = 0.11\ \Omega$ and $R_F = 0.07\ \Omega$, respectively. The diode forward voltage is $V_F = 0.3$ V.

Using (3.2), the duty cycle under ideal operating conditions is $D = 1 - V_I/V_O = 0.4$. The equivalent averaged resistance is $r = 0.116\ \Omega$. The overall efficiency, excluding the switching loss and the loss in the filter capacitor, is $\eta = 0.99 = 99\%$. Therefore, the duty cycle to achieve the rated output voltage is $D = 0.41$. The output current is $I_O = V_O/R_L = 0.5$ A. The average inductor current is $I_L = I_O/(1-D) = 0.83$ A.

3.2 Large-Signal and Small-Signal Models of PWM Boost Converter for CCM

The DC model is perturbed about the steady-state operating point to obtain a large-signal nonlinear model. The large-signal nonlinear model comprises slow time-varying AC quantities superimposed on DC quantities that are expressed as follows:

$$i_S = I_S + i_s,$$
$$i_L = I_L + i_l,$$
$$v_D = V_D + v_d,$$
$$v_I = V_I + v_i,$$
$$v_O = V_O + v_o,$$

and

$$d_T = D + d, \tag{3.14}$$

where i_S, i_L, v_D, v_I, v_O, and d_T are the large-signal switch current, inductor current, diode voltage, input voltage, output voltage, and duty cycle, respectively. The quantities i_s, i_l, v_d, v_i, v_o, and d are the small-signal switch current, inductor current, diode voltage, input voltage, output voltage, and duty cycle, respectively. The averaged switch current and averaged diode voltage given in (3.4) and (3.6) can be represented in terms of their large-signal equivalents as

$$i_S = d_T i_L = (D + d)(I_L + i_l) = DI_L + Di_l + I_L d + i_l d \tag{3.15}$$

and

$$v_D = d_T v_O = (D + d)(V_O + v_o) = DV_O + Dv_o + V_O d + v_o d. \tag{3.16}$$

The large-signal nonlinear model of a boost DC–DC converter in CCM is shown in Figure 3.4. The following small-signal conditions can be applied to obtain a linearized large-signal model:

$$V_O \gg v_o,$$
$$I_L \gg i_l,$$

and

$$D \gg d. \tag{3.17}$$

Applying small-signal conditions to (3.15) and (3.16), the small-signal nonlinear products $i_l d$ and $v_o d$ are eliminated, yielding a linearized large-signal model [46, 48]. Figure 3.5 shows the large-signal linear model of the boost DC–DC converter. The AC terms are separated from DC components to obtain the linear, averaged, small-signal model shown in Figure 3.6.

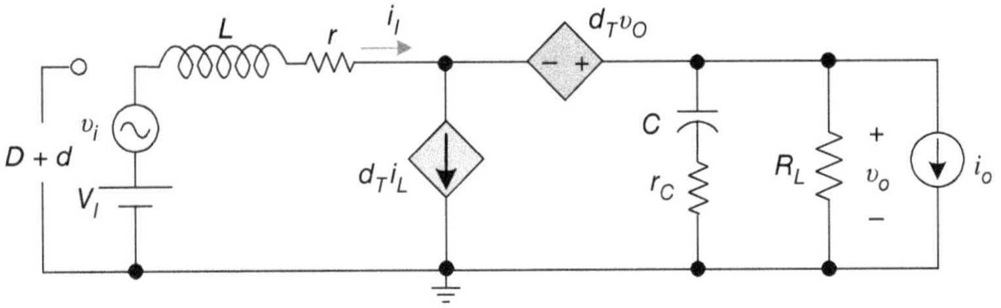

Figure 3.4 Large-signal nonlinear model of the PWM boost DC–DC converter in CCM.

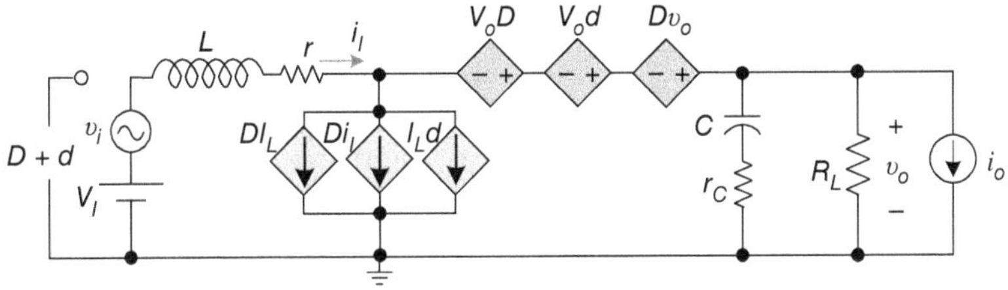

Figure 3.5 Large-signal linear model of the PWM boost DC–DC converter in CCM.

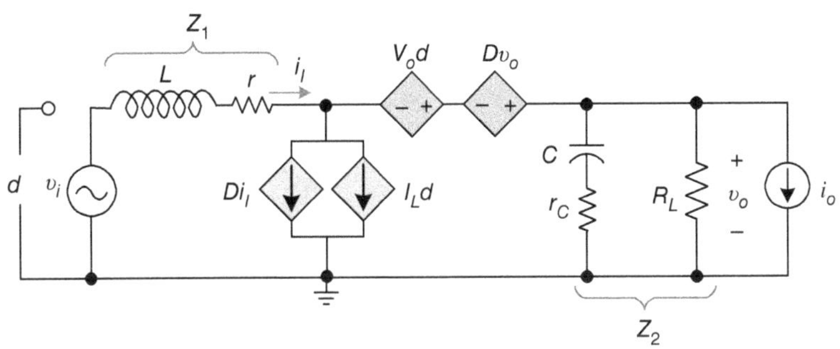

Figure 3.6 Small-signal model of the PWM boost DC–DC converter in CCM.

3.3 Power-Stage Transfer Functions

The dynamic performance of the boost converter power stage is determined using the small-signal dynamic model obtained by the circuit averaging technique developed in Section 3.2. The transfer functions are categorized into *control* transfer functions and *disturbance* transfer functions. In open-loop systems, the control transfer functions are between the duty cycle and the output voltage or the inductor current. The disturbance transfer functions are between the supply voltage or output current and the output voltage or inductor current. The following current-loop and voltage-loop transfer functions for the boost converter are determined in this section:

1. Duty cycle-to-output voltage transfer function T_p (control).
2. Duty cycle-to-inductor current transfer function T_{pi} (control).

3. Input voltage-to-output voltage transfer function M_v (disturbance).
4. Input voltage-to-inductor current transfer function M_{vi} (disturbance).
5. Reverse current gain A_i (disturbance).
6. Input impedance Z_i (disturbance).
7. Output impedance Z_o (disturbance).

The small-signal model of the boost DC–DC converter is shown in Figure 3.6. Using this model, the governing equations to derive the transfer functions are developed in this section.

The impedances Z_1 and Z_2 are

$$Z_1 = r + sL \tag{3.18}$$

and

$$Z_2 = R_L || \left(r_C + \frac{1}{sC}\right) = \frac{R_L\left(r_C + \frac{1}{sC}\right)}{R_L + r_C + \frac{1}{sC}}. \tag{3.19}$$

Using the model, the following governing state equations are derived. The current through the parallel combination of the load resistor and the filter capacitor is

$$i_{Z_2} = \frac{v_o}{Z_2}. \tag{3.20}$$

The current through the inductor is

$$i_l = Di_l + I_L d + i_{Z2} + i_o = Di_l + I_L d + \frac{v_o}{Z_2} + i_o = \frac{I_L d}{1-D} + \frac{v_o}{Z_2(1-D)} + \frac{i_o}{1-D}. \tag{3.21}$$

The input current is

$$i_i = i_l. \tag{3.22}$$

Applying KVL, the output voltage is

$$v_o = v_i - i_l Z_1 + Dv_o + V_o d. \tag{3.23}$$

3.3.1 Duty Cycle-to-Output Voltage Transfer Function T_p

The disturbances v_i and i_o shown in Figure 3.6 are set to zero. Using the governing equations, the duty cycle-to-output voltage transfer function in the impedance form is [45]

$$T_p(s) = \left.\frac{v_o(s)}{d(s)}\right|_{v_i=i_o=0} = \frac{V_O}{1-D}\frac{1-\frac{Z_1}{(1-D)^2R_L}}{1+\frac{Z_1}{(1-D)^2Z_2}} = \frac{V_O}{1-D}\frac{Z_2[(1-D)^2R_L - Z_1]}{R_L[(1-D)^2Z_2 + Z_1]}. \tag{3.24}$$

Substituting (3.18) and (3.19) into (3.24) gives the duty cycle-to-output voltage transfer function in the s domain as

$$T_p(s) = \left.\frac{v_o(s)}{d(s)}\right|_{v_i=i_o=0} = T_{px}\frac{(s+\omega_{zn})(s-\omega_{zp})}{s^2+2\xi\omega_o s+\omega_o^2} = T_{po}\frac{\left(1+\frac{s}{\omega_{zn}}\right)\left(1-\frac{s}{\omega_{zp}}\right)}{\left(\frac{s}{\omega_o}\right)^2+\frac{2\xi s}{\omega_o}+1}, \tag{3.25}$$

where

$$T_{px} = -\frac{V_O}{1-D}\frac{r_C}{R_L+r_C}, \tag{3.26}$$

the DC gain T_{po} is

$$T_{po} = T_p(0) = \frac{V_O}{1-D}\frac{R_L(1-D)^2-r}{R_L(1-D)^2+r}, \tag{3.27}$$

the angular corner frequency or the angular undamped natural frequency is

$$\omega_o = \sqrt{\frac{r+(1-D)^2R_L}{LC(R_L+r_C)}}, \tag{3.28}$$

and the damping ratio is

$$\xi = \frac{L+C[r(R_L+r_C)+(1-D)^2R_Lr_C]}{2\sqrt{LC(R_L+r_C)[r+(1-D)^2R_L]}}. \tag{3.29}$$

The angular frequency of the left-half plane (LHP) zero is

$$\omega_{zn} = \frac{1}{Cr_C}, \tag{3.30}$$

and the angular frequency of the right-half plane (RHP) zero is

$$\omega_{zp} = \frac{R_L(1-D)^2-r}{L} = \frac{R_L(1-D)^2-Dr_{DS}-(1-D)R_F-r_L}{L}. \tag{3.31}$$

The magnitude of the transfer function is

$$|T_p| = T_{po}\sqrt{\frac{\left[1+\left(\frac{\omega}{\omega_{zn}}\right)^2\right]\left[1+\left(\frac{\omega}{\omega_{zp}}\right)^2\right]}{\left[1-\left(\frac{\omega}{\omega_o}\right)^2\right]^2+4\xi^2\left(\frac{\omega}{\omega_o}\right)^2}}, \tag{3.32}$$

and the phase of the transfer function is

$$\phi_{Tp} = -180^\circ + \tan^{-1}\left(\frac{\omega}{\omega_{zn}}\right) - \tan^{-1}\left(\frac{\omega}{\omega_{zp}}\right) - \tan^{-1}\left[\frac{2\xi\left(\frac{\omega}{\omega_o}\right)}{1-\left(\frac{\omega}{\omega_o}\right)^2}\right]. \tag{3.33}$$

For the boost converter designed in Section 3.1.3, the following are determined for the duty cycle-to-output voltage transfer function T_p:

1. DC gain, corner frequency, frequency of the RHP zero, frequency of the LHP zero, damping coefficient, and the bandwidth of the transfer function.

2. Location of the poles and zeros.
3. Bode magnitude and phase plot of the transfer function.
4. Response of the output voltage for a step change in the duty cycle by $\Delta D = 0.1$.
5. Impact of the load resistance and the duty cycle on the location of the poles and zeros on the s-plane.

From the design example, $R_L = 40\ \Omega$, $L = 50\ \mu\text{H}$, $C = 100\ \mu\text{F}$, $r_L = 30\ \text{m}\Omega$, $r_C = 25\ \text{m}\Omega$, $r_{DS} = 0.11\ \Omega$, $R_F = 0.07\ \Omega$, and $V_F = 0.3$ V. The duty cycle is $D = 0.41$ to achieve a rated output voltage $V_O = 20$ V.

For the given specifications, the DC gain $T_{po} = 33.05$ V = 30.38 dBV, natural corner frequency $f_o = \omega_o/2\pi = 1.34$ kHz, frequency of the RHP zero $f_{zp} = \omega_{zp}/2\pi = 45$ kHz, and frequency of the LHP zero $f_{zn} = \omega_{zn}/2\pi = 63.66$ kHz. The damping coefficient is $\xi = 0.1628$. The gain of the transfer function at the 3-dB bandwidth is $T_p(BW) = T_{po}/\sqrt{2} = 23.36$ V = 27.38 dBV.

The roots of the characteristic equation of the transfer function T_p are the poles of the converter power stage. Equating the denominator in (3.25) to zero results in the poles of the transfer function

$$p_1, p_2 = \sigma \pm j\omega_d = -\xi\omega_o \pm j\omega_o\sqrt{1-\xi^2} = \omega_o(-\xi \pm j\sqrt{1-\xi^2}). \quad (3.34)$$

The poles are complex conjugate and are located on the left-half of the s-plane. The real part of the poles is dependent on the damping coefficient. At $\omega_o = 8.4493$ krad/s and $\xi = 0.1628$, the poles are located at

$$p_1, p_2 = -1375.5 \pm j8336.5 \text{ rad/s}. \quad (3.35)$$

The RHP zero is located at

$$z_p = \omega_{zp} = 281.05 \times 10^3 \text{ rad/s}. \quad (3.36)$$

The LHP zero is at

$$z_n = -\omega_{zn} = -400 \times 10^3 \text{ rad/s}. \quad (3.37)$$

Figure 3.7 shows the theoretically obtained magnitude and phase plots of the duty cycle-to-output voltage transfer function. The bandwidth of the converter is higher than the corner frequency f_o since the damping coefficient is low. Beyond the 3-dB bandwidth, the magnitude plot rolls off by −40 dB/dec up to the frequency of the RHP zero. The duty cycle-to-output voltage transfer function is a good representation of non-minimum phase properties of the converter. The RHP zero produces a positive slope for the magnitude and a negative slope for the phase, i.e. the RHP zero acts as a pole for the phase. Therefore, the slope of the magnitude plot beyond f_{zp} is −20 dB/dec and further increases to 0 dB/dec as the frequency approaches that of the LHP zero f_{zn}. Because of the RHP zero, the phase decreases by an extra 90°/dec in addition to the −180°/dec caused by the complex-conjugate pole pair. As a result, the phase crosses −180° at nearly the 3-dB bandwidth frequency.

Figure 3.8 shows the output voltage response to a step change in the duty cycle by $\Delta D = 0.1$. The output voltage increased from 20 to 23.3 V in nearly 4 ms. The step change in the output voltage is $\Delta V_O = 23.3 - 20 = 3.3$ V. Therefore, the ratio $\Delta V_O/\Delta D = 3.3/0.1 = 33$ V/V, which is nearly equal to the DC gain of the duty

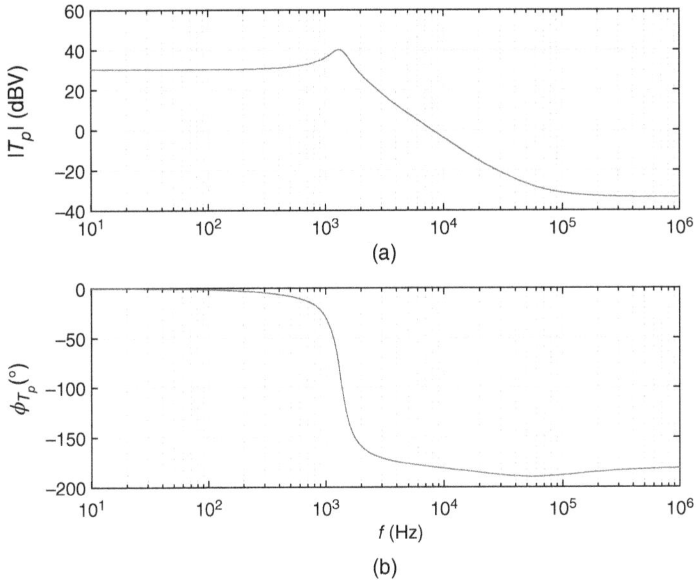

Figure 3.7 Theoretically obtained plots of duty cycle-to-output voltage transfer function T_p. (a) Magnitude. (b) Phase.

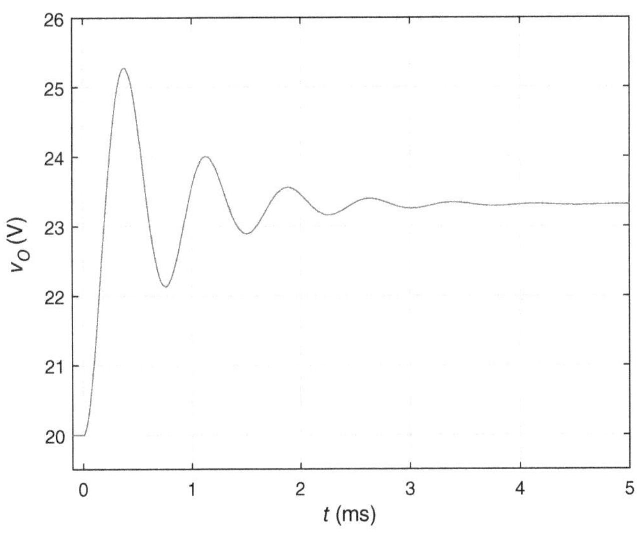

Figure 3.8 Response of the output voltage to step change in the duty cycle by $\Delta D = 0.1$.

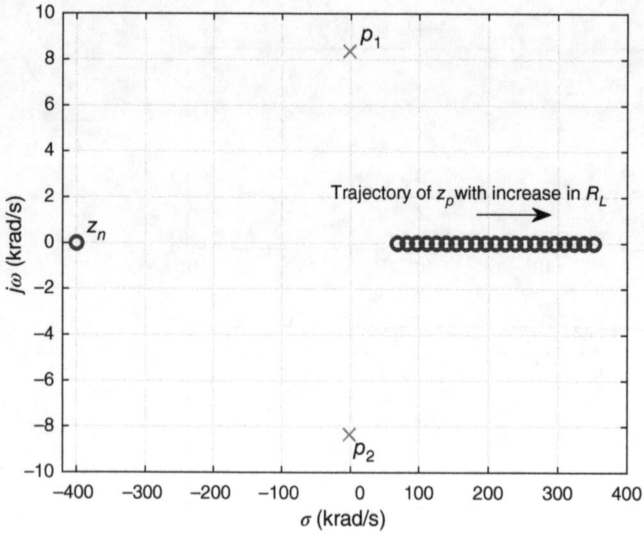

Figure 3.9 Trajectory of the poles and zeros of T_p transfer function as functions of the load resistance.

cycle-to-output voltage transfer function T_{po}. The presence of the RHP zero also causes a drop in the output voltage at the instant of step change. The RHP zero also reduces the speed of the converter in a closed-loop environment, thereby increasing the settling time.

Figure 3.9 shows the location of the power-stage poles p_1, p_2, LHP zero z_n, and the RHP zero z_p at different values of the load resistance R_L varying from 10 to 50 Ω. The location of the LHP zero z_n caused by the filter capacitor C and its equivalent series resistance r_C is independent of the load resistance. It is also observed that the complex-conjugate poles do not relocate with change in the load resistance. The RHP zero moves toward ∞ with increase in the load resistance. This means that the effect of the RHP zero is not significant at high load resistance values. Figure 3.10 shows the Bode magnitude and phase plots of the T_p transfer function at selected values of the load resistance. The DC gain is nearly independent of the load resistance. At low load resistance, the RHP zero is closer to the origin, and the phase plot crosses −180° in the vicinity of the bandwidth f_{BW} = 2 kHz. However, as the load resistance is increased, the phase near f_{BW} also increases. In the transient domain, a high load resistance corresponds to a larger damping coefficient and a reduction in ringing in the output voltage.

Figure 3.11 shows the trajectories of the power-stage poles p_1, p_2, LHP zero z_n, and the RHP zero z_p at different values of the duty cycle varying from 0.3 to 0.95. The location of the LHP zero z_n caused by the filter capacitor C and its equivalent series resistance r_C is independent of the duty cycle. At duty cycles larger than a critical value, the poles become real with p_1 moving toward the origin and p_2 moving toward −∞. An interesting phenomenon occurs with the RHP zero. As the duty cycle increases, the RHP zero moves toward the origin. At a critical duty cycle, the RHP

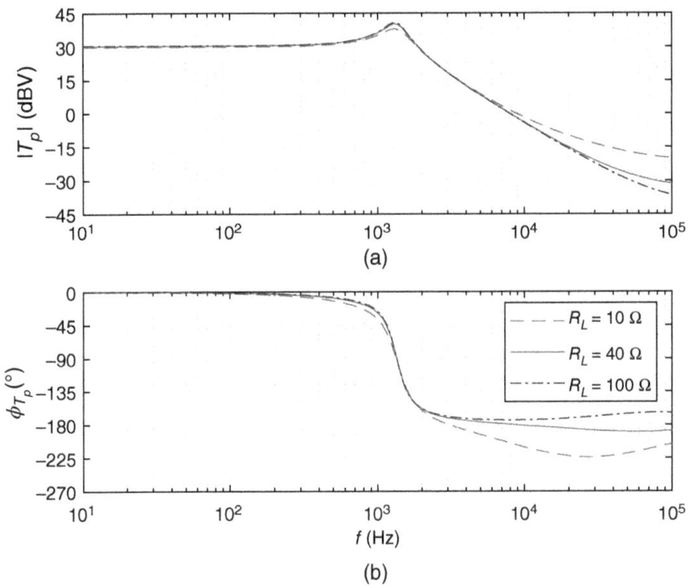

Figure 3.10 Theoretically obtained plots of duty cycle-to-output voltage transfer function T_p at selected load resistances. (a) Magnitude. (b) Phase.

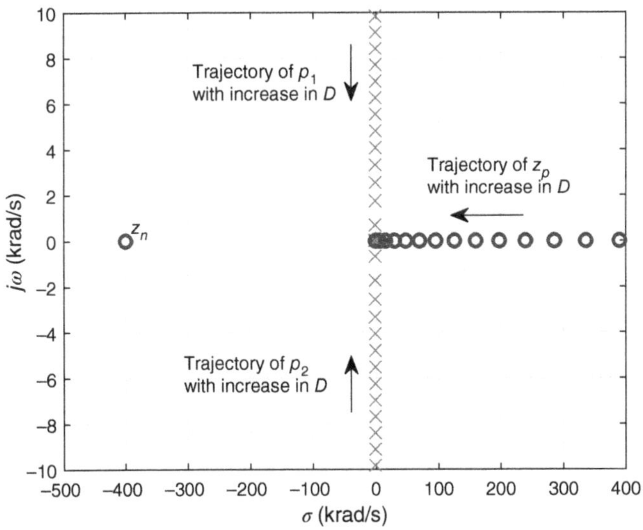

Figure 3.11 Trajectory of the poles and zeros of T_p transfer function as functions of the duty cycle.

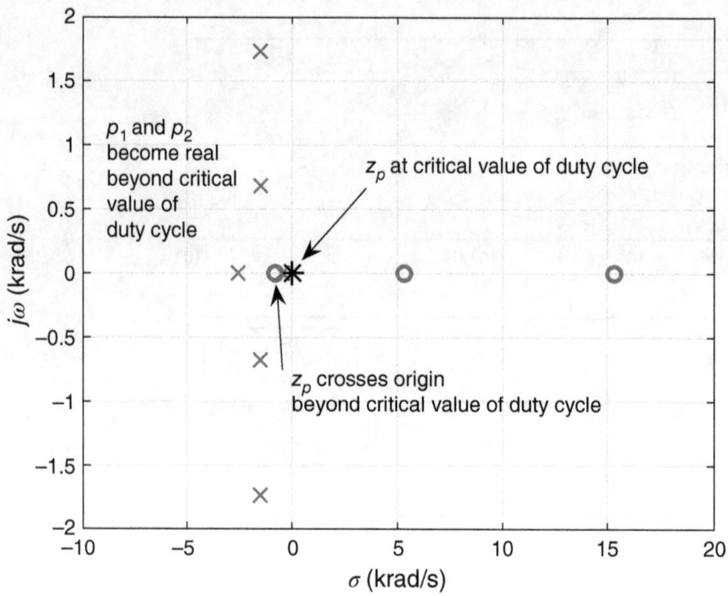

Figure 3.12 The poles and zeros of T_p transfer function beyond the critical value of the duty cycle.

zero z_p is at the origin. Hence, the expression for the critical duty cycle can be derived by setting $\omega_{zp} = 0$. The critical duty cycle is expressed and calculated as

$$D_{Cr} = \frac{2R_L + r_{DS} - R_F - \sqrt{4R_L(r_L + r_{DS}) + R_F(R_F - 2r_{DS}) + r_{DS}^2}}{2R_L} = 0.9413. \tag{3.38}$$

Beyond the critical duty cycle D_{Cr}, by the influence of losses in the converter, the RHP zero enters the left-half of the s-plane. Figure 3.12 shows the root locus of the poles p_1 and p_2 and zero z_p of T_p closer to the origin to illustrate the behavior and the relocation beyond D_{Cr}. The stability of the boost converter can also be determined using the Bode plot of the T_p transfer function analyzed at different duty cycles. Figure 3.13 shows the Bode magnitude and phase plot of the T_p transfer function at selected values of the duty cycle. As the duty cycle increases, the DC gain T_{po} increases, and the corner frequency f_o decreases. At larger duty cycles, the phase plot crosses −180°, and the phase margin (PM) is negative. Any perturbation in the duty cycle around this operating point can cause the output voltage to endure sustained oscillations. Such undesirable effects can be avoided, and the full range of the duty ratio can be utilized with an inner-current loop.

3.3.2 Duty Cycle-to-Inductor Current Transfer Function T_{pi}

The duty cycle-to-inductor current power-stage transfer function is required to design and characterize the inner-current loop. By controlling the inductor current,

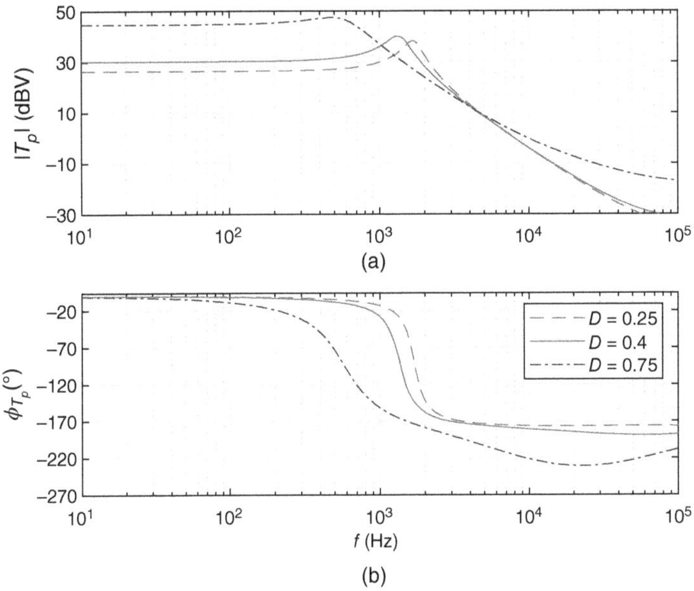

Figure 3.13 Theoretically obtained plots of duty cycle-to-output voltage transfer function T_p at selected duty ratios. (a) Magnitude. (b) Phase.

whose energy states change at the rate of the switching period, an inner-current loop with bandwidth much larger than the outer-voltage loop can be designed. The disturbances v_i and i_o shown in Figure 3.6 are set to zero to derive the duty cycle-to-inductor current transfer function. Using KVL and KCL, the duty cycle-to-inductor current transfer function is

$$T_{pi}(s) = \left.\frac{i_l(s)}{d(s)}\right|_{v_i=i_o=0} = V_O \frac{1+\frac{Z_2}{R_L}}{Z_1+(1-D)^2 Z_2}. \tag{3.39}$$

Substituting (3.18) and (3.19) into (3.39),

$$T_{pi}(s) = \left.\frac{i_l(s)}{d(s)}\right|_{v_i=i_o=0} = T_{pix}\frac{s+\omega_{zi}}{s^2+2\xi\omega_o s+\omega_o^2} = T_{pio}\frac{1+\frac{s}{\omega_{zi}}}{\left(\frac{s}{\omega_o}\right)^2+\frac{2\xi s}{\omega_o}+1}, \tag{3.40}$$

where T_{pix} is

$$T_{pix} = \frac{V_O(R_L+2r_C)}{L(R_L+r_C)}, \tag{3.41}$$

the DC gain T_{pio} is

$$T_{pio} = \frac{2V_O}{(1-D)^2 R_L + r}, \tag{3.42}$$

and the angular frequency of the LHP zero is

$$\omega_{zi} = \frac{1}{C\left(\frac{R_L}{2}+r_C\right)}. \tag{3.43}$$

The magnitude of the duty cycle-to-inductor current transfer function is

$$|T_{pi}| = T_{pio}\sqrt{\frac{1+\left(\frac{\omega}{\omega_{zi}}\right)^2}{\left[1-\left(\frac{\omega}{\omega_o}\right)^2\right]^2+4\xi^2\left(\frac{\omega}{\omega_o}\right)^2}}, \tag{3.44}$$

and the phase of the transfer function is

$$\phi_{Tpi} = -180° + \tan^{-1}\left(\frac{\omega}{\omega_{zi}}\right) - \tan^{-1}\left[\frac{2\xi\left(\frac{\omega}{\omega_o}\right)}{1-\left(\frac{\omega}{\omega_o}\right)^2}\right]. \tag{3.45}$$

The duty cycle-to-inductor current transfer function exhibits a dominant zero response. At $f = f_{BWi}$, where f_{BWi} is the bandwidth of the duty cycle-to-output voltage transfer function, we have

$$\frac{|T_{pi}(f_{BW})|}{T_{pio}} = \frac{1}{\sqrt{2}}. \tag{3.46}$$

The bandwidth of the T_{pi} transfer function is governed by the dominant converter zero z_i. Using the condition (3.46) in (3.44) gives

$$\frac{T_{pi}(f_{BWi})}{T_{pio}} = \frac{1}{\sqrt{2}} = \sqrt{\frac{1+\left(\frac{\omega}{\omega_{zi}}\right)^2}{\left[1-\left(\frac{\omega_{BW}}{\omega_o}\right)^2\right]^2+4\xi^2\left(\frac{\omega_{BW}}{\omega_o}\right)^2}}, \tag{3.47}$$

which yields

$$\frac{1+\left(\frac{f_{BWi}}{f_{zi}}\right)^2}{\left[1-\left(\frac{f_{BWi}}{f_o}\right)^2\right]^2+4\xi^2\left(\frac{f_{BWi}}{f_o}\right)^2} = \frac{1}{2} \tag{3.48}$$

and produces the bandwidth of the duty cycle-to-inductor current transfer function as

$$f_{BWi} = f_o\sqrt{1-2\xi^2+\left(\frac{f_o}{f_{zi}}\right)^2+\frac{1}{2}\sqrt{\left[4\xi^2-2\left(1+\frac{f_o}{f_{zi}}\right)^2\right]^2-1}}. \tag{3.49}$$

The expression (3.49) is valid for transfer functions with dominant zero response and is not applicable for converters that exhibit a dominant pole effect.

For the boost converter designed in Section 3.1.3, the following results can be determined for the duty cycle-to-inductor current transfer function T_{pi}:

1. DC gain, corner frequency, frequency of the LHP zero, damping coefficient, and the bandwidth of the transfer function.
2. Location of the poles and zeros.
3. Bode magnitude and phase plot of the transfer function.

4. Response of the inductor current for a step change in the duty cycle by $\Delta D = 0.1$.
5. Impact of the load resistance and the duty cycle on the location of the poles and zeros on the s-plane.

From Section 3.1.3, $R_L = 40\ \Omega$, $L = 50\ \mu\text{H}$, $C = 100\ \mu\text{F}$, $r_L = 30\ \text{m}\Omega$, $r_C = 25\ \text{m}\Omega$, $r_{DS} = 0.11\ \Omega$, $R_F = 0.07\ \Omega$, and $V_F = 0.3$ V. The duty cycle is $D = 0.41$ to achieve a rated output voltage $V_O = 20$ V.

Using the given specifications, the DC gain $T_{pio} = 2.8\ \text{A} = 8.94$ dBA, the natural corner frequency $f_o = \omega_o/2\pi = 1.34$ kHz, and the frequency of the LHP zero $f_{zi} = \omega_{zi}/2\pi = 79.48$ Hz. The damping coefficient is $\xi = 0.1628$. The gain of the transfer function at the 3-dB bandwidth is $T_{pi}(BW) = T_{pio}/\sqrt{2} = 1.98\ \text{A} = 5.94$ dBA. The bandwidth is $f_{BWi} = 32.23$ kHz, i.e. $f_{BWi}/f_o = 24$.

The poles are complex conjugate and are located on the left-half of the s-plane. The real part of the poles is dependent on the damping coefficient. At $\omega_o = 8.4493$ krad/s and $\xi = 0.1628$, the poles are located at

$$p_1, p_2 = -1375.5 \pm j8336.5\ \text{rad/s}. \tag{3.50}$$

The LHP zero is at

$$z_i = -\omega_{zi} = -499.4\ \text{rad/s}. \tag{3.51}$$

Figure 3.14 shows the theoretically obtained magnitude and phase plots of the duty cycle-to-inductor current transfer function. The bandwidth of the converter is nearly 16 times wider than the corner frequency f_o. The duty cycle-to-inductor current transfer function for the boost converter exhibits a low-frequency zero f_{zi} and

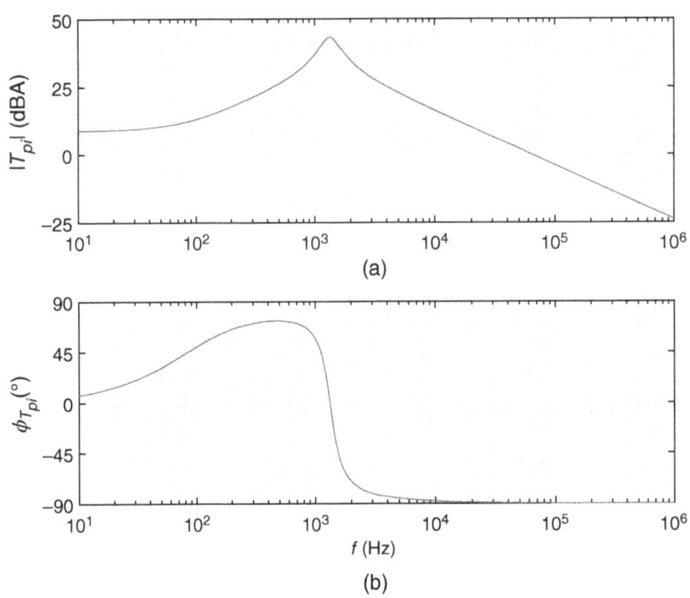

Figure 3.14 Theoretically obtained plots of duty cycle-to-inductor current transfer function T_{pi}. (a) Magnitude. (b) Phase.

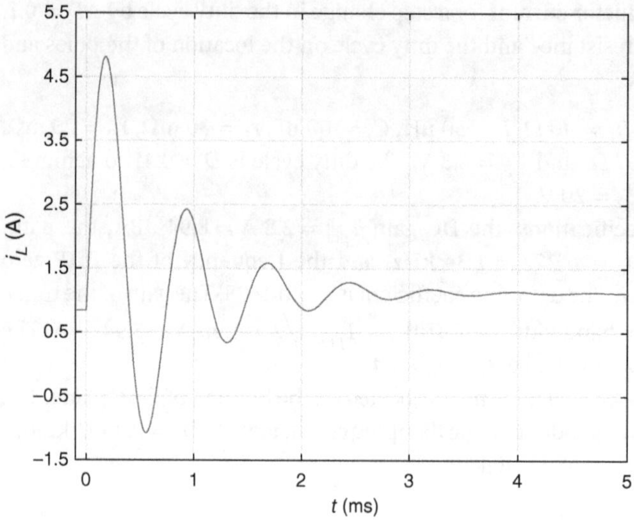

Figure 3.15 Response of the inductor current to step change in the duty cycle by $\Delta D = 0.1$.

a complex-conjugate pole-pair at f_o. The phase starts at zero and rises to 45° due to f_{zi}. This means that at low frequencies, the duty cycle signal leads the inductor current by a particular phase angle. The bandwidth of T_{pi} is larger than that of T_p. Therefore, the inner-current loop presents a faster response in inductor current for changes in duty cycle. The low-frequency zero f_{zi} depends on the load resistance R_L and the parameters of the filter capacitance, namely C and r_C. As the value of R_L is decreased, f_{zi} moves toward f_o. The roots of the characteristic equation (denominator of T_{pi}) or the poles of T_{pi} transition from complex conjugate to real.

Figure 3.15 shows the inductor current response to a step change in the duty cycle by $\Delta D = 0.1$. The inductor current increased from 0.83 to 1.11 A in nearly 4 ms. The step change in the inductor current is $\Delta I_L = 1.11 - 0.83 = 0.28$ A. Therefore, the ratio $\Delta I_L/\Delta D = 0.28/0.1 = 2.8$ A, which is equal to the DC gain of the duty cycle-to-inductor current transfer function T_{pio}.

Figure 3.16 shows the trajectories of the power-stage poles p_1, p_2 and the LHP zero z_i at different values of the load resistance R_L varying from 10 to 50 Ω. It is observed that the real components of the complex-conjugate poles move toward the origin with increase in the load resistance. The LHP zero also moves toward the origin with increase in the load resistance. Figure 3.17 shows the Bode magnitude and phase plots of the T_{pi} transfer function at selected values of the load resistance. The DC gain decreases with the increase in the load resistance. As the load resistance increases, the frequency of the LHP zero decreases. As a result, a large phase boost is observed at lower frequencies for higher values of R_L. The phase plot reaches −90° for all values of the load resistance.

Figure 3.18 shows the trajectories of the power-stage poles p_1, p_2 and the LHP zero z_i at different values of the duty cycle varying from 0.3 to 0.95. In accordance with (3.34), the imaginary components of the poles p_1 and p_2 move toward the origin with the increase in the duty cycle. At duty cycles larger than a critical value, the

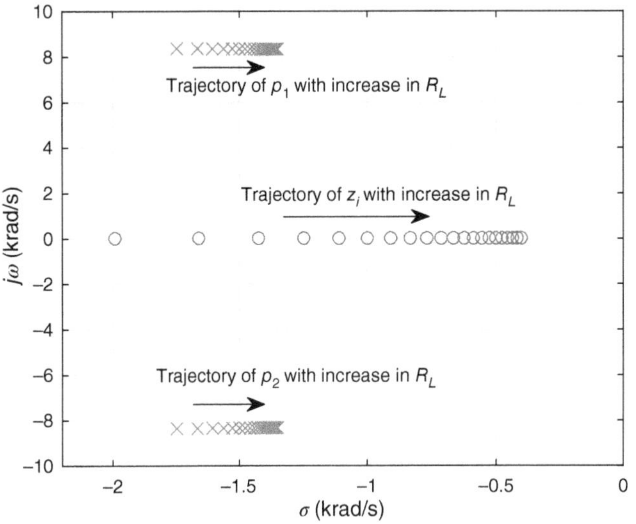

Figure 3.16 Trajectory of the poles and zeros of T_{pi} transfer function as functions of the load resistance.

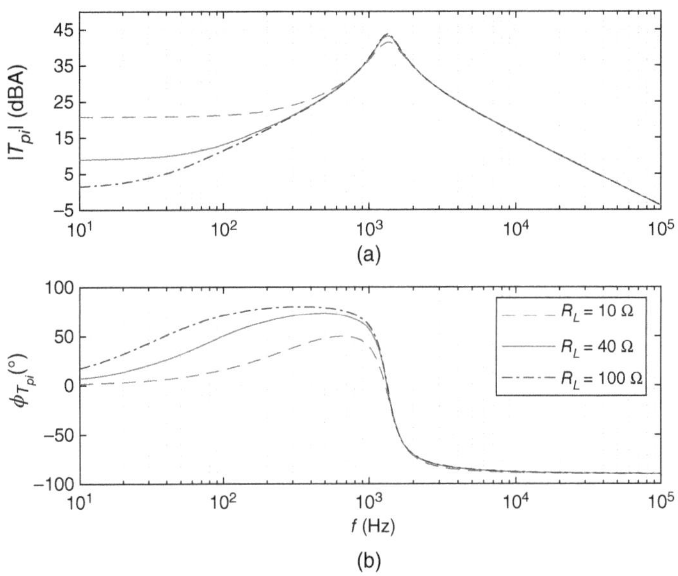

Figure 3.17 Theoretically obtained plots of duty cycle-to-inductor current transfer function T_{pi} at selected load resistances. (a) Magnitude. (b) Phase.

poles become real with p_1 moving toward the origin and p_2 moving toward $-\infty$. The location of the LHP zero z_i is independent of the duty cycle. Figure 3.19 shows the Bode magnitude and phase plot of the T_{pi} transfer function at selected values of the duty cycle. As the duty cycle increases, the DC gain T_{pio} also increases rapidly, and the corner frequency f_o decreases. The phase starts at 0° and saturates at −90° at

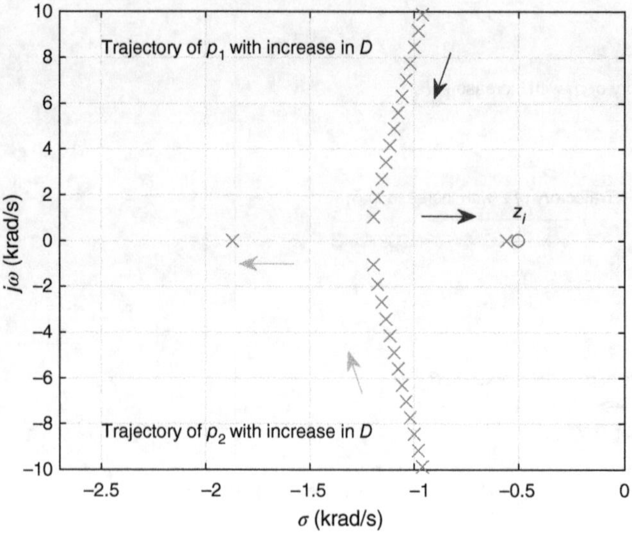

Figure 3.18 Trajectory of the poles and zeros of T_{pi} transfer function as functions of the duty cycle.

high frequencies for all values of duty cycle. As the duty cycle increases, the phase boost decreases. Since the phase at high frequencies is −90°, the phase margin of the inner-current loop is nearly fixed for all the duty cycles. This offers two advantages: (i) a wide range of duty cycle can be achieved for current control, and (ii) simpler controller techniques can be used.

3.3.3 Input Voltage-to-Output Voltage Transfer Function M_v

The input voltage-to-output voltage transfer function is obtained by setting $d = 0$ and $i_o = 0$ in Figure 3.6. The input voltage-to-output voltage transfer function M_v is

$$M_v(s) = \left.\frac{v_o(s)}{v_i(s)}\right|_{d=i_o=0} = \frac{Z_2(1-D)}{Z_2(1-D)^2 + Z_1}. \tag{3.52}$$

Substituting (3.18) and (3.19) into (3.52) gives the input voltage-to-output voltage transfer function in the s-domain as (Figure 3.19)

$$M_v(s) = \left.\frac{v_o(s)}{v_i(s)}\right|_{d=i_o=0} = M_{vx}\frac{s+\omega_{zn}}{s^2 + 2\xi\omega_o s + \omega_o^2} = M_{vo}\frac{1+\frac{s}{\omega_{zn}}}{\left(\frac{s}{\omega_o}\right)^2 + \frac{2\xi s}{\omega_o} + 1}, \tag{3.53}$$

where

$$M_{vx} = \frac{(1-D)R_L r_C}{L(R_L + r_C)} \tag{3.54}$$

and the DC gain M_{vo} is

$$M_{vo} = \frac{(1-D)R_L}{(1-D)^2 R_L + r}. \tag{3.55}$$

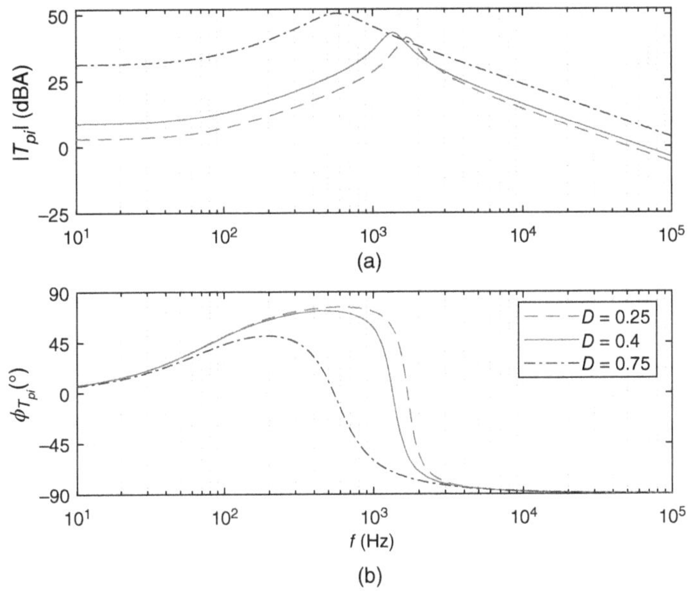

Figure 3.19 Theoretically obtained plots of duty cycle-to-inductor current transfer function T_{pi} at selected duty ratios. (a) Magnitude. (b) Phase.

The angular frequency of the LHP zero is given in (3.30).

Figure 3.20 shows the theoretically obtained magnitude and phase plots of input voltage-to-output voltage transfer function. The gain at DC is $M_{vo} = 1.67$ A/V = 4.43 dBA/V. The phase starts at 0°. At high frequencies the phase is −90°.

3.3.4 Input Voltage-to-Inductor Current Transfer Function M_{vi}

The input voltage-to-inductor current transfer function is obtained by setting $d = 0$ and $i_o = 0$ in Figure 3.6. Setting $d = 0$ in (3.21) and rearranging, we get

$$i_l = \frac{Dv_i}{Z_1 + Z_2}. \tag{3.56}$$

Hence, the input voltage-to-inductor current transfer function M_{vi} is

$$M_{vi}(s) = \left.\frac{i_l(s)}{v_i(s)}\right|_{d=i_o=0} = \frac{1}{Z_1 + (1-D)^2 Z_2}. \tag{3.57}$$

Substituting (3.18) and (3.19) into (3.57) gives the input voltage-to-inductor current transfer function in the s-domain as

$$M_{vi}(s) = \left.\frac{i_l(s)}{v_i(s)}\right|_{d=i_o=0} = M_{vix}\frac{s + \omega_{zvi}}{s^2 + 2\xi\omega_o s + \omega_o^2} = M_{vio}\frac{1 + \frac{s}{\omega_{zvi}}}{\left(\frac{s}{\omega_o}\right)^2 + \frac{2\xi s}{\omega_o} + 1}, \tag{3.58}$$

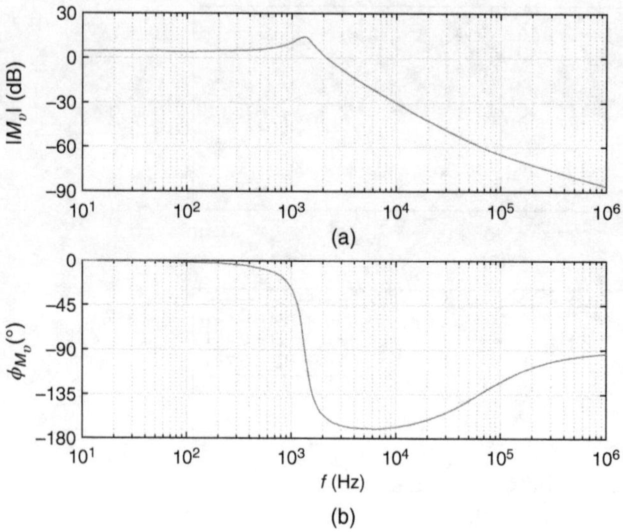

Figure 3.20 Theoretically obtained plots of input-to-output voltage transfer function M_v. (a) Magnitude. (b) Phase.

where

$$M_{vix} = \frac{1}{L}, \tag{3.59}$$

the DC gain M_{vio} is

$$M_{vio} = \frac{1}{(1-D)^2 R_L + r}, \tag{3.60}$$

and the angular frequency of the LHP zero is

$$\omega_{zvi} = \frac{1}{C(R_L + r_C)}. \tag{3.61}$$

Figure 3.21 shows the theoretically obtained magnitude and phase plots of input voltage-to-inductor current transfer function. The gain at DC is $M_{vio} = 0.07$ A/V $= -23$ dBA/V. Through the M_{vi} current-loop relevant transfer function, one can observe that the current loop offers a lower audio-susceptibility at DC than that provided by the voltage-loop transfer function M_v. All the other frequency-domain properties remain identical to T_{pi}.

3.3.5 Reverse Current Gain A_i

The small-signal model to derive the output-to-inductor current transfer function is shown in Figure 3.22. This model is obtained by setting $v_i = 0$ and $d = 0$. An independent voltage source v_t is applied at the output, which forces a current i_t. The output current-to-inductor current transfer function, i.e. reverse current transfer function A_i, is

$$A_i(s) = \left.\frac{i_l(s)}{i_o(s)}\right|_{d=v_i=0} = \frac{(1-D)Z_2}{Z_1 + (1-D)^2 Z_2}. \tag{3.62}$$

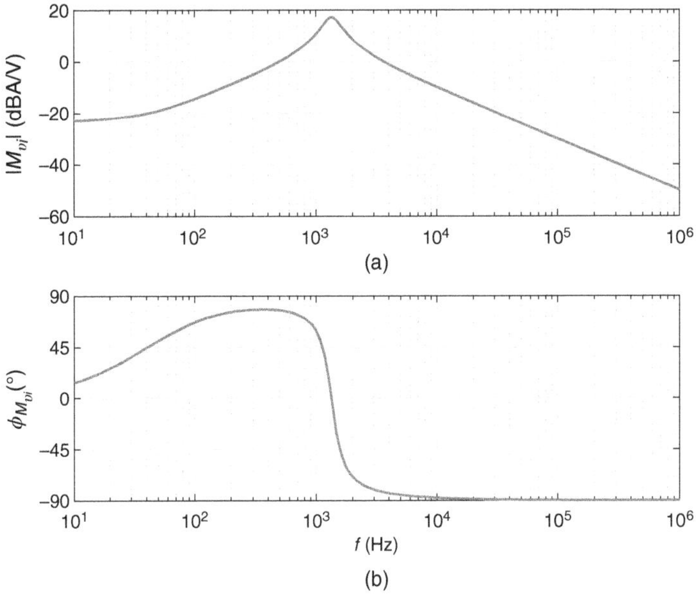

Figure 3.21 Theoretically obtained plots of input voltage-to-inductor current transfer function M_{vi}. (a) Magnitude. (b) Phase.

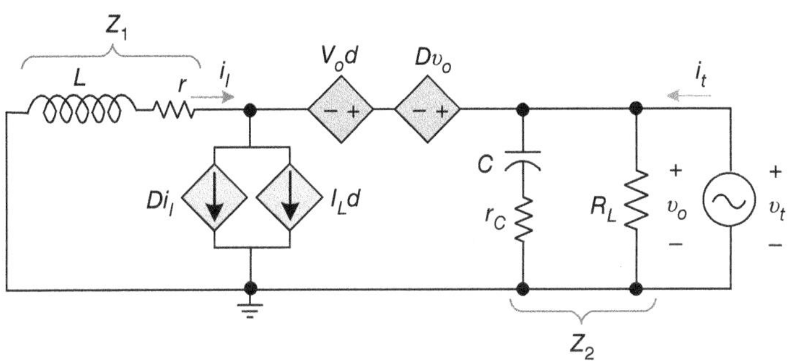

Figure 3.22 Small-signal model of the PWM boost DC–DC converter in CCM to derive the output current-to-inductor current transfer function A_i.

Substituting (3.18) and (3.19) into (3.62) gives the output current-to-inductor current transfer function in the s domain as

$$A_i(s) = \left.\frac{i_l(s)}{i_o(s)}\right|_{d=v_i=0} = A_{ix}\frac{s+\omega_{zn}}{s^2+2\xi\omega_o s+\omega_o^2} = A_{io}\frac{1+\frac{s}{\omega_{zn}}}{\left(\frac{s}{\omega_o}\right)^2+\frac{2\xi s}{\omega_o}+1}, \tag{3.63}$$

where

$$A_{ix} = \frac{(1-D)R_L r_C}{L(R_L+r_C)}, \tag{3.64}$$

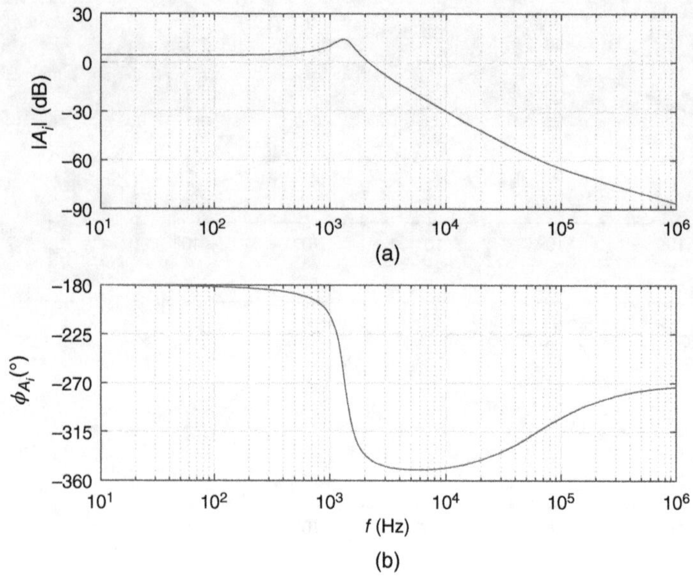

Figure 3.23 Theoretically obtained plots of output current-to-inductor current transfer function A_i. (a) Magnitude. (b) Phase.

the DC gain A_{io} is

$$A_{io} = \frac{(1-D)R_L}{(1-D)^2 R_L + r}, \tag{3.65}$$

and the angular frequency of the LHP zero ω_{zn} is given in (3.30).

Figure 3.23 shows the theoretically obtained magnitude and phase plots of output-to-inductor current transfer function. The gain at DC is $A_{io} = 1.67$ A/A = 4.45 dB, i.e. the reverse current gain is unity at DC and low frequencies up to nearly f_o. Beyond f_o, the impedance offered by the capacitor branch is lower than the inductor, and the small-signal output current begins to flow into the capacitor. Therefore, the gain reduces by −40 dB per decade.

3.3.6 Open-Loop Input Impedance Z_i

The open-loop input impedance is obtained by setting $d = 0$ and $i_o = 0$ in Figure 3.6. The open-loop input impedance is

$$Z_i(s) = \left.\frac{v_i(s)}{i_i(s)}\right|_{d=i_o=0} = Z_1 + Z_2(1-D)^2. \tag{3.66}$$

Substituting (3.18) and (3.19) into (3.66) gives input impedance in the s-domain as

$$Z_i(s) = \left.\frac{v_i(s)}{i_i(s)}\right|_{d=i_o=0} = L_i \frac{s^2 + 2\xi\omega_o s + \omega_o^2}{s + \omega_{zvi}} = R_{io} \frac{\left(\frac{s}{\omega_o}\right)^2 + \frac{2\xi s}{\omega_o} + 1}{1 + \frac{s}{\omega_{zvi}}}, \tag{3.67}$$

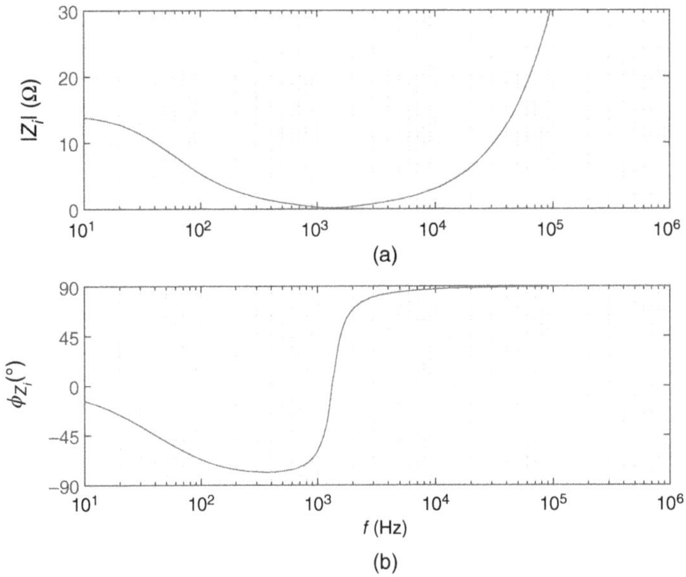

Figure 3.24 Theoretically obtained plots of input impedance Z_i. (a) Magnitude. (b) Phase.

where L_i is the high-frequency inductance and R_{io} is the DC input resistance. At high frequencies,

$$Z_i(\infty) = sL_i = L. \tag{3.68}$$

At DC,

$$R_{io} = Z_i(0) = r + R_L(1 - D)^2. \tag{3.69}$$

The angular frequency of the LHP pole ω_{zvi} is

$$\omega_{zvi} = \frac{1}{C(R_L + r_C)}. \tag{3.70}$$

The gain Z_{ix} increases with the frequency. Figure 3.24 shows the theoretically obtained magnitude and phase plots of the input impedance. The gain at DC is $R_{io} = 14.28\ \Omega$, and the high-frequency inductance is $L_i = 50\ \mu\text{H}$. The gain moves toward zero around f_o and increases rapidly at high frequencies.

3.3.7 Open-Loop Output Impedance Z_o

The small-signal model to derive the output impedance is shown in Figure 3.22. This model is obtained by setting $v_i = 0$ and $d = 0$. The open-loop output impedance is

$$Z_o(s) = \left.\frac{v_o(s)}{i_o(s)}\right|_{d=i_o=0} = \frac{Z_1}{(1-D)^2}||Z_2 = \frac{Z_1 Z_2}{Z_1 + Z_2(1-D)^2}. \tag{3.71}$$

Table 3.1 Summary of calculated values for open-loop transfer functions

Variable	Value
$\lvert T_{po}\rvert$	33.05 V = 30.38 dBV
$\lvert T_{pio}\rvert$	2.8 A = 9 dBA
$\lvert M_{vo}\rvert$	1.67 V/V = 4.5 dB
$\lvert M_{vio}\rvert$	0.07 A/V = −23 dBA/V
$\lvert Z_{io}\rvert$	14.28 Ω
$\lvert Z_{oo}\rvert$	0.325 Ω
ω_o	8.45 krad/s
f_o	1.345 kHz
ξ	0.1628
ω_{zi}	500 rad/s
f_{zi}	80 Hz
ω_{zvi}	250 rad/s
f_{zvi}	40 Hz
ω_{zn}	400 krad/s
f_{zn}	63.7 kHz
ω_{zp}	281 krad/s
f_{zp}	45 kHz
ω_{rl}	2.32 krad/s
f_{rl}	370 kHz

Substituting (3.18) and (3.19) into (3.71) gives the open-loop output impedance in the s domain as

$$Z_o(s) = \left.\frac{v_o(s)}{i_o(s)}\right|_{d=i_o=0} = R_{ox}\frac{(s+\omega_{zn})(s+\omega_{rl})}{s^2+2\xi\omega_o s+\omega_o^2} = R_{oo}\frac{\left(1+\frac{s}{\omega_{zn}}\right)\left(1+\frac{s}{\omega_{rl}}\right)}{\left(\frac{s}{\omega_o}\right)^2+\frac{2\xi s}{\omega_o}+1}, \tag{3.72}$$

where the high-frequency output resistance R_{ox} is

$$R_{ox} = Z_o(\infty) = \frac{R_L r_C}{R_L + r_C}, \tag{3.73}$$

the DC gain R_{oo} is

$$R_{oo} = Z_o(0) = \frac{rR_L}{r + R_L(1-D)^2}, \tag{3.74}$$

and the angular frequency of the LHP zero ω_{rl} is

$$\omega_{rl} = \frac{r}{L}. \tag{3.75}$$

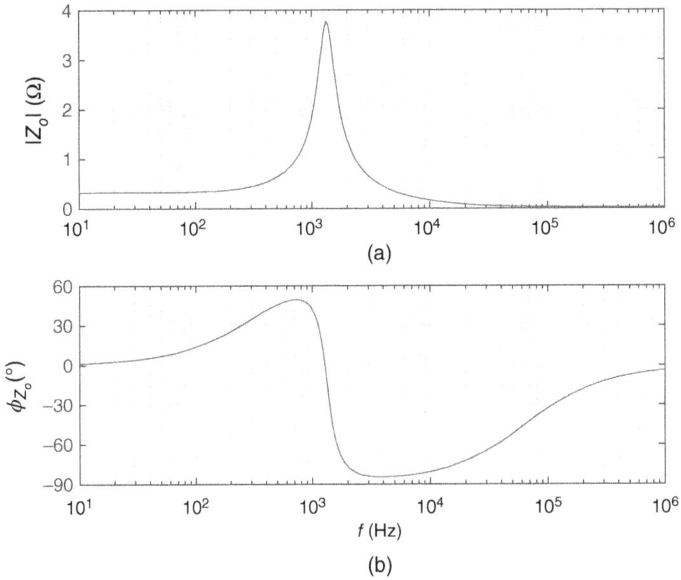

Figure 3.25 Theoretically obtained plots of output impedance Z_o. (a) Magnitude. (b) Phase.

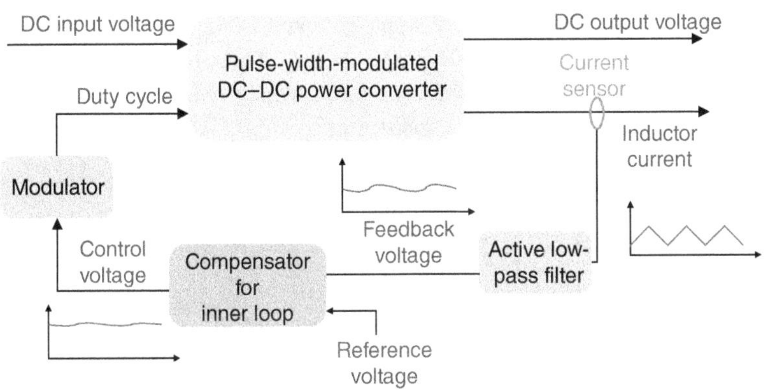

Figure 3.26 Architecture of the inner-current loop with low-pass filter in the feedback path.

The angular frequency of the LHP zero ω_{zn} is given in (3.30). Figure 3.25 shows the theoretically obtained magnitude and phase plots of the output impedance.

The gain at DC is nearly $R_{oo} = 0.325\ \Omega$. The high-frequency output resistance is $R_{ox} = 25\ \text{m}\Omega$. At DC, the inductor is a short circuit, and the filter capacitor is an open circuit. The DC resistance is due to the parallel combination of the load resistance R_L and the equivalent averaged resistance r. With the increase in frequency, the capacitive reactance is high, and the reactance is mostly offered by the inductive reactance. At a frequency f_o, the inductive and capacitive reactances are equal, and the converter filter network undergoes resonance. Beyond f_o, the capacitive reactance dominates and reduces the output impedance to nearly zero. The phase starts at zero and increases toward 90° due to the zero at frequency ω_{rl}. The phase decreases toward −90° around the frequency of poles. However, the LHP zero at ω_{zn} increases the phase back to 0°. Table 21 provides a summary of the characteristics values for open-loop transfer functions (Figure 3.26). A summary of calculated parameters for the open-loop transfer functions for the boost converter are listed in Table 3.1.

3.4 Inner-Current Loop

Architecture of the inner-current loop with low-pass filter in the feedback path is shown in Figure 3.26. Figure 3.27 shows the circuit of the boost DC–DC converter with inner-current loop. A sense resistor R_s is placed in the inductor branch to sense the inductor current. At low duty cycles, the sensed inductor current ripple is high, which can cause the control voltage V_{CI} to intersect with the modulator ramp voltage twice in each cycle, hence resulting in the subharmonic instability. The sensed voltage $v_{RS} = R_s i_L$ contains the switching frequency components and its harmonics, which are attenuated using the loss-pass filter. The theoretically obtained output voltage of the low-pass filter is $V_{FI} = V_{Rs} = R_s I_L$ and is governed by the DC component of the sensed inductor current. At steady state, the feedback voltage V_{FI} is equal to the reference voltage V_{RI} of the inner-current loop, i.e. $V_{FI} = V_{Rs} \approx V_{RI} = R_s I_L$. As a result, the control voltage V_{CI} is nearly ripple-free and is equal to the reference voltage at steady state, i.e. $V_{CI} = V_{FI} = V_{Rs} \approx V_{RI} = R_s I_L$. Thus, in this method, the *average* component of the inductor current is sensed, which reduces the problems due to the subharmonic instability and transistor misgating. The transfer functions

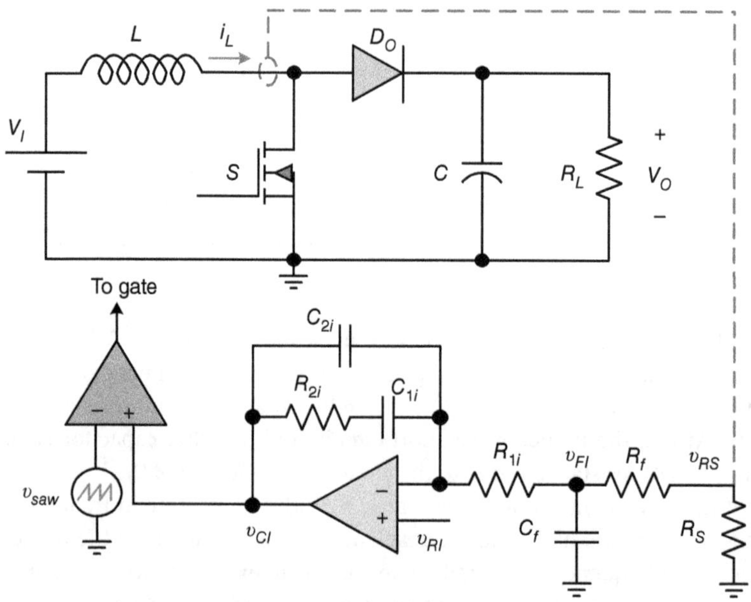

Figure 3.27 Circuit of the boost DC–DC converter with inner-current loop.

related to the inner-current loop, which includes the filter in the feedback path, are derived and discussed in the following sections.

3.4.1 Transfer Function of Filter and Non-inverting Amplifier T_f

The low-pass filter is composed of a resistor R_f and a capacitor C_f. The transfer function of two stages is

$$\begin{aligned} T_f(s) &= \frac{v_{fi}(s)}{v_{rs}(s)} = \frac{Z_{Cf}}{R_f + Z_{Cf}} = \frac{\frac{1}{sC_f}}{R_f + \frac{1}{sC_f}} = \frac{1}{sR_fC_f + 1} \\ &= \frac{1}{R_fC_f}\frac{1}{\frac{1}{R_fC_f} + s} = \frac{\omega_{pf}}{\omega_{pf} + s} = \frac{1}{1 + \frac{s}{\omega_{pf}}}. \end{aligned} \tag{3.76}$$

Setting $s = j\omega$,

$$T_f(j\omega) = \frac{1}{1 + j\frac{\omega}{\omega_{pf}}}, \tag{3.77}$$

which yields

$$T_f(f) = \frac{1}{1 + j\frac{f}{f_{pf}}} = |T_f(f)|e^{j\phi(f)}, \tag{3.78}$$

where

$$|T_f(f)| = \frac{1}{\sqrt{1 + \left(\frac{f}{f_{pf}}\right)^2}} \tag{3.79}$$

and

$$\phi(f) = -\tan^{-1}\left(\frac{f}{f_{pf}}\right). \tag{3.80}$$

Using (3.79),

$$|T_f(f)|^2 = \frac{1}{1 + \left(\frac{f}{f_{pf}}\right)^2}. \tag{3.81}$$

This yields

$$\left(\frac{f}{f_{pf}}\right)^2 = \frac{1}{|T_f(f)|^2} - 1 = \frac{1 - |T_f(f)|^2}{|T_f(f)|^2} \tag{3.82}$$

and

$$\frac{f_{pf}}{f} = \sqrt{\frac{|T_f(f)|^2}{1 - |T_f(f)|^2}}. \tag{3.83}$$

Hence,

$$f_{pf} = f\sqrt{\frac{|T_f(f)|^2}{1 - |T_f(f)|^2}}. \tag{3.84}$$

The magnitude of the transfer function T_f is also given as

$$|T_f(f)| = \frac{\Delta V_{FI}}{\Delta V_{RS}}. \tag{3.85}$$

The filter frequency can be set by the amount of attenuation desired. As an example, to achieve an attenuation of 50% for $f = f_s = 100$ kHz, the filter upper cutoff frequency can be calculated using (3.84) as

$$f_{pf} = f_s\sqrt{\frac{|T_f(f)|^2}{1-|T_f(f)|^2}} = 100\times 10^3 \times \sqrt{\frac{0.50^2}{1-0.50^2}} = 57\text{ kHz}. \tag{3.86}$$

Assuming $R_f = 1\text{ k}\Omega/0.25\ W/2\%$, the value of the filter capacitor can be calculated as $C_f = 1/(2\pi f_{pf} R_f) \approx 2.8$ nF. A 3.2 nF/25 V capacitor is selected.

3.4.2 Transfer Function of Pulse-Width Modulator T_m

The control voltage-to-duty cycle transfer function of the pulse-width modulator is

$$T_m = \frac{d}{v_{ci}} = \frac{D}{V_{CI}} = \frac{1}{V_{Tm}}, \tag{3.87}$$

where V_{Tm} is the amplitude of the sawtooth waveform.

At a duty cycle $D = 0.41$ and assuming the peak sawtooth voltage of the pulse-width modulator is $V_{Tm} = 3$ V, the DC value of the control voltage and the reference voltage required for the rated duty cycle are $V_{CI} = V_{RI} = DV_{Tm} = 1.2$ V. The steady-state average inductor current is $I_L = V_O/R_L(1-D) = 0.83$ A. The value of the sense resistor is

$$R_s = \frac{V_{RI}}{I_L} = \frac{V_{Rs}}{I_L} = \frac{1.2}{0.83} = 1.5\ \Omega. \tag{3.88}$$

A standard resistor $R_s = 1.5\ \Omega/0.25\ W/2\%$ is selected. From Section 3.4.1, $R_f = 1\text{ k}\Omega/0.25\ W/2\%$ and $C_f = 3.2$ nF/25 V.

3.4.3 Uncompensated Loop Gain T_{ki}

The natural behavior of the inner-current loop can be determined using the uncompensated loop gain as

$$T_{ki} = \frac{v_{fi}}{v_{ei}} = T_m T_{pi} R_s T_f = T_{kio}\frac{\left(1+\frac{s}{\omega_{zi}}\right)}{\left(1+\frac{s}{\omega_{pf}}\right)\left(1+\frac{2\xi s}{\omega_o}+\frac{s^2}{\omega_o^2}\right)}, \tag{3.89}$$

where T_m, T_{pi}, and T_f are given in (3.87), (3.40), and (3.76), respectively. The DC gain T_{kio} is given by

$$T_{kio} = \frac{R_s}{V_{Tm}}\frac{2V_O}{R_L(1-D)^2+r} = 1.36\text{ V/V}. \tag{3.90}$$

Figure 3.28 shows the theoretically obtained magnitude and phase plots of the uncompensated loop gain of the inner-current loop.

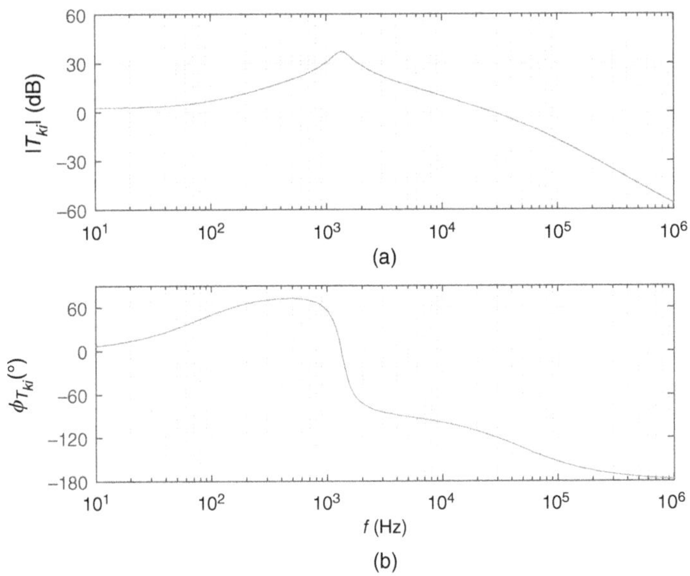

Figure 3.28 Theoretically obtained plots of uncompensated loop gain T_{ki}. (a) Magnitude. (b) Phase.

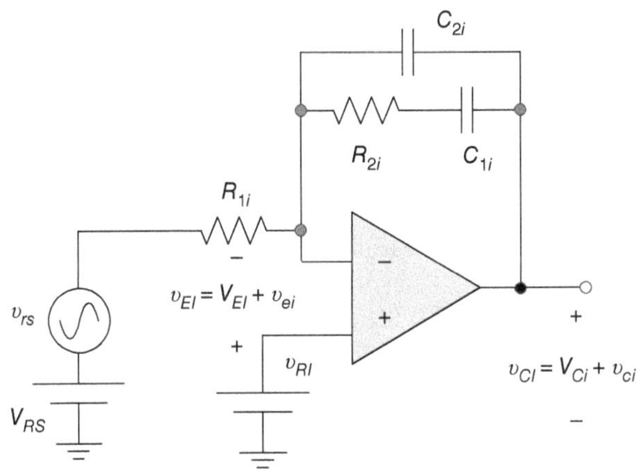

Figure 3.29 Circuit of Type-II controller.

3.4.4 Transfer Function of Control Circuit T_{ci}

Figure 3.29 shows the circuit of a single-lead integral controller, which is also called the Type-II controller. A single-lead integral control circuit introduces a pole at the origin, which is required to boost the DC gain and a pole–zero pair to adjust the crossover frequency f_c and maintain required phase margin (PM). A detailed procedure to determine the transfer function of the controller to improve the DC gain (>20 dB) and to achieve an inner-current loop phase margin $PM = 60°$ is explained in [45]. The remainder of this section details the Type-II controller circuit design.

The magnitude of the uncompensated loop gain transfer function can be determined using

$$|T_{ki}(f)| = T_{kio}\frac{\sqrt{1+\left(\frac{f}{f_{zi}}\right)^2}}{\sqrt{1+\left(\frac{f}{f_{pf}}\right)^2}\sqrt{1-\left(\frac{f}{f_o}\right)^2+(4\xi^2\left(\frac{f}{f_o}\right)^2}}, \tag{3.91}$$

and the phase is

$$\phi_{Tk}(f) = -180^\circ + \tan^{-1}\left(\frac{f}{f_{zi}}\right) - \tan^{-1}\left(\frac{f}{f_{pf}}\right) - \tan^{-1}\frac{2\xi\frac{f}{f_o}}{1-\left(\frac{f}{f_o}\right)^2}. \tag{3.92}$$

The desired crossover frequency for the loop gain is chosen as $f_c = 10$ kHz. At this frequency, the magnitude of the uncompensated loop gain transfer function is $|T_{ki}(f_c)| = 3.087 = 9.79$ dB, and its phase is $\phi_{Tk}(f_c) = -100^\circ$. The required phase margin is $PM = 60^\circ$. Therefore, the phase boost required at f_c is

$$\phi_m = PM - \phi_{Tk}(f_c) - 90^\circ = 69^\circ, \tag{3.93}$$

giving the geometric mean between the controller corner frequencies as

$$K = \tan\left(\frac{\phi_m}{2} + 45^\circ\right) = 5.45. \tag{3.94}$$

Assuming $R_{1i} = 1\ \text{k}\Omega/0.25\ W/2\%$, the value of C_{2i} is

$$C_{2i} = \frac{|T_k(f_c)|}{2\pi f_c K R_{1i}} = 9\ \text{nF}. \tag{3.95}$$

A 9 nF/50 V standard capacitor is selected. The expression for C_{1i} is

$$C_{1i} = C_{2i}(K^2 - 1) = 259\ \text{nF}. \tag{3.96}$$

A 270 nF/25 V standard capacitor is selected. The expression for R_{2i} is

$$R_{2i} = \frac{K}{2\pi f_c C_{1i}} = 335\ \Omega. \tag{3.97}$$

A 336 $\Omega/0.25\ W/1\%$ standard resistor is selected. The transfer function T_{ci} of the current-loop control circuit is

$$T_{ci}(s) = \frac{v_{ci}(s)}{v_{ei}(s)} = T_{cio}\frac{1+\frac{s}{\omega_{zci}}}{s\left(1+\frac{s}{\omega_{pci}}\right)}, \tag{3.98}$$

where

$$T_{cio} = \frac{1}{R_{1i}(C_{1i}+C_{2i})} = 3.6\ \text{kV/V}, \tag{3.99}$$

$$\omega_{zci} = \frac{1}{R_{2i}C_{1i}} = 11\ \text{krad/s}, \tag{3.100}$$

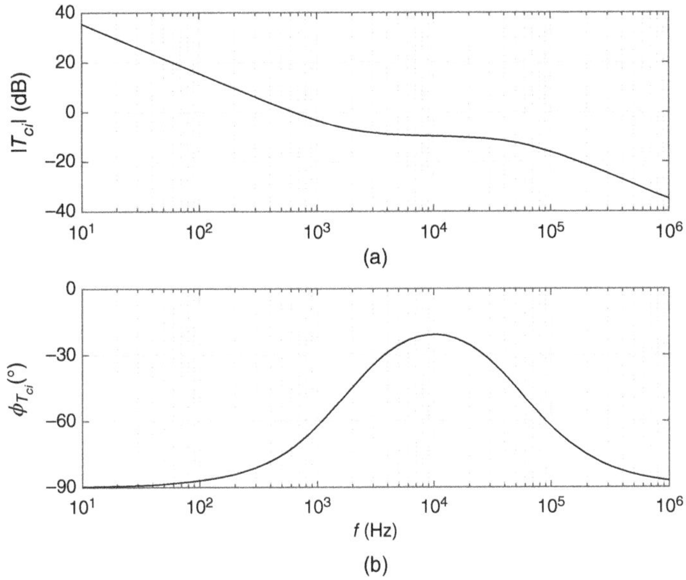

Figure 3.30 Theoretically obtained plots of controller transfer function T_{ci}. (a) Magnitude. (b) Phase.

and

$$\omega_{pci} = \frac{C_{1i} + C_{2i}}{R_{2i}C_{1i}C_{2i}} = 342 \text{ krad/s}. \tag{3.101}$$

Figure 3.30 shows the theoretically obtained magnitude and phase plots of the controller transfer function used in the inner-current loop. The control transfer function T_{ci} is incorporated into the critical path to determine the inner-loop gain and is discussed in Section 3.4.5.

3.4.5 Loop Gain of Inner-Current Loop T_i

The loop gain of the compensated inner-current loop is

$$T_i = \frac{v_{ei}}{v_{fi}} = T_{ki}T_{ci} = T_m T_{pi} R_s T_f T_{ci}, \tag{3.102}$$

which yields

$$T_i = T_{io}\frac{\left(1 + \frac{s}{\omega_{zi}}\right)\left(1 + \frac{s}{\omega_{zci}}\right)}{s\left(1 + \frac{s}{\omega_{pf}}\right)\left(1 + \frac{s}{\omega_{pci}}\right)\left(1 + \frac{2\xi s}{\omega_o} + \frac{s^2}{\omega_o^2}\right)}, \tag{3.103}$$

where T_{io} is the gain at $s = 0$ given as

$$T_{io} = T_{kio}T_{cio} = \frac{R_s}{V_{Tm}}\frac{2V_O}{R_L(1-D)^2 + r}\frac{1}{R_{1i}(C_{1i} + C_{2i})}. \tag{3.104}$$

Figure 3.31 shows the theoretically obtained magnitude and phase plots of compensated loop gain of the inner-current loop. Using the controller designed

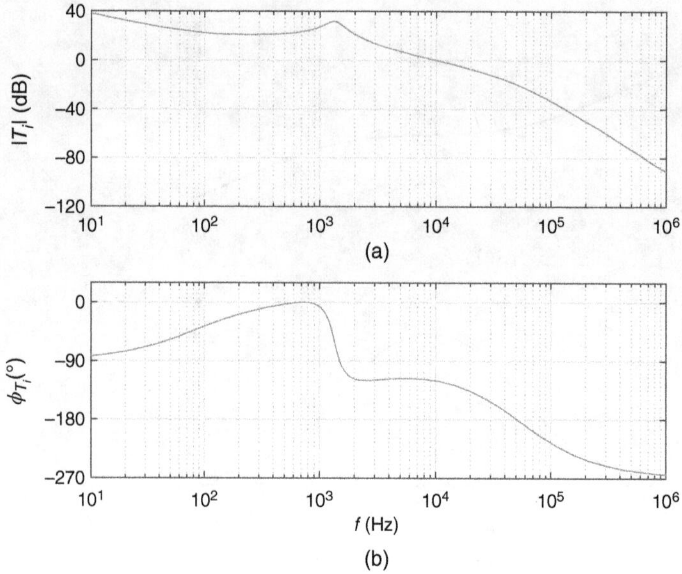

Figure 3.31 Theoretically obtained plots of compensated loop gain T_i of the inner current loop. (a) Magnitude. (b) Phase.

in Section 3.4.4, the calculated value of compensated loop gain at DC is $T_{io} = 5\ kV/V = 74\ dB$. The phase margin is $PM = 60°$, and the gain margin is $20\ dB$. The closed-loop transfer functions for the inner-current loop are derived and discussed in Section 3.5.

3.5 Closed-Loop Transfer Functions for Inner-Current Loop

The following closed-inner-current loop transfer functions are derived:

1. Reference voltage-to-inductor current transfer function T_{icl} (control).
2. Reference voltage-to-output voltage transfer function T_{picl} (control).
3. Input voltage-to-inductor current transfer function M_{icl} (disturbance).
4. Input voltage-to-output voltage transfer function M_{vicl} (disturbance).
5. Input voltage-to-duty cycle transfer function M_{di} (disturbance).
6. Input impedance Z_{iicl} (disturbance).
7. Output impedance Z_{oicl} (disturbance).

3.5.1 Reference Voltage-to-Inductor Current Transfer Function T_{icl}

Using the block diagram shown in Figure 3.32, the closed-loop reference voltage-to-inductor current transfer function is

$$T_{icl}(s) = \left.\frac{i_l(s)}{v_{ri}(s)}\right|_{v_i=i_o=0} = \frac{T_{ci}T_mT_{pi}}{1+T_i}. \tag{3.105}$$

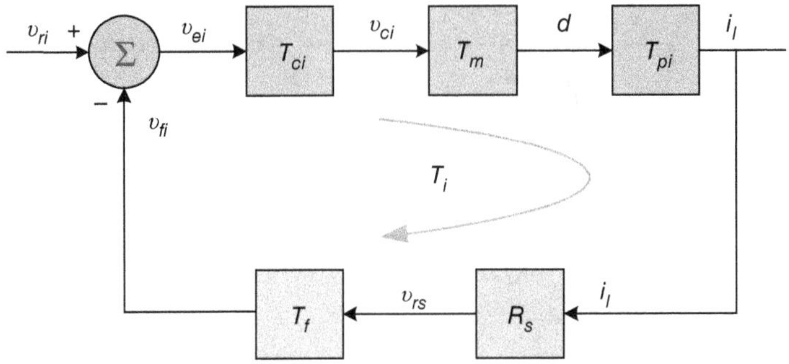

Figure 3.32 Block diagram used to derive the inner-loop control-to-inductor current transfer function T_{icl}.

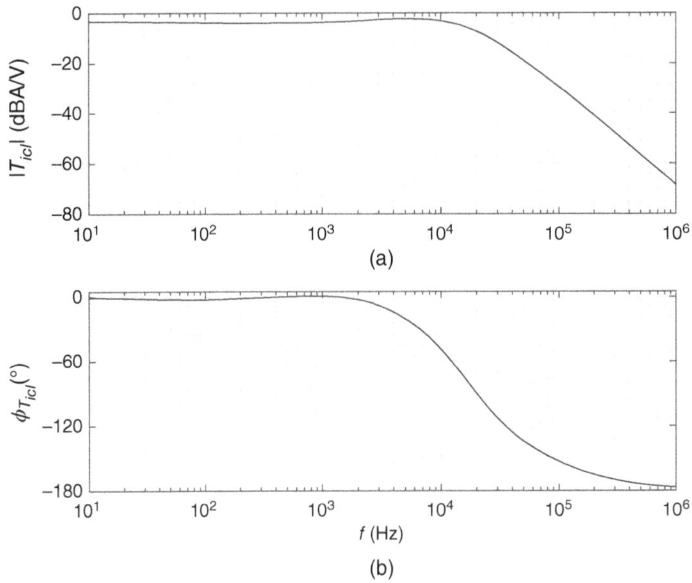

Figure 3.33 Theoretically obtained plots of inner-current loop reference voltage-to-inductor current transfer function T_{icl}. (a) Magnitude. (b) Phase.

Figure 3.33 shows the theoretically obtained magnitude and phase plots of the inner-current loop reference voltage-to-inductor current transfer function.

3.5.2 Reference Voltage-to-Output Voltage Transfer Function T_{picl}

Using the block diagram shown in Figure 3.34, the closed-loop reference voltage-to-output voltage transfer function is

$$T_{picl}(s) = \left.\frac{v_o(s)}{v_{ri}(s)}\right|_{v_i=i_o=0} = \frac{T_{ci}T_mT_p}{1+T_i}. \tag{3.106}$$

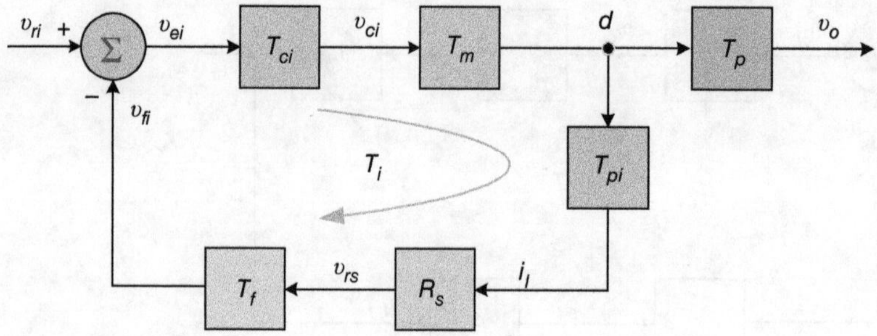

Figure 3.34 Block diagram required to derive the inner-loop control-to-output voltage transfer function T_{picl}.

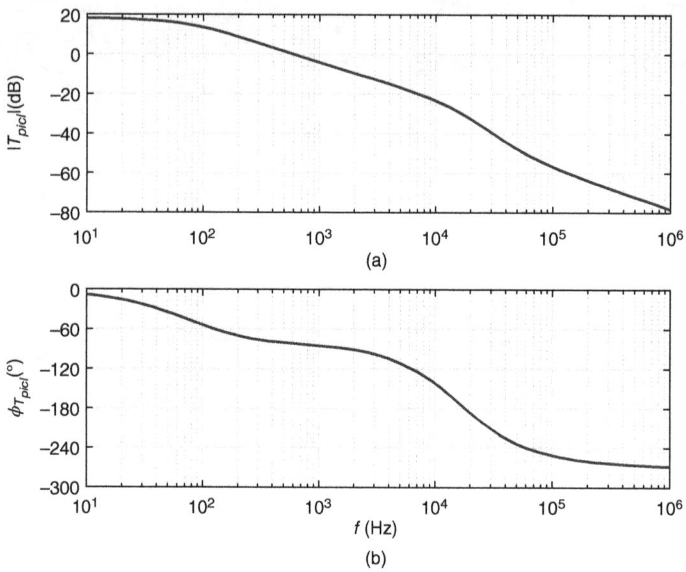

Figure 3.35 Theoretically obtained plots of reference voltage-to-output voltage transfer function T_{picl}. (a) Magnitude. (b) Phase.

Figure 3.35 shows the theoretically obtained magnitude and phase plots of the inner-current loop reference voltage-to-output voltage transfer function.

3.5.3 Input Voltage-to-Inductor Current Transfer Function M_{icl}

Using the block diagram shown in Figure 3.36, the closed-loop input voltage-to-inductor current transfer function is

$$M_{icl}(s) = \left.\frac{i_l(s)}{v_i(s)}\right|_{v_{ri}=i_o=0}. \tag{3.107}$$

From the block diagram,

$$i_l = i_l' + i_l'' = -T_{ci}T_mT_{pi}T_fR_si_l + M_{vi}v_i = (-T_i)i_l + M_{vi}v_i, \tag{3.108}$$

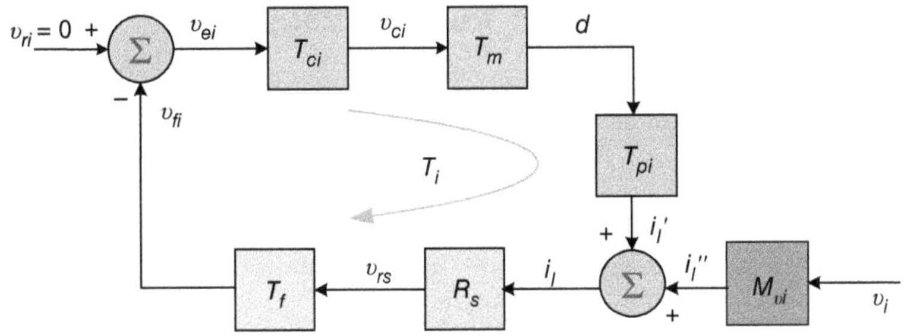

Figure 3.36 Block diagram required to derive the inner-loop input voltage-to-inductor current transfer function M_{icl}.

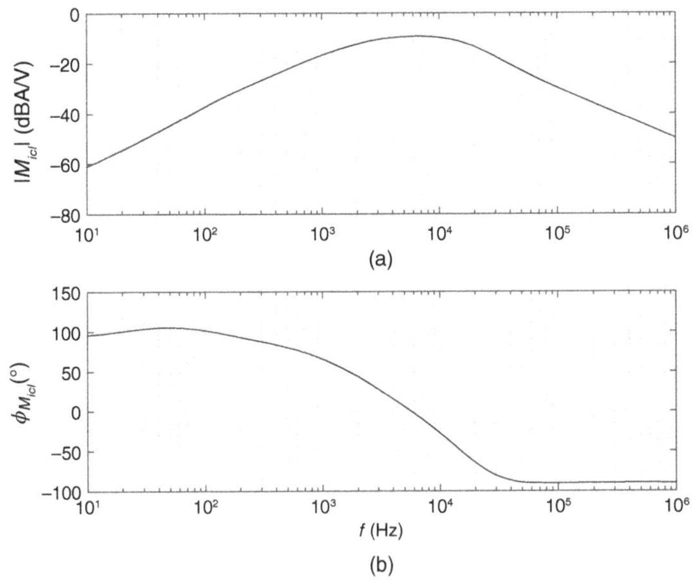

Figure 3.37 Theoretically obtained plots of input voltage-to-inductor current transfer function M_{icl}. (a) Magnitude. (b) Phase.

and

$$i_l(1 + T_i) = M_{vi} v_i. \tag{3.109}$$

Hence,

$$M_{icl}(s) = \frac{i_l(s)}{v_i(s)} = \frac{M_{vi}}{1 + T_i}. \tag{3.110}$$

Figure 3.37 shows the theoretically obtained magnitude and phase plots of the inner-current loop input voltage-to-inductor current transfer function.

3.5.4 Input Voltage-to-Output Voltage Transfer Function M_{vicl}

Using the block diagram shown in Figure 3.38, the closed-loop input voltage-to-output voltage transfer function is

$$M_{vicl}(s) = \left.\frac{v_o(s)}{v_i(s)}\right|_{v_{ri}=i_o=0}. \tag{3.111}$$

From the block diagram,

$$i_l = i_l' + i_l'' = T_{pi}d + M_{vi}v_i \tag{3.112}$$

and

$$\frac{-d}{T_{ix}} = T_{pi}d + M_{vi}v_i, \tag{3.113}$$

where

$$T_{ix} = \frac{T_i}{T_{pi}}. \tag{3.114}$$

Rearranging (3.113) gives

$$-d\left(\frac{1}{T_{ix}} + T_{pi}\right) = M_{vi}v_i \tag{3.115}$$

and

$$-d\left(\frac{1 + T_{ix}T_{pi}}{T_{ix}}\right) = M_{vi}v_i, \tag{3.116}$$

where solving for d provides

$$d = -\frac{M_{vi}T_{ix}}{1 + T_i}v_i. \tag{3.117}$$

From the block diagram in Figure 3.38,

$$v_o = v_o' + v_o'' = T_p d + M_v v_i. \tag{3.118}$$

Substituting d from (3.117) into (3.118), we get

$$v_o = T_p\left(-\frac{M_{vi}T_{ix}}{1 + T_i}v_i\right) + M_v v_i = \left(M_v - \frac{T_p M_{vi} T_{ix}}{1 + T_i}\right)v_i. \tag{3.119}$$

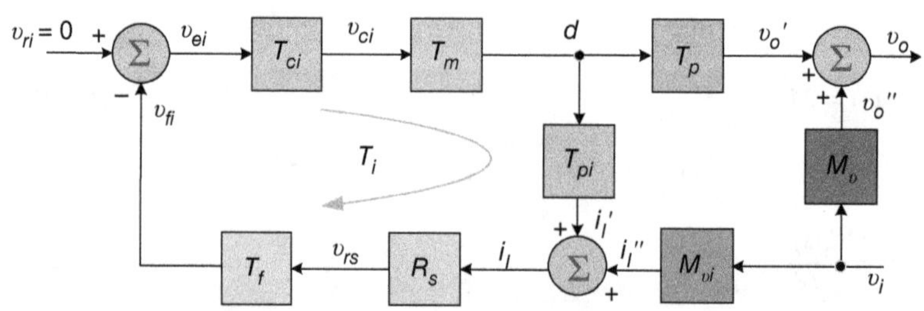

Figure 3.38 Block diagram required to derive the inner-loop input voltage-to-output voltage transfer function M_{vicl}.

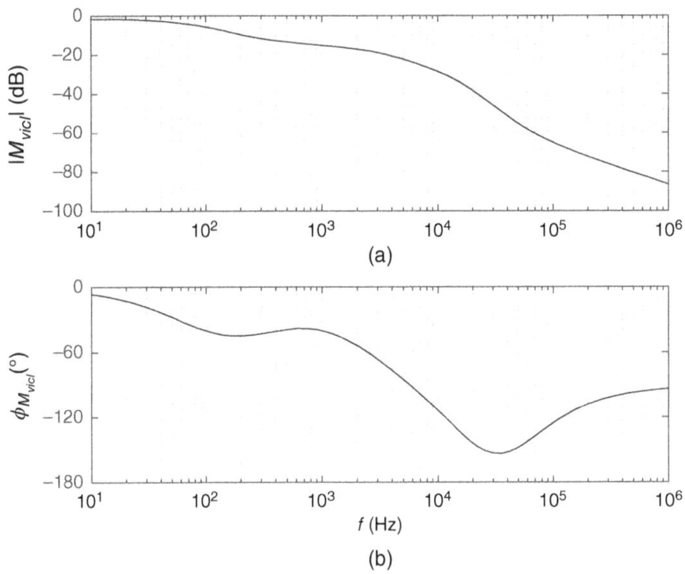

Figure 3.39 Theoretically obtained magnitude and phase plots of input voltage-to-output voltage transfer function M_{vicl}.

The closed-loop input-to-output voltage transfer function is

$$M_{vicl}(s) = \frac{v_o(s)}{v_i(s)} = M_v(1 + T_i) - \frac{T_p M_{vi} T_{ix}}{1 + T_i}. \tag{3.120}$$

Figure 3.39 shows the theoretically obtained magnitude and phase plots of the inner-current loop input voltage-to-output voltage transfer function.

3.5.5 Input Voltage-to-Duty Cycle Transfer Function M_{di}

Using the block diagram shown in Figure 3.38, the closed-loop input voltage-to-duty cycle transfer function is

$$M_{di}(s) = \left.\frac{d(s)}{v_i(s)}\right|_{v_{ri}=i_o=0}. \tag{3.121}$$

From (3.117),

$$d = -\frac{M_{vi} T_{ix}}{1 + T_i} v_i. \tag{3.122}$$

Hence, the closed-loop input voltage-to-duty cycle transfer function is

$$M_{di}(s) = \frac{d(s)}{v_i(s)} = -\frac{M_{vi} T_{ix}}{1 + T_i} = -\frac{M_{vi}}{T_{pi}} \frac{T_i}{1 + T_i}. \tag{3.123}$$

Figure 3.40 shows the theoretically obtained magnitude and phase plots of the inner-current loop input voltage-to-duty cycle transfer function.

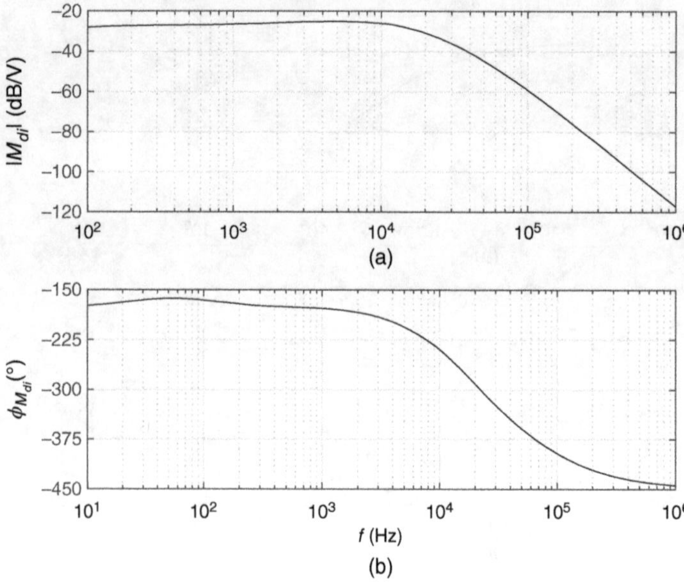

Figure 3.40 Theoretically obtained plots of input voltage-to-duty cycle transfer function M_{di}. (a) Magnitude. (b) Phase.

3.5.6 Input Impedance Z_{iicl}

The closed-inner-loop input impedance is

$$Z_{iicl}(s) = \left.\frac{v_i(s)}{i_i(s)}\right|_{i_o=v_{ri}=0}. \tag{3.124}$$

From Figure 3.41, the inductor current is

$$i_l = i_l' + i_l'' = T_{pi}d + M_{vi}v_i, \tag{3.125}$$

and the duty cycle is

$$d = T_{ci}T_m v_{ei} = T_{ci}T_m(-T_f R_s i_l). \tag{3.126}$$

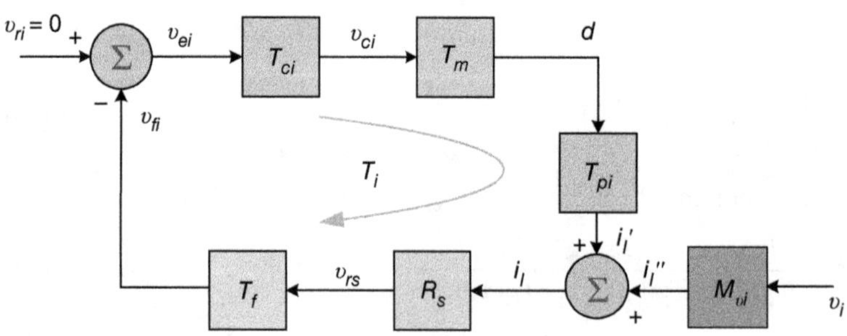

Figure 3.41 Block diagram used to derive the closed-inner-loop input impedance Z_{iicl}.

Substituting (3.126) into (3.125),

$$i_l = -T_{pi}T_{ci}T_mT_fR_si_l + M_{vi}v_i = -T_ii_l + M_{vi}v_i. \tag{3.127}$$

Rearranging (3.127),

$$i_l = \frac{M_{vi}}{1+T_i}v_i. \tag{3.128}$$

The input current is

$$i_i = i_l = \frac{M_{vi}}{1+T_i}v_i. \tag{3.129}$$

The closed-loop input admittance is

$$Y_{iicl}(s) = \frac{i_i(s)}{v_i(s)} = \frac{M_{vi}}{1+T_i}. \tag{3.130}$$

For the small-signal model of the boost converter, we have

$$M_{vi} = \frac{i_i}{v_i} = \frac{i_l}{v_i} = \frac{1}{Z_i} = Y_i. \tag{3.131}$$

Hence, the closed-inner-loop input impedance is

$$Z_{iicl}(s) = \left.\frac{v_i(s)}{i_i(s)}\right|_{i_o=v_{ri}=0} = \frac{1}{Y_{iicl}} = Z_i(1+T_i). \tag{3.132}$$

Figure 3.42 shows the theoretically obtained magnitude and phase plots of the closed-inner-loop input impedance.

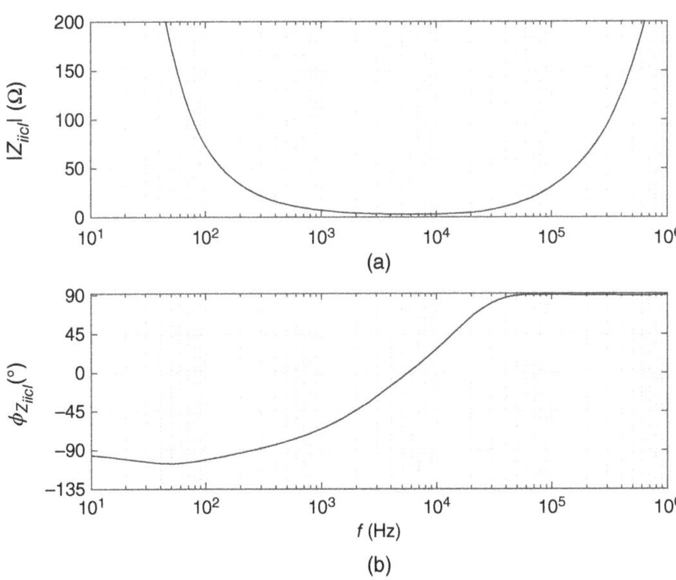

Figure 3.42 Theoretically obtained plots of closed-inner-loop input impedance Z_{iicl}. (a) Magnitude. (b) Phase.

3.5.7 Output Impedance Z_{oicl}

Using the block diagram shown in Figure 3.43, the closed-inner-loop output impedance is

$$Z_{oicl}(s) = \left.\frac{v_o(s)}{i_o(s)}\right|_{v_{ri}=v_i=0}. \tag{3.133}$$

From the block diagram,

$$i_l = i_l' + i_l'' = T_{pi}d + A_i i_o \tag{3.134}$$

and

$$\frac{-d}{T_{ix}} = T_{pi}d + A_i i_o. \tag{3.135}$$

Rearranging further, we get

$$-d\left(\frac{1}{T_{ix}} + T_{pi}\right) = A_i i_o, \tag{3.136}$$

where

$$d = -\frac{A_i T_{ix}}{1+T_i} i_o. \tag{3.137}$$

From the block diagram in Figure 3.43,

$$v_o = v_o' + v_o'' = T_p d + Z_o i_o. \tag{3.138}$$

Substituting d from (3.117), we get

$$v_o = T_p\left(-\frac{A_i T_{ix}}{1+T_i} i_o\right) + Z_o i_o. = \left(Z_o - \frac{A_i T_p T_{ix}}{1+T_i}\right) i_o. \tag{3.139}$$

The closed-inner-loop output impedance is

$$Z_{oicl}(s) = \frac{v_o(s)}{i_o(s)} = Z_o - \frac{A_i T_p T_{ix}}{1+T_i}. \tag{3.140}$$

Figure 3.44 shows the theoretically obtained magnitude and phase plots of the inner-current-loop output impedance.

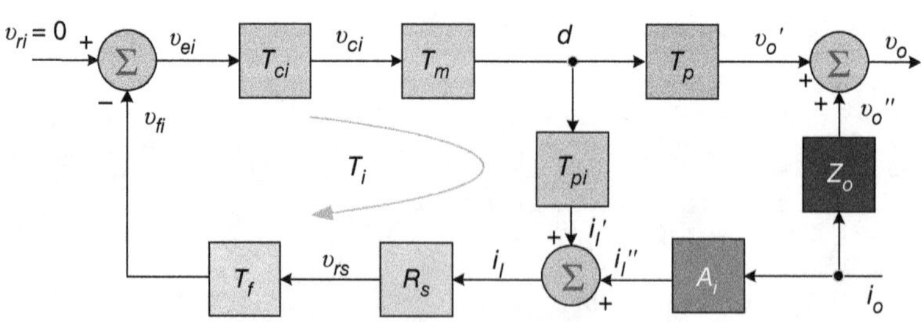

Figure 3.43 Block diagram required to derive the closed-inner-loop output impedance Z_{oicl}.

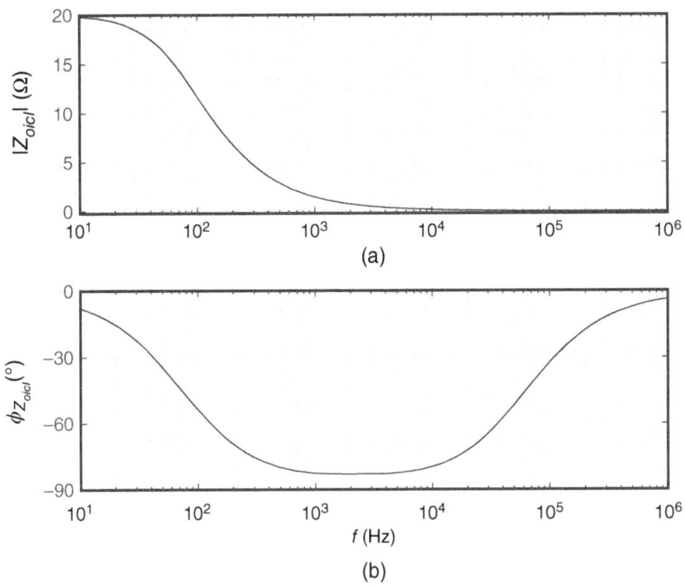

Figure 3.44 Theoretically obtained plots of closed-inner-loop output impedance Z_{oicl}. (a) Magnitude. (b) Phase.

3.6 Outer-Voltage Loop

Figure 3.45 shows the architecture of the average current-mode controlled buck converter with outer-voltage loop. Figure 3.46 shows the circuit of the average current-mode controlled buck converter with outer-voltage loop. To maintain the output voltage constant, it is essential to introduce the outer-voltage loop set with a

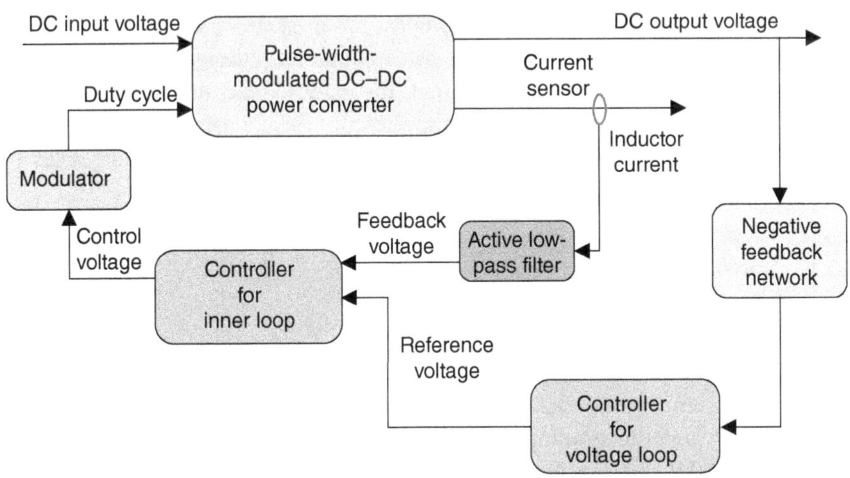

Figure 3.45 Architecture of the average current-mode controlled boost converter with outer-voltage loop.

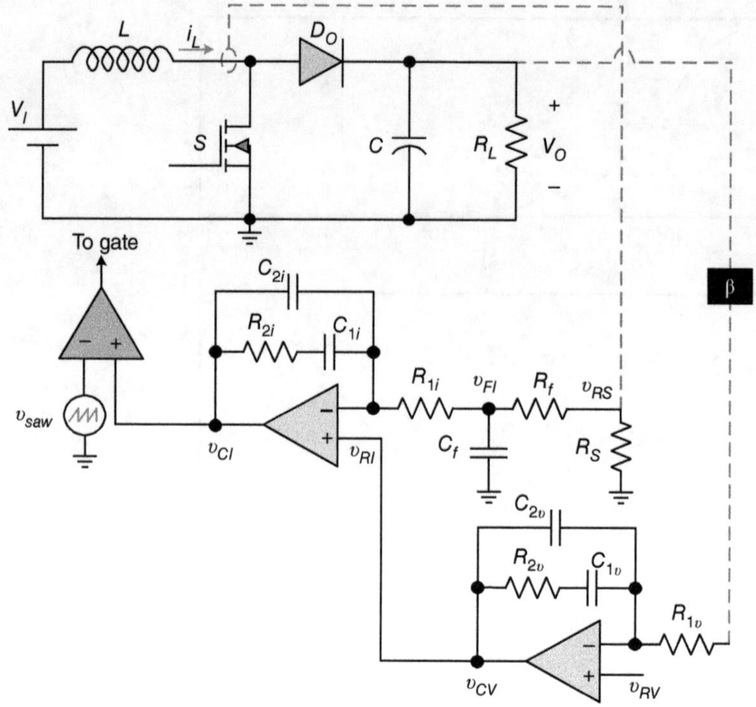

Figure 3.46 Circuit of the average current-mode controlled boost converter with outer-voltage loop.

reference voltage V_{RV} that regulates the output voltage. In a typical two-loop system, the reference voltage to the outer loop is a fraction of the rated output voltage. The fraction is determined by the feedback network, whose voltage gain β is determined by a network of resistors. The procedure to design the outer-voltage loop is the same as designing the inner-current loop. The boost power stage served as the plant for the inner-current loop. However, in the case of outer-loop design, the power stage incorporated within the inner loop becomes the plant to be controlled. Hence, the path between the current reference v_{ri} and the duty cycle d serves as the plant for the outer-voltage loop. The reference voltage V_{RI} for the inner-current loop is set by the outer loop.

3.6.1 Transfer Function of Feedback Network β

The transfer function of the feedback network is given as

$$\beta = \frac{V_{RV}}{V_O}, \tag{3.141}$$

where V_{RV} is the reference voltage to the outer-voltage loop. Assuming the reference voltage $V_{RV} = 10$ V, the transfer function of the feedback network is

$$\beta = \frac{V_{RV}}{V_O} = \frac{10}{20} = 0.5. \tag{3.142}$$

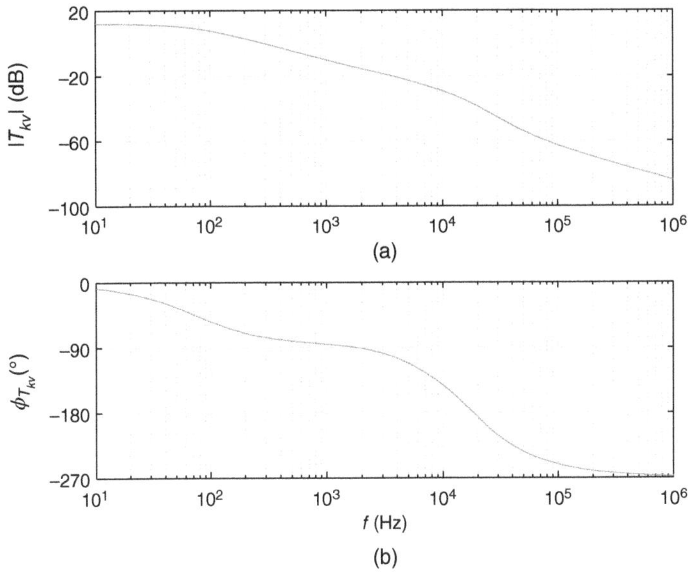

Figure 3.47 Theoretically obtained plots of uncompensated loop gain T_{kv}. (a) Magnitude. (b) Phase.

3.6.2 Uncompensated Loop Gain for Outer-Voltage Loop T_{kv}

The natural behavior of the outer-voltage loop is determined using the uncompensated loop gain as

$$T_{kv} = \frac{v_{fv}}{v_{ev}} = \frac{\beta v_o}{v_{ri}} = \beta T_{picl} = \frac{\beta T_{ci} T_m T_p}{1 + T_i}. \tag{3.143}$$

Figure 3.47 shows the theoretically obtained magnitude and phase plots of the uncompensated loop gain of the outer-voltage loop. The DC gain is nearly $T_{kvo} = 5$ V/V = 14 dB, and the crossover frequency is 286 Hz.

3.6.3 Transfer Function of Control Circuit for Outer-Voltage Loop T_{cv}

Figure 3.48 shows circuit of a single-lead integral (Type-II) controller. A single-lead integral control circuit introduces a pole at the origin, which is required to boost the DC gain, and a pole–zero pair to adjust the crossover frequency f_c to maintain the required PM. A detailed procedure to determine the transfer function of the controller to improve the DC gain (> 20 dB) and achieve an inner-current-loop phase margin $PM = 60°$ is explained in [45]. The remaining section details the Type-II controller design for the outer-voltage loop.

The desired crossover frequency for the loop gain is chosen as $f_c = 2.14$ kHz. At this frequency, the magnitude of the uncompensated loop gain transfer function is $|T_{kv}(f_c)| = 0.158$ V/V = −16 dB, and its phase is $\phi_{Tkv}(f_c) = -91.7°$. The required phase margin is $PM = 60°$. Therefore, the phase boost required at f_c is

$$\phi_m = PM - \phi_{Tkv}(f_c) - 90° = 61.7°, \tag{3.144}$$

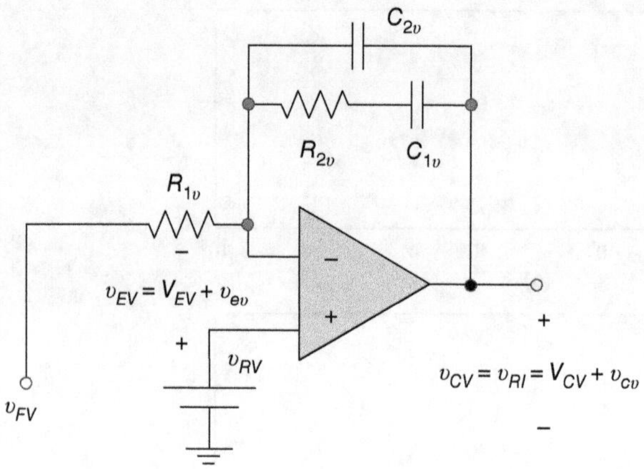

Figure 3.48 Circuit of the Type-II controller used in the outer-voltage loop.

yielding the geometric mean between the controller corner frequencies as

$$K = \tan\left(\frac{\phi_m}{2} + 45^\circ\right) = 3.97. \tag{3.145}$$

Assuming $R_{1v} = 1\ \text{k}\Omega/0.25\ W/2\%$, the value of C_{2v} is

$$C_{2v} = \frac{|T_{kv}(f_c)|}{2\pi f_c K R_{1v}} = 3\ \text{nF}. \tag{3.146}$$

A 3 nF/25 V standard capacitor is selected. The expression for C_{1v} is

$$C_{1v} = C_{2v}(K^2 - 1) = 44\ \text{nF}. \tag{3.147}$$

A 47 nF/20 V standard capacitor is selected. The expression for R_{2v} is

$$R_{2v} = \frac{K}{2\pi f_c C_{1v}} = 6.7\ \text{k}\Omega. \tag{3.148}$$

A 6.8 kΩ/0.25 W/5% standard resistor is selected. The transfer function T_{cv} of the voltage-loop control circuit is

$$T_{cv}(s) = \frac{v_{cv}(s)}{v_{ev}(s)} = T_{cvo}\frac{1 + \frac{s}{\omega_{zcv}}}{s\left(1 + \frac{s}{\omega_{pcv}}\right)}, \tag{3.149}$$

where

$$T_{cvo} = \frac{1}{R_{1v}(C_{1v} + C_{2v})} = 20\ \text{kV/V}, \tag{3.150}$$

$$\omega_{zcv} = \frac{1}{R_{2v}C_{1v}} = 3.2\ \text{krad/s}, \tag{3.151}$$

and

$$\omega_{pcv} = \frac{C_{1v} + C_{2v}}{R_{2v}C_{1v}C_{2v}} = 52\ \text{krad/s}. \tag{3.152}$$

Figure 3.49 shows the theoretically obtained magnitude and phase plots of the controller transfer function used in the inner-current loop. The control transfer function T_{cv} is incorporated into the critical path to determine the outer-voltage loop gain and is discussed in Section 3.6.4.

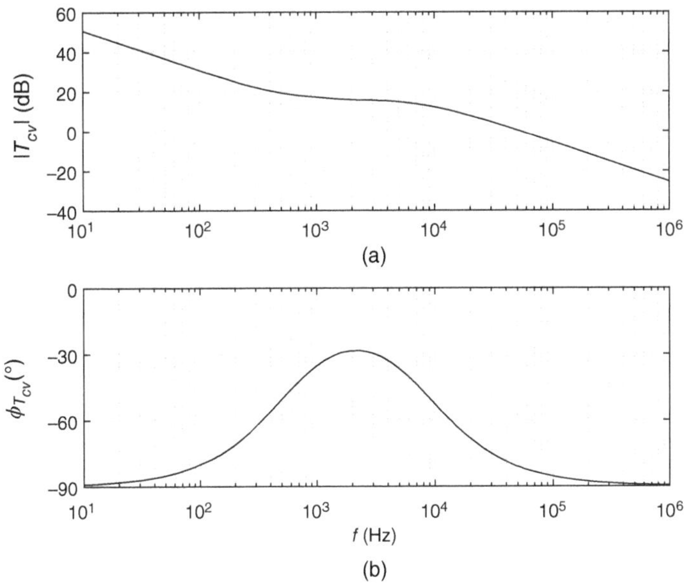

Figure 3.49 Theoretically obtained plots of controller transfer function T_{cv} of the outer-voltage loop. (a) Magnitude. (b) Phase.

3.6.4 Compensated Loop Gain of Outer-Voltage Loop T_v

The loop gain of the outer-voltage loop is

$$T_v = \frac{v_{ev}}{v_{fv}} = T_{kv}T_{cv} = \frac{\beta T_{ci}T_m T_p T_{cv}}{1+T_i} = \frac{T_o}{1+T_i}, \tag{3.153}$$

where $T_o = \beta T_{ci}T_m T_p T_{cv}$. Figure 3.50 shows the theoretically obtained magnitude and phase plots of the loop gain of the outer-voltage loop. Using the controller designed in Section 3.6.4, the value of the compensated loop gain at DC is $T_{vo} = 14\ kV/V = 83\ dB$. The phase margin is $PM = 60°$ and the gain margin is $20\ dB$. The closed-loop transfer functions for the outer-voltage loop are derived and discussed in Section 3.7.

3.7 Closed-Loop Transfer Functions for Outer-Voltage Loop

The following closed-loop control transfer functions are derived:

1. Reference voltage-to-output voltage transfer function T_{pcl} (control).
2. Input voltage-to-output voltage transfer function M_{vcl} (disturbance).
3. Input voltage-to-duty cycle transfer function M_{dv} (disturbance).
4. Input impedance Z_{ivcl} (disturbance).
5. Output impedance Z_{ovcl} (disturbance).

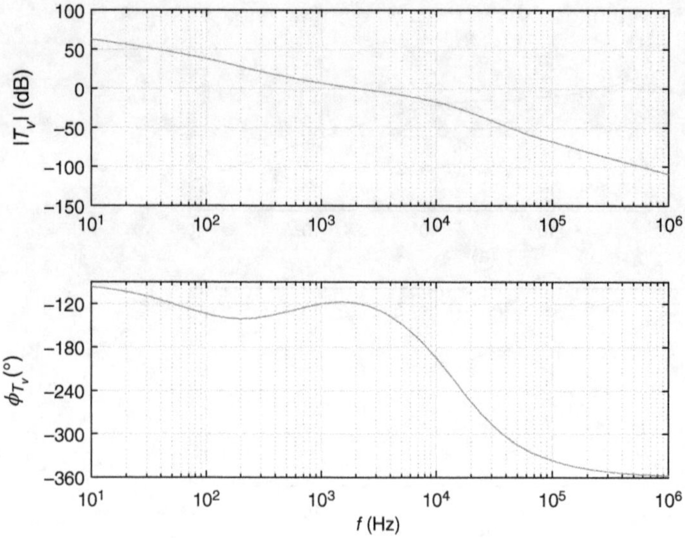

Figure 3.50 Theoretically obtained plots of loop gain T_v of the compensated outer-voltage loop. (a) Magnitude. (b) Phase.

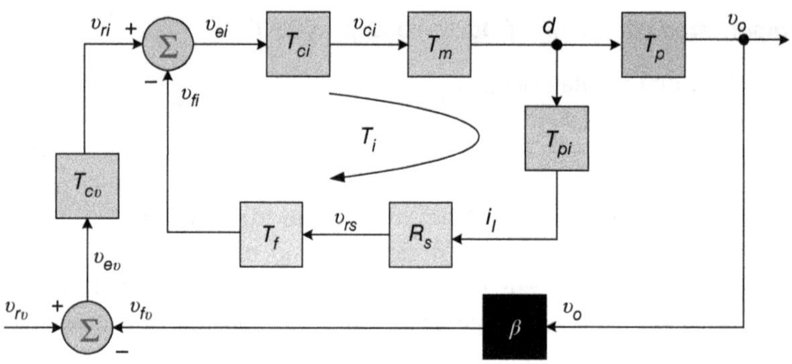

Figure 3.51 Block diagram used to derive the control-to-output voltage transfer function T_{pcl}.

3.7.1 Reference Voltage-to-Output Voltage Transfer Function T_{pcl}

Using the block diagram shown in Figure 3.51, the closed-loop reference voltage-to-output voltage transfer function is

$$T_{pcl}(s) = \left.\frac{v_o(s)}{v_{rv}(s)}\right|_{v_i=i_o=0} = \frac{T_{cv}T_{picl}}{1+T_v}. \tag{3.154}$$

Figure 3.52 shows the theoretically obtained magnitude and phase plots of the reference voltage-to-output voltage transfer function.

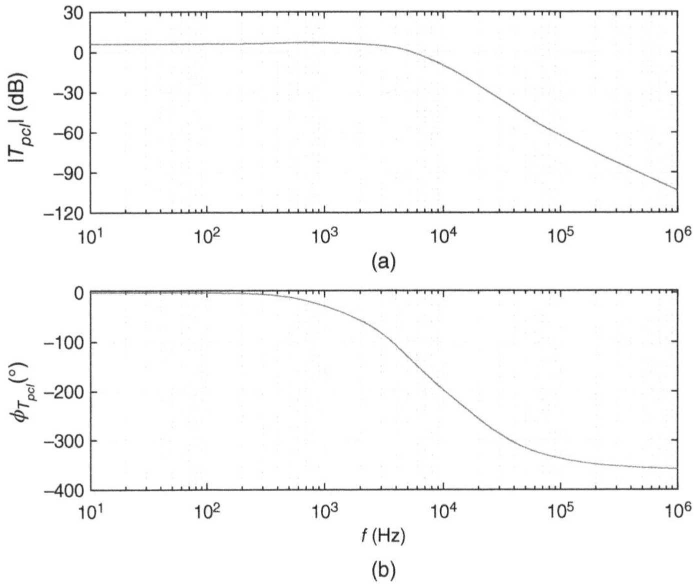

Figure 3.52 Theoretically obtained plots of the reference voltage-to-output voltage T_{pcl}. (a) Magnitude. (b) Phase.

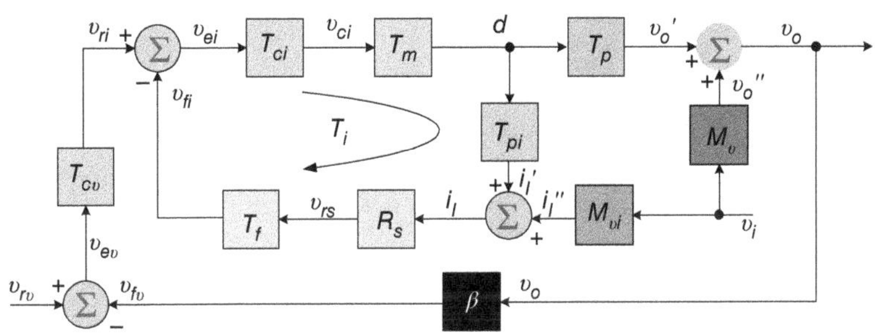

Figure 3.53 Block diagram used to derive the input voltage-to-duty cycle transfer function M_{dv}.

3.7.2 Input Voltage-to-Duty Cycle Transfer Function M_{dv}

Using the block diagram shown in Figure 3.53, the closed-loop input voltage-to-duty cycle transfer function is

$$M_{dv}(s) = \left.\frac{d(s)}{v_i(s)}\right|_{v_{rv} = i_o = 0}. \tag{3.155}$$

From the block diagram,

$$v_o = v_o' + v_o'' = T_p d + M_v v_i. \tag{3.156}$$

Also,

$$v_{fv} = -\beta v_o, \tag{3.157}$$

and

$$v_{cv} = v_{ri} = -\beta T_{cv} v_o. \tag{3.158}$$

The error voltage v_{ei} is

$$v_{ei} = v_{ri} - v_{fi} = -T_{cv}\beta v_o - R_s T_f i_l \tag{3.159}$$

and

$$v_{ei} = \frac{d}{T_m T_{ci}}. \tag{3.160}$$

The inductor current is

$$i_l = T_{pi} d + M_{vi} v_i. \tag{3.161}$$

Substituting (3.161) and (3.160) into (3.159) yields

$$\frac{d}{T_m T_{ci}} = -T_{cv}\beta v_o - R_s T_f (T_{pi} d + M_{vi} v_i). \tag{3.162}$$

Rearranging (3.162) gives

$$v_o = -\frac{d}{T_{cv}\beta}\left(\frac{1}{T_m T_{ci}} + R_s T_f T_{pi}\right) - \frac{v_i}{T_{cv}\beta}\left(R_s T_f M_{vi}\right). \tag{3.163}$$

Equating (3.156) and (3.163) and rearranging,

$$v_i\left(-\frac{M_{vi} R_s T_f}{T_{cv}\beta} - M_v\right) = d\left(T_p + \frac{1}{T_m T_{ci} T_{cv}\beta} + \frac{R_s T_f T_{pi}}{T_{cv}\beta}\right). \tag{3.164}$$

Equating (3.156) and (3.163) and rearranging,

$$-\frac{v_i}{T_{cv}\beta}\left(M_{vi} R_s T_f + M_v T_{cv}\beta\right) = \frac{d}{T_{cv}\beta}\left(T_p T_{cv}\beta + \frac{1 + R_s T_f T_{pi} T_m T_{ci}}{T_m T_{ci}}\right). \tag{3.165}$$

Rearranging further,

$$-v_i\left(M_{vi} R_s T_f + M_v T_{cv}\beta\right) = d\left(T_p T_{cv}\beta + \frac{1 + R_s T_f T_{pi} T_m T_{ci}}{T_m T_{ci}}\right), \tag{3.166}$$

and

$$M_{dv}(s) = \frac{d(s)}{v_i(s)} = -\frac{M_{vi} R_s T_f + \beta M_v T_{cv}}{\beta T_p T_{cv} + \frac{1+T_i}{T_m T_{ci}}}. \tag{3.167}$$

Figure 3.54 shows the theoretically obtained magnitude and phase plots of the input voltage-to-duty cycle transfer function.

3.7.3 Input Voltage-to-Output Voltage Transfer Function M_{vcl}

Using the block diagram shown in Figure 3.55, the closed-loop input voltage-to-output voltage transfer function with outer-voltage loop is

$$M_{vcl}(s) = \left.\frac{v_o(s)}{v_i(s)}\right|_{v_{rv}=i_o=0}. \tag{3.168}$$

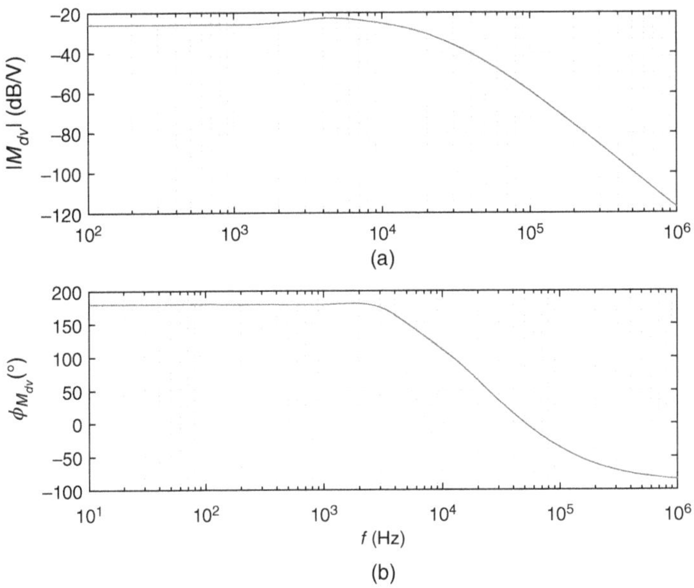

Figure 3.54 Theoretically obtained plots of the input voltage-to-duty cycle transfer function M_{dv}. (a) Magnitude. (b) Phase.

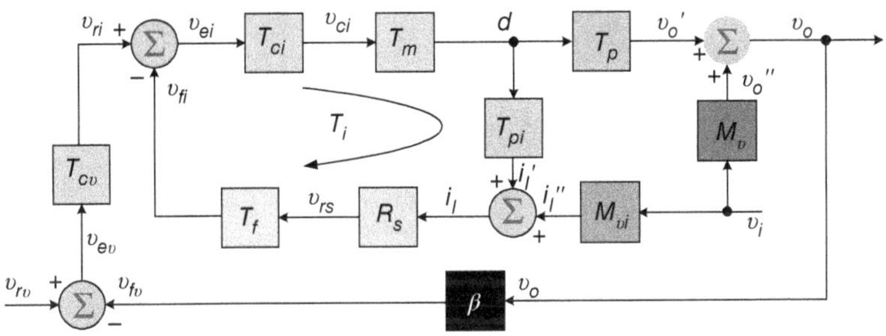

Figure 3.55 Block diagram used to derive the input voltage-to-output voltage transfer function M_{vcl}.

From previous analysis,

$$d(s) = M_{dv} v_i(s). \tag{3.169}$$

Substituting (3.169) into (3.156),

$$v_o = v_o' + v_o'' = T_p d + M_v v_i = T_p M_{dv} v_i + M_v v_i. \tag{3.170}$$

Rearrangement of (3.170) yields

$$M_{vcl}(s) = \frac{v_o(s)}{v_i(s)} = T_p M_{dv} + M_v. \tag{3.171}$$

Figure 3.56 shows the theoretically obtained magnitude and phase plots of the input voltage-to-output voltage transfer function M_{vcl}.

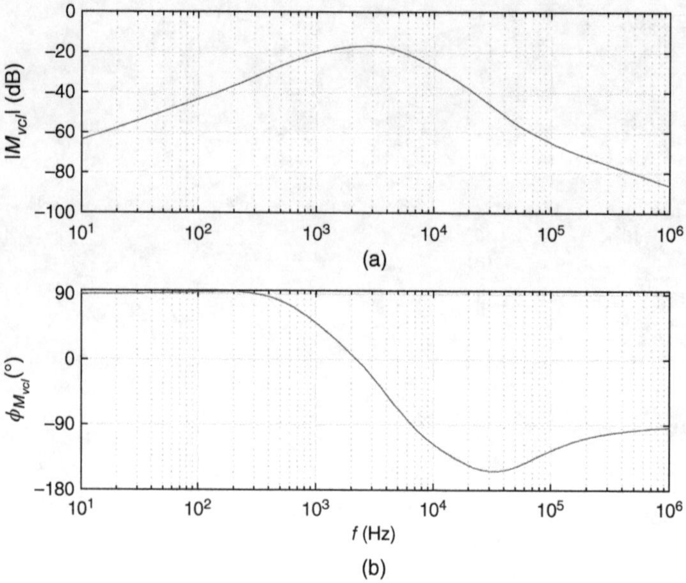

Figure 3.56 Theoretically obtained plots of the input voltage-to-output voltage transfer function M_{vcl}. (a) Magnitude. (b) Phase.

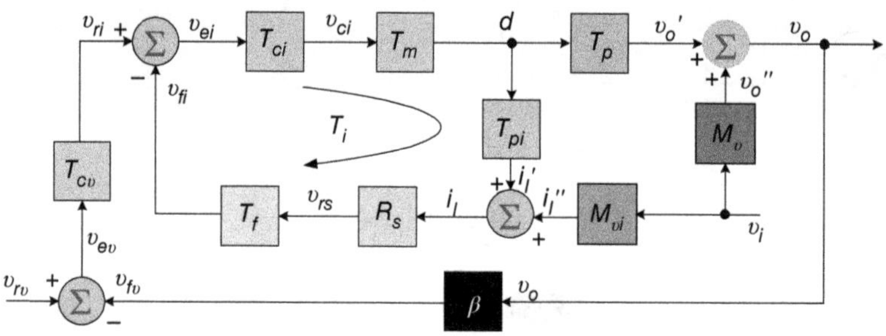

Figure 3.57 Block diagram used to derive the closed-outer-voltage loop input impedance Z_{ivcl}.

3.7.4 Input Impedance Z_{ivcl}

Using the block diagram shown in Figure 3.57, the closed-loop input impedance with outer-voltage loop is

$$Z_{ivcl}(s) = \left. \frac{v_i(s)}{i_i(s)} \right|_{v_{rv}=i_o=0}. \tag{3.172}$$

From the Figure 3.57,

$$v_o = v_o' + v_o'' = T_p d + M_v v_i \tag{3.173}$$

and

$$i_i = i_l = i_l' + i_l'' = T_{pi} d + M_{vi} v_i. \tag{3.174}$$

Also,

$$v_{fv} = \beta v_o, \tag{3.175}$$

$$v_{ev} = v_{rv} - v_{fv} = 0 - \beta v_o = -\beta v_o, \tag{3.176}$$

and

$$v_{cv} = v_{ri} = T_{cv} v_{ev} = -\beta T_{cv} v_o. \tag{3.177}$$

The error voltage v_{ei} is

$$v_{ei} = \frac{d}{T_{ci} T_m}. \tag{3.178}$$

The feedback voltage v_{fi} is

$$v_{fi} = T_f R_s i_l. \tag{3.179}$$

The reference voltage to the inner-current loop is

$$v_{ri} = v_{ei} + v_{fi}. \tag{3.180}$$

Substituting (3.177), (3.178), and (3.179) into (3.180),

$$-\beta T_{cv} v_o = \frac{d}{T_{ci} T_m} + T_f R_s i_l. \tag{3.181}$$

Rearranging gives

$$v_o = -\frac{d}{\beta T_{cv} T_{ci} T_m} - \frac{T_f R_s i_l}{\beta T_{cv}}. \tag{3.182}$$

Substituting (3.173) into (3.182),

$$T_p d + M_v v_i = -\frac{d}{\beta T_{cv} T_{ci} T_m} - \frac{T_f R_s i_l}{\beta T_{cv}}. \tag{3.183}$$

Rearranging (3.183) gives

$$d\left(T_p + \frac{1}{\beta T_{cv} T_{ci} T_m}\right) = -\frac{T_f R_s i_l}{\beta T_{cv}} - M_v v_i. \tag{3.184}$$

Substituting (3.174) into (3.184),

$$d\left(T_p + \frac{1}{\beta T_{cv} T_{ci} T_m}\right) = -\frac{T_f R_s (T_{pi} d + M_{vi} v_i)}{\beta T_{cv}} - M_v v_i, \tag{3.185}$$

$$d\left(T_p + \frac{1}{\beta T_{cv} T_{ci} T_m} + \frac{T_f R_s T_{pi}}{\beta T_{cv}}\right) = \left(-\frac{T_f R_s M_{vi}}{\beta T_{cv}} - M_v\right), and \tag{3.186}$$

$$d\left(\frac{\beta T_p T_{ci} T_m T_{cv} + 1 + T_f R_s T_{pi} T_{ci} T_m}{\beta T_{ci} T_m T_{cv}}\right) = \left(\frac{-T_f R_s M_{vi} - \beta T_{cv} M_v}{\beta T_{cv}}\right), \tag{3.187}$$

and solving for d yields

$$d = -\frac{T_{ci} T_m (T_f R_s M_{vi} + \beta M_v T_{cv})}{\beta T_p T_{ci} T_m T_{cv} + 1 + T_f R_s T_{pi} T_{ci} T_m} v_i. \tag{3.188}$$

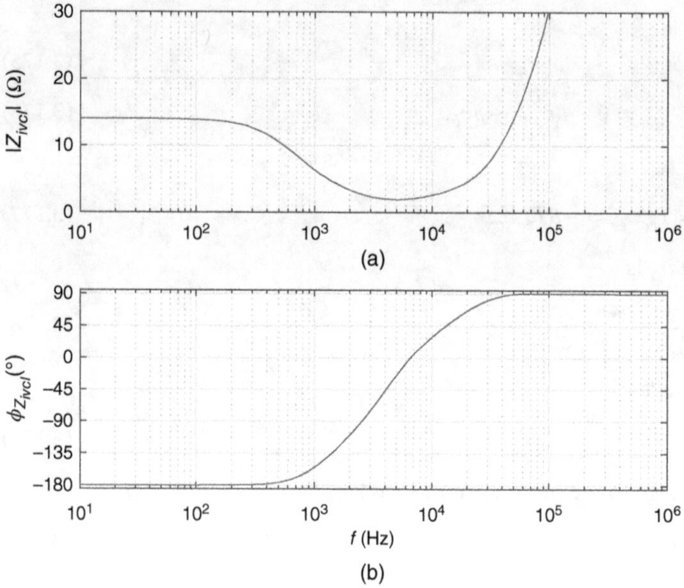

Figure 3.58 Theoretically obtained plots of the closed-outer-voltage loop input impedance Z_{ivcl}. (a) Magnitude. (b) Phase.

Substituting (3.188) into (3.174),

$$i_i = \left[-\frac{T_{ci} T_m (T_f R_s M_{vi} + \beta M_v T_{cv})}{\beta T_p T_{ci} T_m T_{cv} + 1 + T_f R_s T_{pi} T_{ci} T_m} T_{pi} + M_{vi} \right] v_i. \tag{3.189}$$

The closed-loop input admittance with outer-voltage loop is

$$Y_{ivcl}(s) = \frac{i_i(s)}{v_i(s)} = -\frac{T_{ci} T_m (T_f R_s M_{vi} + \beta M_v T_{cv})}{\beta T_p T_{ci} T_m T_{cv} + 1 + T_f R_s T_{pi} T_{ci} T_m} T_{pi} + M_{vi}. \tag{3.190}$$

The closed-loop input impedance with outer-voltage loop is

$$Z_{ivcl}(s) = \frac{v_i(s)}{i_i(s)} = \frac{1}{Y_{ivcl}(s)}. \tag{3.191}$$

Figure 3.58 shows the theoretically obtained magnitude and phase plots of the closed-loop input impedance Z_{ivcl}.

3.7.5 Output Impedance Z_{ovcl}

Using the block diagram shown in Figure 3.59, the closed-loop output impedance with outer-voltage loop is

$$Z_{ovcl}(s) = \left. \frac{v_o(s)}{i_o(s)} \right|_{v_{rv} = v_i = 0}. \tag{3.192}$$

From Figure 3.59,

$$v_o = v_o' + v_o'' = T_p d + Z_o i_o, \tag{3.193}$$

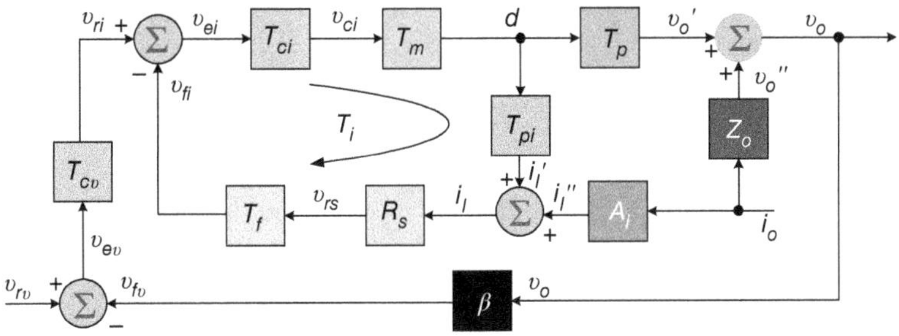

Figure 3.59 Block diagram used to derive the closed-outer-voltage-loop output impedance Z_{ovcl}.

and

$$i_l = i_l' + i_l'' = A_i i_o + T_{pi} d. \tag{3.194}$$

Also,

$$v_{fv} = \beta v_o, \tag{3.195}$$

and

$$v_{cv} = v_{ri} = -\beta T_{cv} v_o. \tag{3.196}$$

The error voltage v_{ei} is

$$v_{ei} = v_{ri} - v_{fi} = -T_{cv}\beta v_o - R_s T_f i_l \tag{3.197}$$

and

$$v_{ei} = \frac{d}{T_m T_{ci}}. \tag{3.198}$$

Substituting (3.194) and (3.198) into (3.197) yields

$$\frac{d}{T_m T_{ci}} = -T_{cv}\beta v_o - R_s T_f (A_i i_o + T_{pi} d). \tag{3.199}$$

Rearranging (3.199),

$$d = -\frac{(T_{cv}\beta v_o + R_s T_f A_i i_o) T_{ci} T_m}{1 + T_i}. \tag{3.200}$$

Substituting (3.200) into (3.193) yields

$$v_o = T_p \left[-\frac{(T_{cv}\beta v_o + R_s T_f A_i i_o) T_{ci} T_m}{1 + T_i} \right] + Z_o i_o, \tag{3.201}$$

and rearranging further, we get

$$v_o \left(1 + \frac{\beta T_p T_{cv} T_{ci} T_m}{1 + T_i} \right) = i_o \left(Z_o - \frac{T_p R_s T_f A_i T_{ci} T_m}{1 + T_i} \right). \tag{3.202}$$

Hence, the closed-loop output impedance with outer-voltage loop is

$$Z_{ovcl}(s) = \frac{v_o(s)}{i_o(s)} = \frac{Z_o(1 + T_i) - A_i T_p T_{ix}}{1 + T_i + T_p T_{cv} \beta T_{ci} T_m} = \frac{Z_o(1 + T_i) - A_i T_p T_{ix}}{1 + T_i + T_o}. \tag{3.203}$$

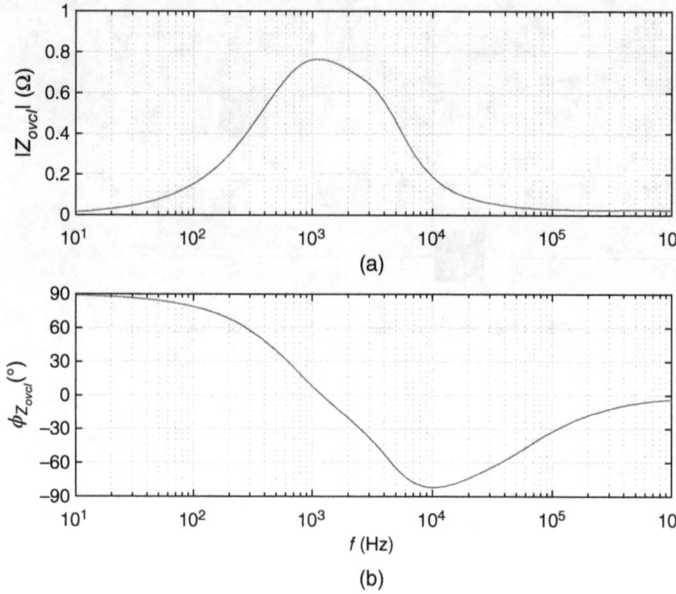

Figure 3.60 Theoretically obtained plots of the closed-outer-voltage loop output impedance Z_{ovcl}. (a) Magnitude. (b) Phase.

Figure 3.60 shows the theoretically obtained magnitude and phase plots of the closed-loop output impedance Z_{ovcl}.

3.8 Comparison of Closed-Loop and Open-Loop Step Responses

3.8.1 Response of Output Voltage to Step Change in Input Voltage

Figure 3.61 compares the responses of the output voltage for unit step changes in input voltage by 1 V for (i) open loop (M_v), (ii) closed-inner-current loop (M_{vicl}), and (iii) closed-outer-voltage loop in the presence of the inner-current loop (M_{vcl}). In the open-loop configuration, a step change $\Delta V_I = 1$ V in the input voltage causes the output voltage to achieve a new steady-state value $V_O(\infty) = 21.67$ V. Therefore, the step change in the output voltage is $\Delta V_O = V_O(\infty) - V_O(0) = 21.67 - 20 = 1.67$ V, yielding $M_{vo} = (\Delta V_O)/(\Delta V_I) = 1.67/1 = 1.67$ V/V, which was also observed in the magnitude response of M_v. The settling time is nearly 4 ms. For the closed-loop response determined by the current-loop transfer function M_{vicl}, the step change in input voltage causes the output voltage to rise to its new steady-state value $V_O(\infty) = 20.8$ V in a settling time of 6 ms. The step change in the output voltage is $\Delta V_O = V_O(\infty) - V_O(0) = 20.8 - 20 = 0.8$ V, yielding $M_{viclo} = (\Delta V_O)/(\Delta V_I) = 0.8/1 = 0.8$ V/V, which can be confirmed with magnitude plot of M_{vicl} in Figure 3.39. However, for the closed-loop response

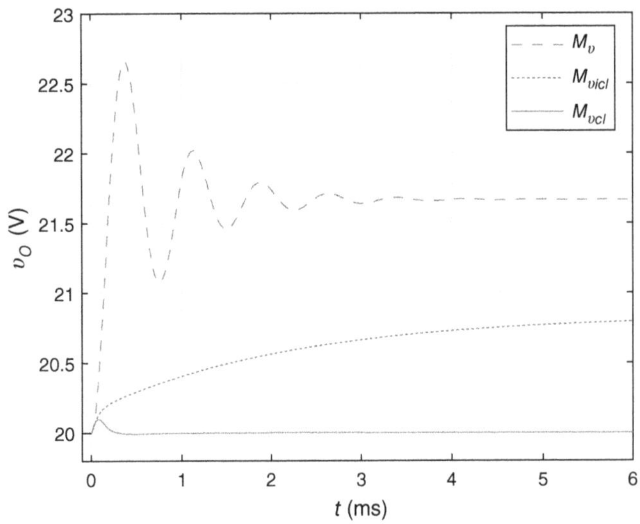

Figure 3.61 Comparison of responses in output voltage for step changes in input voltage by 1 V, obtained using M_v, M_{vicl}, and M_{vcl} transfer functions.

determined by the voltage-loop transfer function M_{vcl}, the step change in input voltage causes the output voltage to return to steady-state value $V_O(\infty) = 20$ V in 0.25 ms.

3.8.2 Response of Output Voltage to Step Change in Duty Cycle, Current-Loop Reference Voltage, and Voltage-Loop Reference Voltage

Figure 3.62 compares the responses of the output voltage for step changes in (i) duty cycle for open-loop by 0.1 (T_p), (ii) current-loop reference voltage for the closed-inner-current loop by 1 V (T_{picl}), and (iii) voltage-loop reference voltage for the closed-outer-voltage loop by 1 V (T_{pcl}). In the open-loop configuration, a step change in duty cycle $\Delta D = 0.1$ causes the output voltage to achieve a new steady-state value $V_O(\infty) = 23.3$ V. Therefore, the step change in the output voltage is $\Delta V_O = V_O(\infty) - V_O(0) = 23.305 - 20 = 3.305$ V, yielding $T_{po} = (\Delta V_O)/\Delta D = 3.305/0.1 = 33.05$, which was also observed in the magnitude plot of T_p. The step response shows a settling time of ≈3 ms. The output voltage response for the closed-inner-current loop achieves a steady-state value $V_O(\infty) = 27.9$ V in a settling time of 8 ms. The step change in the output voltage is $\Delta V_O = V_O(\infty) - V_O(0) = 27.9 - 20 = 7.9$ V, yielding $T_{piclo} = (\Delta V_O)/(\Delta V_I) = 7.9/1 = 7.9$ V/V, which can be confirmed with magnitude plot of T_{picl} in Figure 3.35. The output voltage response for the closed-outer-voltage loop achieves a steady-state value $V_O(\infty) = 22$ V in a rapid settling time of nearly 1 ms. The step change in the output voltage is $\Delta V_O = V_O(\infty) - V_O(0) = 22 - 20 = 2$ V, yielding $T_{pclo} = (\Delta V_O)/(\Delta V_I) = 2/1 = 2$ V/V, which can be confirmed with magnitude plot of T_{pcl} in Figure 3.52.

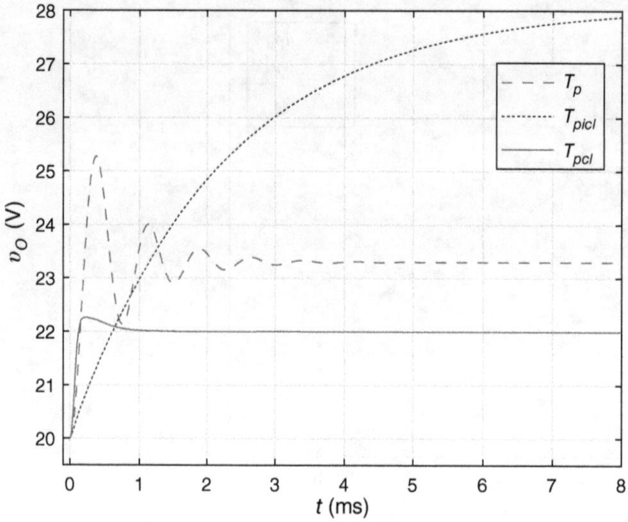

Figure 3.62 Comparison of responses in output voltage for step changes in duty cycle by 0.1, current-loop reference voltage by 1 V, and voltage-loop reference voltage by 1 V, obtained using T_p, T_{picl}, and T_{pcl} transfer functions, respectively.

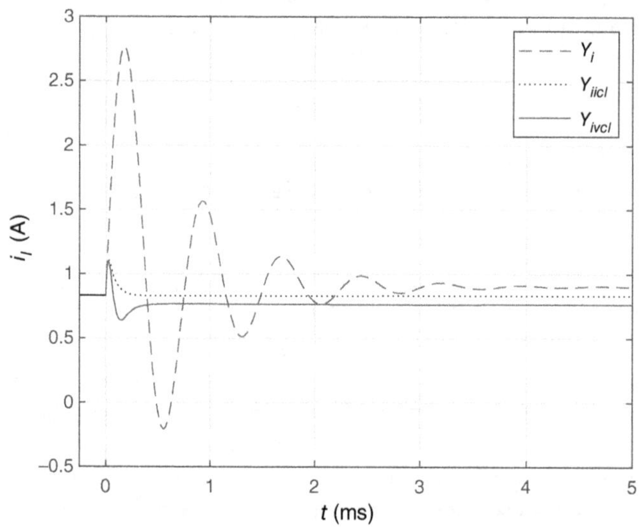

Figure 3.63 Comparison of responses in input current for step changes in the input voltage by 1 V, obtained using Y_i, Y_{iicl}, and Y_{ivcl} transfer functions.

3.8.3 Response of Input Current to Step Change in Input Voltage

Figure 3.63 compares the responses of the input current for unit step change in input voltage by 1 V for (i) open loop (Y_i), (ii) closed-inner-current loop (Y_{iicl}), and (iii) closed-outer-voltage loop (Y_{ivcl}). In the open-loop configuration, a step change in input voltage $\Delta V_I = 1$ V causes the input current to achieve a new steady-state value $I_I(\infty) = 0.9$ A. Therefore, the step change in the input current is $\Delta I_I = I_I(\infty) - I_I(0) = 0.9 - 0.833 = 0.07$ A, yielding $Z_{io} = R_i = (\Delta V_I)/(\Delta I_I) = 1/0.07 = 14.28\ \Omega$, which was also observed in the magnitude plot of Z_i. The step response achieves steady state in 4 ms. The input current response for the closed-inner-current loop achieves a steady-state value $I_I(\infty) = 0.83$ A in a settling time of 0.25 ms. The input current response for the closed-outer-voltage loop achieves a steady-state value $I_I(\infty) = 0.76$ A in a rapid settling time of nearly 0.25 ms. The reduction in the input current in the negative direction for a positive step change in output voltage is due to the negative input resistance caused at DC by the closed-inner-current and closed-outer-voltage loops.

3.8.4 Response of Output Voltage to Step Change in Load Current

Figure 3.64 compares the responses of the output voltage for unit step changes in output current by 1 A for (i) open loop (Z_o), (ii) closed-inner-current loop (Z_{oicl}), and (iii) closed-outer-voltage loop (Z_{ovcl}). In the open-loop configuration, a step change in output current $\Delta I_O = 1$ A causes the output voltage to achieve a new steady-state value $V_O(\infty) = 20.325$ V. Therefore, the step change in the output voltage is $\Delta V_O = V_O(\infty) - V_O(0) = 20.325 - 20 = 0.325$ V, yielding $Z_{oo} = R_o = (\Delta V_O)/(\Delta I_O) = 0.325/1 = 0.325\ \Omega$. The step response shows a settling time of 3 ms. The output voltage response for the closed-inner-current loop achieves a steady-state value $V_O(\infty) = 39.4$ V in a settling time of 8 ms. The output voltage for the closed-outer-voltage loop returns to the steady-state value $V_O(\infty) = V_O(0) = 20$ V in a settling time of 1 ms.

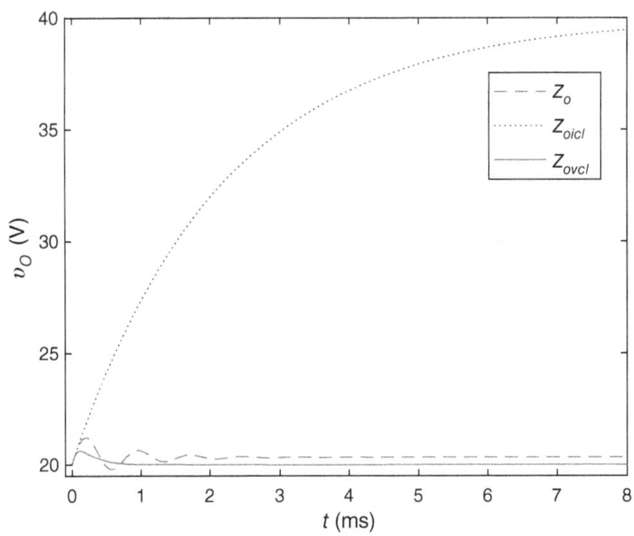

Figure 3.64 Comparison of responses in the output voltage for step changes in the output current by 1 A, obtained using Z_o, Z_{oicl}, and Z_{ovcl} transfer functions.

3.9 Summary

1. The small-signal, linear model of the boost converter in CCM consists of a current-controlled current source representing the transistor and a voltage-controlled voltage source representing the diode, in addition to the input AC voltage source, inductor, and filter capacitor.
2. The small-signal duty cycle-to-output voltage transfer function of a lossy boost converter is a second-order transfer function constituting an LHP real zero, an RHP real zero, and a pair of complex-conjugate poles.
3. The small-signal duty cycle-to-inductor transfer function of a lossy boost converter is a second-order transfer function with a real LHP zero and a pair of complex-conjugate LHP poles.
4. The small-signal input-to-output voltage transfer function of a lossy boost converter is a second-order transfer function exhibiting an LHP zero and a pair of complex-conjugate LHP poles. The location of the LHP zero and complex poles are identical to that of the duty cycle-to-output voltage transfer function.
5. The small-signal input voltage-to-inductor transfer function of a lossy boost converter is a second-order transfer exhibiting an LHP zero and a pair of complex-conjugate LHP poles.
6. The input impedance of a lossy boost converter exhibits a pair of LHP complex-conjugate zeros and an LHP real zero. The input resistance is a function of the load resistance and the duty cycle. The location of the complex zero is identical to the complex pole produced by the duty cycle-to-output voltage transfer function.
7. The output impedance of a lossy boost converter has two LHP real zeros and LHP complex-conjugate poles. The output resistance is a function of the load resistance. The location of the complex pole is identical to that produced by the duty cycle-to-output voltage transfer function.
8. The current-mode control of a boost converter can be achieved using an integral single-lead controller and can effectively reduce the DC steady-state error and improve the closed-loop bandwidth. This is a major advantage of the voltage-mode control, where the presence of RHP zero in the duty cycle-to-output voltage transfer function demands double-lead or higher forms of control schemes for effective compensation.

4

Average Current-Mode Control of Buck-Boost DC–DC Converter

This chapter explains the average current-mode control of pulse-width modulated (PWM) buck-boost DC–DC converter operated in continuous-conduction mode (CCM) [49, 50]. In this chapter, the circuit averaging technique is presented, which is based on the principle of conservation of energy. The detailed operation of the buck-boost converter in CCM is provided in [45]. The main objectives of this chapter are as follows:

1. To describe DC characteristics of the buck-boost converter.
2. To derive and analyze the small-signal linear model of the buck-boost converter.
3. To derive and analyze the open-loop power stage transfer functions of the buck-boost converter.
4. To derive and analyze the open-loop input and output impedances of the buck-boost converter.
5. To design and analyze a controller circuit for the buck-boost converter.
6. To derive and analyze the closed-inner-current-loop transfer functions of the buck-boost converter.
7. To demonstrate the behavior of average current-mode controlled buck-boost converter.
8. To derive and analyze the closed-inner-current-loop transfer functions of the buck-boost converter.
9. To derive and analyze the closed-outer-voltage-loop transfer functions of the buck-boost converter.

As the outcome of this chapter, the reader will gain sufficient knowledge in analysis, design, and characterization of buck-boost DC–DC converter employing average current-mode control. The subsequent analysis in this chapter is based on the following assumptions:

1. The diode capacitance and the lead inductance are zero, and thereby the switching losses are neglected.
2. Passive components are linear, time invariant, and frequency dependent.
3. The output impedance of the supply voltage source is zero for both DC and AC components.
4. The converter operates in steady state.

Average Current-Mode Control of DC-DC Power Converters, First Edition.
Marian K. Kazimierczuk, Dalvir K. Saini, and Agasthya Ayachit.

5. The switching period is much shorter than the time constants of the reactive components designed for CCM operation.
6. The DC output voltage is constant; however, the DC supply voltage and the load resistance are variable.

4.1 Circuit Description, DC Model, and Design

4.1.1 Circuit Description

The circuit of the PWM buck-boost converter is shown in Figure 4.1. The circuit consists of a power MOSFET S, a diode D_o, an inductor L, a filter capacitor C, and a load resistor R_L. The switch is turned ON and OFF at a switching frequency f_s and a duty cycle D. The parasitic resistances of the passive and semiconductor components are considered in the analysis. The equivalent series resistance of the inductor and capacitor are r_L and r_C, respectively. The ON-state resistance of the MOSFET and the ON resistance of the diode are r_{DS} and R_F, respectively, and the diode forward voltage is V_F. The output capacitance of the MOSFET is C_o.

4.1.2 DC Model

The DC voltage transfer functions of the buck-boost converter operated in CCM are provided in detail in [45]. The idealized current and voltage waveforms of the switching network of the buck-boost converter are shown in Figure 4.2. From Figure 4.2, the switch current waveform can be expressed as

$$i_S(t) = \begin{cases} I_L, & \text{for } 0 < t \leq DT \\ 0, & \text{for } DT < t \leq T. \end{cases} \tag{4.1}$$

The DC component of switch current is

$$I_S = \frac{1}{T}\int_0^T i_S(t)dt = \frac{1}{T}\int_0^{DT} i_S(t)dt = \frac{1}{T}\int_0^{DT} I_L dt = DI_L = D(I_I - I_O). \tag{4.2}$$

From the circuit, the diode inverted voltage is expressed by the following relationship:

$$-v_D(t) = v_{LD}(t) = v_{AK}(t), \tag{4.3}$$

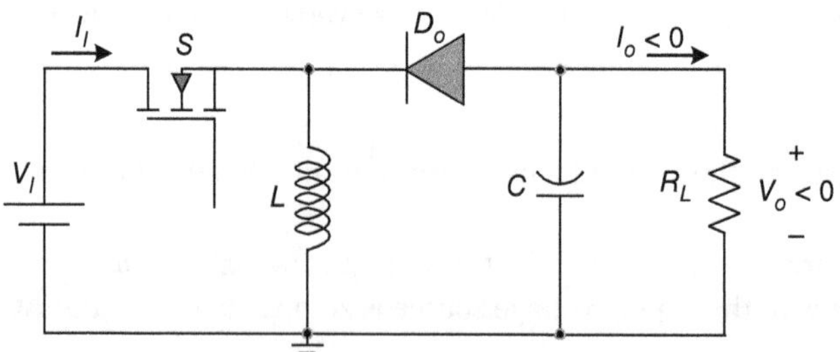

Figure 4.1 Circuit of the PWM buck-boost DC–DC converter.

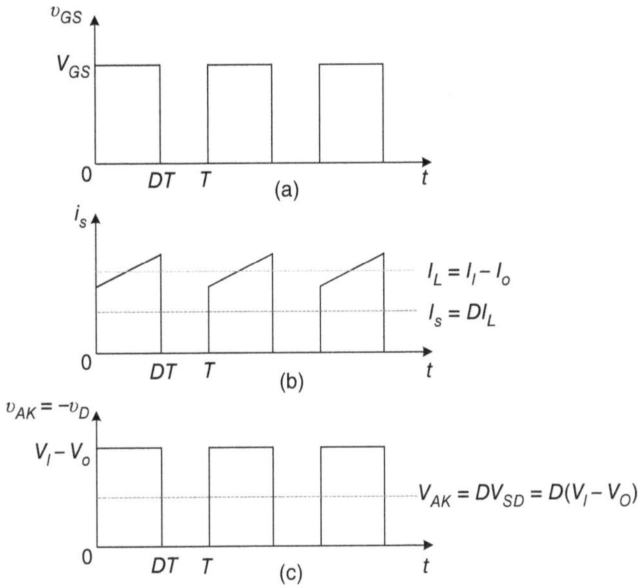

Figure 4.2 Waveforms of ideal switching network. (a) Gate-to-source voltage of switch. (b) Switch current. (c) Diode voltage.

where v_{AK} is the voltage between diode anode and cathode and v_{LD} is the voltage across the node connecting the inductor, switch, and diode. The DC component of the diode inverted voltage is

$$-V_D = V_{LD} = V_{AK}. \tag{4.4}$$

From Figure 4.2, the diode inverted voltage waveform can be expressed as

$$v_{LD}(t) = -v_D(t) = v_{AK}(t) = \begin{cases} V_{SD} = V_I - V_O, & \text{for } 0 < t \leq DT \\ 0, & \text{for } DT < t \leq T. \end{cases} \tag{4.5}$$

The DC component of diode inverted voltage in terms of the input and output voltage is

$$V_{LD} = -V_D = V_{AK} = \frac{1}{T}\int_0^T v_{LD}dt = \frac{1}{T}\int_0^{DT} v_{LD}dt = \frac{1}{T}\int_0^{DT} V_{SD}dt = DV_{SD}$$
$$= D(V_I - V_O). \tag{4.6}$$

As a result, the switch can be replaced by a current-controlled DC source represented by (4.2) and the diode by a voltage-controlled voltage source represented by (4.6). The DC model of the buck-boost converter is obtained by replacing the switch with a controlled current source, diode by a controlled voltage source, by treating the capacitor C as an open circuit and the inductor L as a short circuit. Figure 4.3 shows the DC model of the buck-boost converter in CCM.

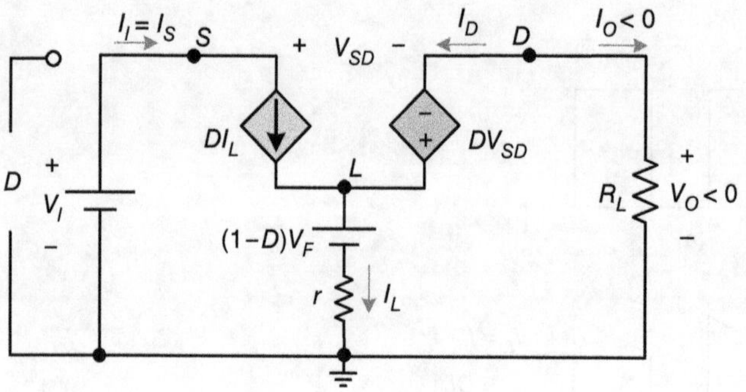

Figure 4.3 DC model of the PWM buck-boost DC–DC converter in CCM.

Applying the principle of conservation of energy [37, 45], the equivalent averaged resistance r placed in series with the inductance L is the sum of the averaged resistance of the switch, diode, and inductor branch given by

$$r = Dr_{DS} + (1 - D)R_F + r_L. \tag{4.7}$$

Using (4.2),

$$I_I = I_S = DI_L = D(I_I - I_O), \tag{4.8}$$

yielding the DC input current as

$$I_I = -\frac{D}{1-D} I_O = -\frac{D}{1-D}\frac{V_O}{R_L}. \tag{4.9}$$

Therefore, the dc current transfer function of the buck-boost converter is

$$M_{IDC} = \frac{I_O}{I_I} = -\frac{1-D}{D}. \tag{4.10}$$

The DC inductor current is

$$I_L = -\frac{I_O}{1-D} = -\frac{V_O}{R_L(1-D)}. \tag{4.11}$$

For a lossless buck-boost converter, $r = 0$ and $V_F = 0$. Applying KVL to the output loop yields

$$D(V_I - V_O) = -V_O, \tag{4.12}$$

producing the DC voltage transfer function for the lossless converter as

$$M_{VDC} = \frac{V_O}{V_I} = -\frac{D}{1-D}. \tag{4.13}$$

For the lossy buck-boost converter, the DC voltage transfer function is dependent on r and V_F and can be derived by applying KVL to the output loop

$$V_O = -DV_{SD} + (1 - D)V_F + rI_L, \tag{4.14}$$

producing

$$V_O = -D(V_I - V_O) + (1 - D)V_F - \frac{rV_O}{(1 - D)R_L}, \tag{4.15}$$

yielding the DC voltage transfer function of a lossy buck-boost converter as

$$M_{VDC(lossy)} = \frac{V_O}{V_I} = -\frac{D}{1-D}\frac{1}{1+\frac{r}{R_L(1-D)^2} - \frac{V_F}{V_O}}. \tag{4.16}$$

The efficiency of the lossy buck-boost converter can be obtained using the DC model as

$$\eta = \frac{P_O}{P_I} = \frac{V_O I_O}{V_I I_I} = \frac{1}{1+\frac{r}{R_L(1-D)^2} - \frac{V_F}{V_O}}. \tag{4.17}$$

The overall efficiency of the buck-boost converter [45] is

$$\begin{aligned}\eta &= \frac{P_O}{P_I} = \frac{V_O I_O}{V_I I_I} = M_{IDC}M_{VDC}\\ &= \frac{1}{1+\frac{r_L + Dr_{DS}}{(1-D)^2 R_L} + \frac{R_F + Dr_C}{(1-D)R_L} - \frac{V_F}{V_O} + \frac{f_s C_o R_L(1+M_{VDC})^2}{M_{VDC}^2}}.\end{aligned} \tag{4.18}$$

4.1.3 Design Example

The subsequent analysis is performed for the buck-boost converter designed for the following specifications:

1. Supply voltage $V_I = 12$ V.
2. Output voltage $V_O = -5$ V.
3. Output power $P_O = 10$ W.
4. Load resistance $R_L = 2.5\ \Omega$.
5. Switching frequency $f_s = 200$ kHz.
6. Inductance to ensure CCM operation $L = 10\ \mu$H.
7. Filter capacitance $C = 120\ \mu$F.

The equivalent series resistances of the inductor is $r_L = 10$ mΩ, and the capacitor is $r_C = 10$ mΩ. The ON-state resistance of the MOSFET and the ON resistance of the diode are $r_{DS} = 0.11\ \Omega$ and $R_F = 0.07\ \Omega$. The diode forward voltage is $V_F = 0.3$ V. The calculated value of converter efficiency is $\eta = 88\%$. The required duty cycle to achieve the rated output voltage is $D = 0.3213$. The equivalent averaged resistance $r \approx 0.1\ \Omega$.

The calculated values of open-loop DC quantities using equations in Section 4.1.2 are input current $I_I = 0.83$ A, output current $I_O = -2$ A, dc current transfer function $M_{IDC} = -2.4$, DC voltage transfer function $M_{VDC} = -0.42$, inductor current $I_L = 2.95$ A, switch current $I_S = 0.83$ A, and diode voltage $|V_D| = 5.46$ V.

4.2 Large-Signal and Small-Signal Models of PWM Buck-Boost Converter in CCM

The DC model is perturbed by small-signal, low-frequency, sinusoidal excitation about the DC operating point to obtain a large-signal nonlinear model. The

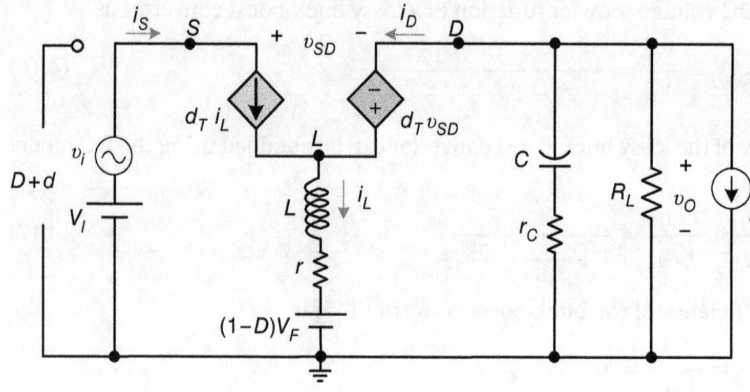

Figure 4.4 Large-signal nonlinear model of the PWM buck-boost DC–DC converter in CCM.

large-signal nonlinear model of a buck-boost DC–DC converter in CCM is shown in Figure 4.4. The large-signal expressions for the switch current and the diode inverted voltage, respectively, are

$$i_S = d_T i_L \tag{4.19}$$

and from (4.6)

$$-v_D = v_{LD} = v_{AK} = d_T v_{SD}, \tag{4.20}$$

where the large-signal quantities are expressed as the sum of DC and AC components

$$i_S = I_S + i_s, \tag{4.21}$$

$$i_L = I_L + i_l, \tag{4.22}$$

$$v_{LD} = V_{LD} + v_{ld}, \tag{4.23}$$

$$v_I = V_I + v_i, \tag{4.24}$$

$$v_O = V_O + v_o, \tag{4.25}$$

and

$$d_T = D + d, \tag{4.26}$$

where i_S, i_L, v_D, v_I, v_O, and d_T are the large-signal switch current, inductor current, diode voltage, input voltage, output voltage, and duty cycle, respectively. The quantities i_s, i_l, v_d, v_i, v_o, and d are the small-signal switch current, inductor current, diode voltage, input voltage, output voltage, and duty cycle, respectively. Hence, from (4.2),

$$i_S = d_T i_L = (D + d)(I_L + i_l) = DI_L + Di_l + I_L d + i_l d \tag{4.27}$$

and from (4.6),

$$\begin{aligned}
-v_D = v_{LD} = d_T v_{SD} &= d_T(v_I - v_O) = (D+d)[(V_I + v_i) - (V_O + v_o)] \\
&= (D+d)(V_I + v_i) - (D+d)(V_O + v_o) \\
&= DV_I + Dv_i + V_I d + v_i d - DV_O - Dv_o - V_O d - v_o d \\
&= DV_I - DV_O + Dv_i - Dv_o + V_I d - V_O d + v_i d - v_o d \\
&= D(V_I - V_O) + D(v_i - v_o) + (V_I - V_O)d + (v_i - v_o)d. \qquad (4.28)
\end{aligned}$$

In (4.27) and (4.28), the nonlinear small-signal products $i_l d$, $v_i d$, and $v_o d$ are eliminated by applying the small-signal conditions ($d << D$, $v_i << V_I$, $i_l << I_L$, $v_o << V_O$), yielding a linearized large-signal model. Hence, the linearized large-signal expressions for the switch current and the diode inverted voltage can be derived by

$$i_S = I_S + i_s = DI_L + Di_l + I_L d \qquad (4.29)$$

and

$$-v_D = v_{LD} = V_{LD} + v_{ld} = D(V_I - V_O) + D(v_i - v_o) + (V_I - V_O)d. \qquad (4.30)$$

Figure 4.5 shows the large-signal linear model of the buck-boost DC–DC converter. Using the principle of superposition, the DC terms are separated to obtain the DC averaged model as shown in Figure 4.3 and is represented by the expressions

$$I_S = DI_L \qquad (4.31)$$

and

$$V_{LD} = D(V_I - V_O). \qquad (4.32)$$

The AC terms are separated to obtain the linear, averaged, small-signal model as shown in Figure 4.6 and is represented by the expressions

$$i_s = Di_l + I_L d \qquad (4.33)$$

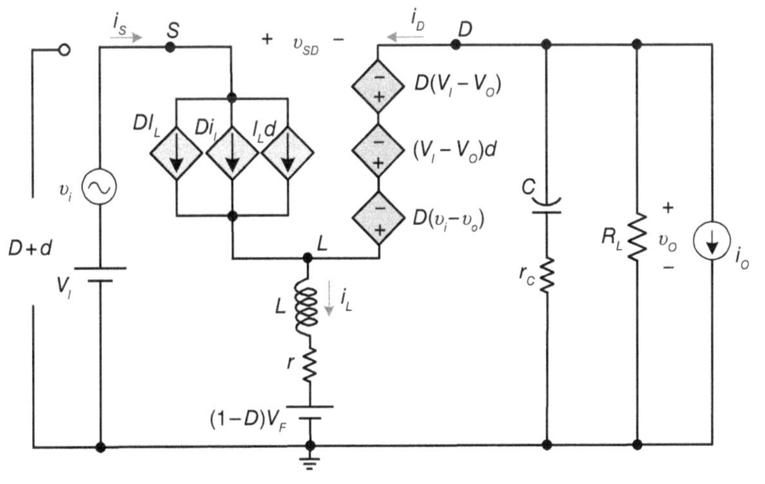

Figure 4.5 Large-signal linear model of the PWM buck-boost DC–DC converter in CCM.

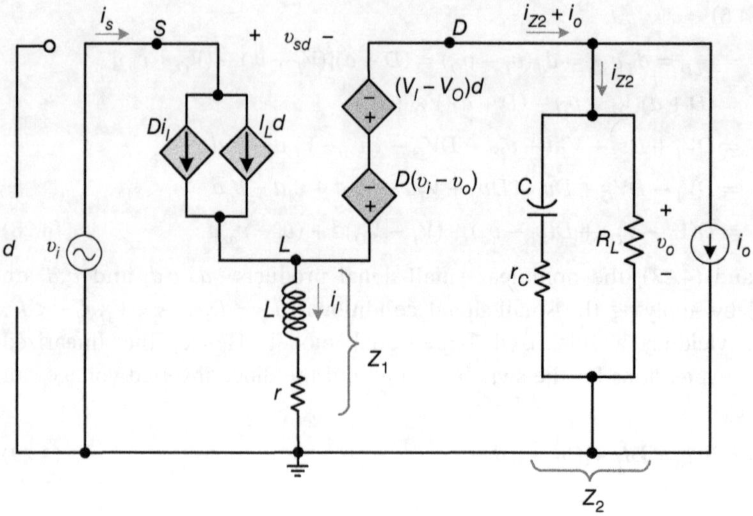

Figure 4.6 Small-signal model of the PWM buck-boost DC–DC converter in CCM.

and

$$v_{ld} = D(v_i - v_o) + (V_I - V_O)d. \tag{4.34}$$

Subsequent analysis in this chapter considers the small-signal model in Figure 4.6 to determine the power-stage transfer functions of the buck-boost DC–DC converter.

4.3 Power-Stage Transfer Functions

The following power-stage transfer functions are discussed in this section:

1. Duty cycle-to-output voltage transfer function T_p (control).
2. Duty cycle-to-inductor current transfer function T_{pi} (control).
3. Input voltage-to-output voltage transfer function M_v (disturbance).
4. Input voltage-to-inductor current transfer function M_{vi} (disturbance).
5. Reverse current gain A_i (disturbance).
6. Input impedance Z_i (disturbance).
7. Output impedance Z_o (disturbance).

The small-signal model of the PWM buck-boost converter is shown in Figure 4.6. The state equations governing the small-signal model are derived. The impedances Z_1 and Z_2 are

$$Z_1 = r + sL \tag{4.35}$$

and

$$Z_2 = R_L || \left(r_C + \frac{1}{sC} \right) = \frac{R_L \left(r_C + \frac{1}{sC} \right)}{R_L + r_C + \frac{1}{sC}}. \tag{4.36}$$

The current through the parallel combination of the load resistance and the filter capacitor is

$$i_{Z_2} = \frac{v_o}{Z_2}. \tag{4.37}$$

The input current is

$$i_i = i_s = Di_l + I_L d. \tag{4.38}$$

The current through the inductor is

$$i_l = i_s - (i_{Z_2} + i_o) = Di_l + I_L d - i_{Z_2} - i_o, \tag{4.39}$$

yielding

$$i_l = \frac{I_L d}{1-D} - \frac{v_o}{Z_2(1-D)} - \frac{i_o}{1-D}. \tag{4.40}$$

Applying Kirchhoff's voltage law to the output loop provides

$$v_o = -(V_I - V_O)d - D(v_i - v_o) + i_l Z_1. \tag{4.41}$$

4.3.1 Duty Cycle-to-Output Voltage Transfer Function T_p

The duty cycle-to-output voltage transfer function is obtained by setting $v_i = 0$ and $i_o = 0$ in Figure 4.7. Applying KVL the output loop with $v_i = 0$ provides

$$v_o = -(V_I - V_O)d + Dv_o + i_l Z_1. \tag{4.42}$$

The current through the inductor with $i_o = 0$ is

$$i_l = i_s - i_{Z_2} = Di_l + I_L d - i_{Z_2}. \tag{4.43}$$

The current through the impedance Z_2 is

$$i_{Z_2} = \frac{v_o}{Z_2}. \tag{4.44}$$

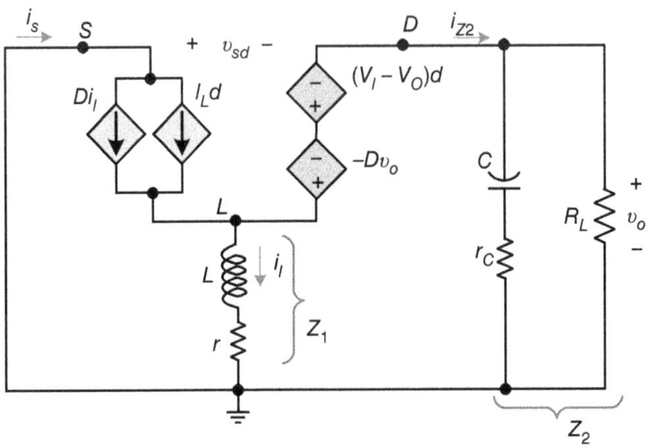

Figure 4.7 Small-signal model of the PWM buck-boost converter to derive the duty cycle-to-output voltage transfer function T_p.

Substituting (4.63) into (4.62) gives

$$i_l = Di_l + I_L d - \frac{v_o}{Z_2} \tag{4.45}$$

or

$$i_l = \frac{I_L d}{1-D} - \frac{v_o}{Z_2(1-D)}. \tag{4.46}$$

Substituting (4.65) into (4.42),

$$v_o = -(V_I - V_O)d + Dv_o + \left[\frac{I_L d}{1-D} - \frac{v_o}{Z_2(1-D)}\right] Z_1, \tag{4.47}$$

yielding

$$v_o(1-D) = -\left(\frac{V_I}{V_O} - 1\right) V_O d + \frac{I_L Z_1}{1-D} d - \frac{Z_1}{Z_2(1-D)} v_o. \tag{4.48}$$

Substituting for $V_I/V_O = 1/M_{VDC} = -(1-D)/D$ and $I_L = -V_O/[R_L(1-D)]$ into (4.48) and rearranging gives

$$(1-D)\left[1 + \frac{Z_1}{Z_2(1-D)^2}\right] v_o = -\frac{V_O}{D}\left[1 - \frac{DZ_1}{R_L(1-D)^2}\right] d. \tag{4.49}$$

Therefore, the duty cycle-to-output transfer function is

$$T_p(s) = \left.\frac{v_o(s)}{d(s)}\right|_{v_i=i_o=0} = -\frac{V_O}{D(1-D)} \frac{1 - \dfrac{DZ_1}{R_L(1-D)^2}}{1 + \dfrac{Z_1}{Z_2(1-D)^2}}. \tag{4.50}$$

Substituting (4.35) and (4.36) into (4.50) gives the duty cycle-to-output transfer function in the s-domain as

$$T_p(s) = \left.\frac{v_o(s)}{d(s)}\right|_{v_i=i_o=0} = T_{px} \frac{(s+\omega_{zn})(s-\omega_{zp})}{s^2 + 2\xi\omega_o s + \omega_o^2} = T_{po} \frac{\left(1 + \dfrac{s}{\omega_{zn}}\right)\left(1 - \dfrac{s}{\omega_{zp}}\right)}{\left(\dfrac{s}{\omega_o}\right)^2 + \dfrac{2\xi s}{\omega_o} + 1}, \tag{4.51}$$

where the DC gain T_{po} is

$$T_{po} = T_p(0) = -\frac{V_O}{D(1-D)} \frac{Dr - (1-D)^2 R_L}{r + (1-D)^2 R_L}; \tag{4.52}$$

the high-frequency gain T_{px} is

$$T_{px} = T_p(\infty) = -\frac{V_O}{1-D} \frac{r_C}{R_L + r_C}; \tag{4.53}$$

the angular corner frequency, i.e. the angular undamped natural frequency, is

$$\omega_o = \sqrt{\frac{r + R_L(1-D)^2}{LC(R_L + r_C)}}; \tag{4.54}$$

and the damping ratio is

$$\xi = \frac{L + C[r(R_L + r_C) + R_L r_C(1 - D)^2]}{2\sqrt{LC(R_L + r_C)[R_L(1 - D)^2 + r]}}. \tag{4.55}$$

The angular frequency of the left-half plane (LHP) zero is

$$\omega_{zn} = \frac{1}{r_C C} \tag{4.56}$$

and the angular frequency of the right-half plane (RHP) zero is

$$\omega_{zp} = \frac{R_L(1 - D)^2 - Dr}{DL} = \frac{R_L(1 - D)^2 - Dr_L - D^2 r_{DS} - D(1 - D)R_F}{DL}. \tag{4.57}$$

The magnitude of the transfer function is

$$|T_p| = T_{po}\sqrt{\frac{\left[1 + \left(\frac{\omega}{\omega_{zn}}\right)^2\right]\left[1 + \left(\frac{\omega}{\omega_{zp}}\right)^2\right]}{\left[1 - \left(\frac{\omega}{\omega_o}\right)^2\right]^2 + 4\xi^2\left(\frac{\omega}{\omega_o}\right)^2}} \tag{4.58}$$

and the phase of the transfer function is

$$\phi_{Tp} = -180^\circ + \tan^{-1}\left(\frac{\omega}{\omega_{zn}}\right) - \tan^{-1}\left(\frac{\omega}{\omega_{zp}}\right) - \tan^{-1}\left[\frac{2\xi\left(\frac{\omega}{\omega_o}\right)}{1 - \left(\frac{\omega}{\omega_o}\right)^2}\right]. \tag{4.59}$$

For the design specification given in Section 4.1.3, the gain at DC and low frequencies is nearly $T_{po} = 20.67$ V = 26.31 dBV. The natural corner frequency of the second-order low-pass filter formed by the inductor, capacitor, and load resistor is $f_o = 3.2$ kHz. The gain decreases by −40 dB/dec beyond f_o. Figure 4.8 shows the theoretically obtained magnitude and phase plots of duty cycle-to-output voltage transfer function. The duty cycle-to-output voltage transfer function of the buck-boost converter exhibits an RHP zero located at $f_{zp} = \omega_{zp}/2\pi = 56$ kHz. The RHP zero is a zero for magnitude, but it behaves like a pole for phase. The RHP zero cancels the effect of one of the complex-conjugate poles reducing the fall-off slope to −20 dB/dec. Beyond f_{zp}, the gain decreases by −20 dB/dec and is finally canceled by the LHP zero $f_{zn} = \omega_{zn}/2\pi = 132.6$ kHz created by the capacitor C and its equivalent series resistance r_C. The gain at high frequencies is constant and equal to −30 dB at high frequencies. The phase is −180° at DC and low frequencies indicating a phase inversion between the duty cycle and the output voltage, i.e. as duty cycle increases or decreases, the output voltage decreases or increases, respectively. With the increase in frequency, the phase rolls off by −180° due to the complex-conjugate poles. In addition, the presence of the RHP zero adds an additional phase of −45° per decade at f_{zp}, thereby having the phase cross −360°. As the perturbation frequency increases, the LHP zero at ω_{zn}, which offers a positive slope of 45° per decade, negates the effect of the RHP zero, thereby increasing the phase to −360°.

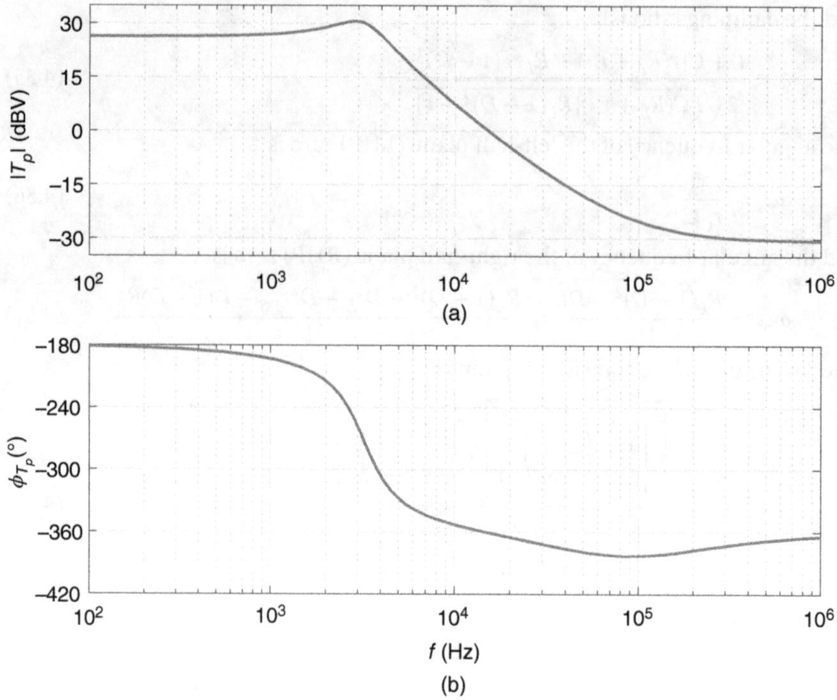

Figure 4.8 Theoretically obtained plots of duty cycle-to-output voltage transfer function T_p. (a) Magnitude. (b) Phase.

Figure 4.9 shows the locations of the power-stage poles p_1, p_2, the LHP zero z_n, and the RHP zero z_p at different values of the load resistance varying from 2.5 Ω to 10 Ω. The location of the LHP zero z_n caused by the filter capacitor C and its equivalent series resistance r_C is independent of the load resistance. It is observed that the complex-conjugate poles do not relocate with change in the load resistance. The RHP zero moves toward ∞ with the increase in the load resistance. This means that the effect of the RHP zero is not significant at high load resistance values. Figure 4.10 shows the Bode magnitude and phase plots of the T_p transfer function at selected values of the load resistance. The DC gain increases slightly with the increase in the load resistance. At low load resistance, the RHP zero is closer to the origin, and the phase plot crosses −180° in the vicinity of the bandwidth frequency $f_{BW} = 2$ kHz. However, as the load resistance increases, the phase near f_{BW} also increases. In the transient domain, a high load resistance corresponds to a larger damping coefficient and a reduction in ringing in the output voltage.

Figure 4.11 shows the trajectories of the power-stage poles p_1, p_2, the LHP zero z_n, and the RHP zero z_p at different values of the duty cycle varying from 0.2 to 0.92. The location of the LHP zero z_n caused by the filter capacitor C and its equivalent series resistance r_C is independent of the duty cycle. The imaginary components of the poles p_1 and p_2 move toward the origin with the increase in duty cycle. The RHP zero z_p also moves toward the origin with the increase in the duty cycle. As D increases from 0 to 1, z_p decreases from $-\infty$ to $-(r_L + r_{DS})/L$. At duty cycles larger

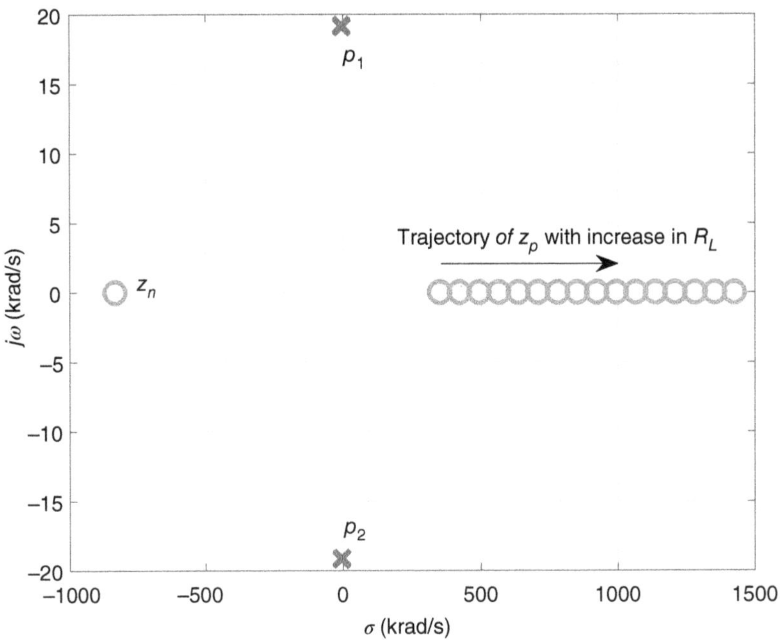

Figure 4.9 Trajectory of the poles and zeros of T_p transfer function as functions of the load resistance.

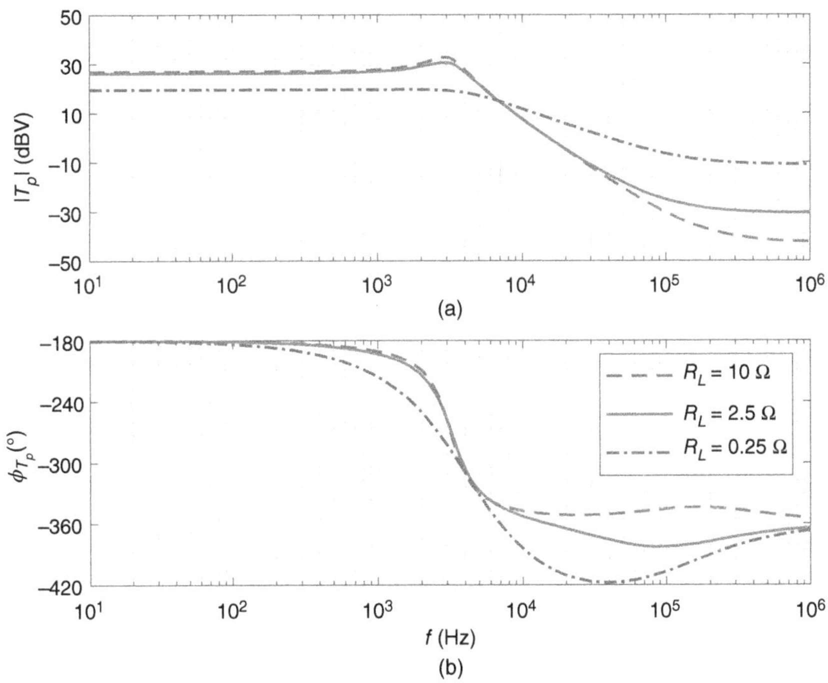

Figure 4.10 Theoretically obtained plots of duty cycle-to-output voltage transfer function T_p at selected load resistances. (a) Magnitude. (b) Phase.

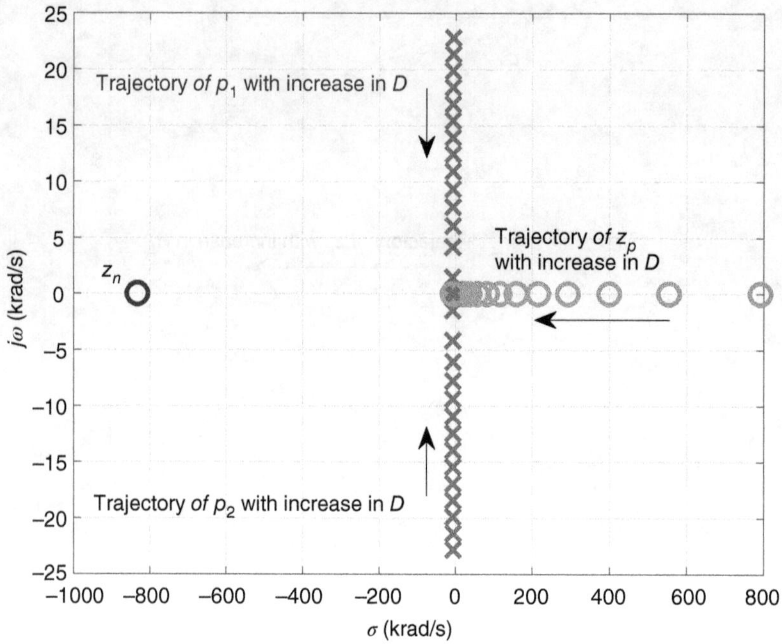

Figure 4.11 Trajectory of the poles and zeros of T_p transfer function as functions of the duty cycle.

than a critical value, the poles become real with p_1 moving toward the origin and p_2 moving toward $-\infty$. An interesting phenomena occurs with the RHP zero. At a critical duty cycle, the RHP zero is at the origin. Hence, the expression for the critical duty cycle can be derived by setting $\omega_{zp} = 0$. The critical duty cycle is calculated as

$$D_{Cr} = \frac{2R_L + r_L + R_F - \sqrt{4R_L(r_L + r_{DS}) + R_F(2r_L + R_F) + r_L^2}}{2(R_L - r_{DS} + R_F)} = 0.81. \tag{4.60}$$

Beyond the critical duty cycle D_{Cr}, by the influence of losses in the converter, the RHP zero enters the left-half of the s-plane. Figure 4.12 shows the root locus of the poles p_1 and p_2 and zero z_p of T_p closer to the origin to illustrate the behavior and the relocation beyond the critical duty cycle value. The stability of the buck-boost converter can also be determined using the Bode plot of T_p transfer function analyzed at different duty cycles. Figure 4.13 shows the Bode magnitude and phase plots of the T_p transfer function at selected values of the duty cycle. As the duty cycle increases, the DC gain T_{po} increases, and the corner frequency f_o decreases.

4.3.2 Duty Cycle-to-Inductor Current Transfer Function T_{pi}

The duty cycle-to-output voltage transfer function is obtained by setting $v_i = 0$ and $i_o = 0$ in Figure 4.6. Setting $v_i = 0$ results in

$$v_o = -(V_I - V_O)d + Dv_o + i_l Z_1 = -\frac{(V_I - V_O)d}{1 - D} + \frac{i_l Z_1}{1 - D}. \tag{4.61}$$

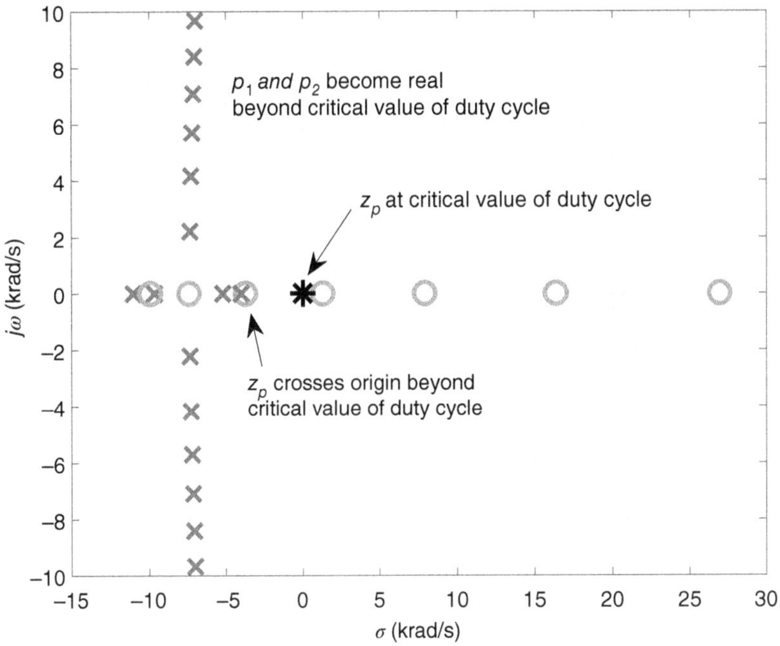

Figure 4.12 The poles and zeros of T_p transfer function beyond the critical value of the duty cycle.

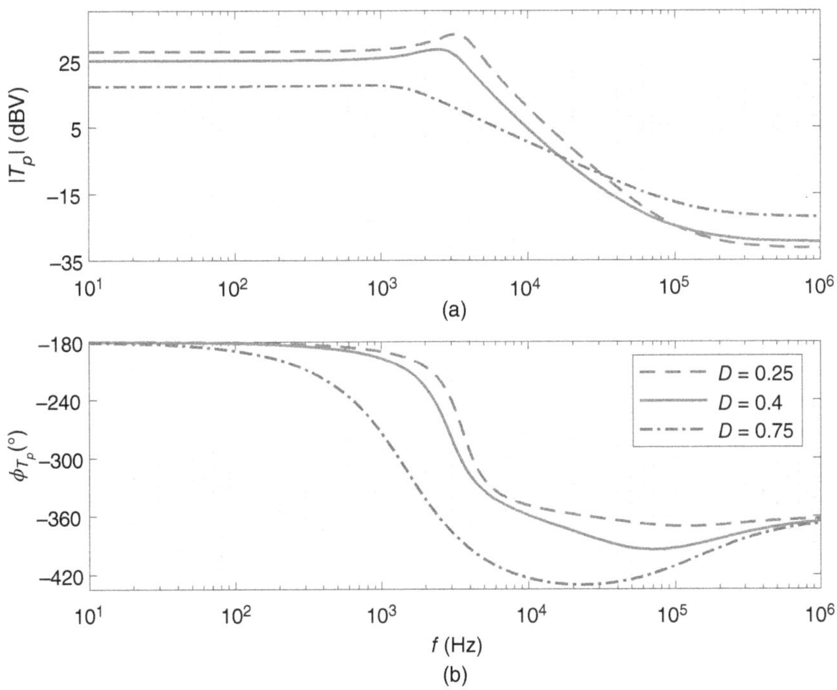

Figure 4.13 Theoretically obtained plots of duty cycle-to-output voltage transfer function T_p at selected duty ratios. (a) Magnitude. (b) Phase.

The current through the inductor is

$$i_l = i_s - i_{Z_2} = Di_l + I_L d - i_{Z_2}. \tag{4.62}$$

The current through the impedance Z_2 is

$$i_{Z_2} = \frac{v_o}{Z_2}. \tag{4.63}$$

Substituting (4.63) into (4.62) gives

$$i_l = Di_l + I_L d - \frac{v_o}{Z_2} \tag{4.64}$$

or

$$i_l = \frac{I_L d}{1-D} - \frac{v_o}{Z_2(1-D)}. \tag{4.65}$$

Substituting (4.61) into (4.65) yields

$$i_l = \frac{I_L d}{1-D} + \frac{(V_I - V_O)d}{Z_2(1-D)^2} - \frac{i_l Z_1}{Z_2(1-D)^2}. \tag{4.66}$$

Rearranging (4.66), we get

$$i_l \left[1 + \frac{Z_1}{Z_2(1-D)^2}\right] = \frac{I_L}{1-D} d + \frac{(V_I - V_O)d}{Z_2(1-D)^2}. \tag{4.67}$$

Substituting for $V_I/V_O = 1/M_{VDC} = -(1-D)/D$ and $I_L = -V_O/[R_L(1-D)]$ into (4.67) and rearranging gives

$$i_l \left[1 + \frac{Z_1}{Z_2(1-D)^2}\right] = -\frac{V_O}{(1-D)^2 R_L}\left(1 + \frac{R_L}{DZ_2}\right). \tag{4.68}$$

Hence, the duty cycle-to-inductor current transfer function is

$$T_{pi}(s) = \left.\frac{i_l(s)}{d(s)}\right|_{v_i=i_o=0} = -\frac{V_O}{(1-D)^2 R_L} \frac{1 + \dfrac{R_L}{DZ_2}}{1 + \dfrac{Z_1}{Z_2(1-D)^2}}. \tag{4.69}$$

Substituting (4.35) and (4.36) into (4.69) gives the duty cycle-to-inductor current transfer function in the s-domain as

$$T_{pi}(s) = \left.\frac{i_l(s)}{d(s)}\right|_{v_i=i_o=0} = T_{pix}\frac{s + \omega_{zi}}{s^2 + 2\xi\omega_o s + \omega_o^2} = T_{pio}\frac{1 + \dfrac{s}{\omega_{zi}}}{\left(\dfrac{s}{\omega_o}\right)^2 + \dfrac{2\xi s}{\omega_o} + 1}, \tag{4.70}$$

where

$$T_{pix} = -\frac{V_O(R_L + r_C + Dr_C)}{DL(R_L + r_C)}, \tag{4.71}$$

the DC gain T_{pio} is

$$T_{pio} = T_{pi}(0) = -\frac{V_O(1+D)}{D[r + R_L(1-D)^2]}, \tag{4.72}$$

and the angular frequency of the LHP zero is

$$\omega_{zi} = \frac{1 + D}{C(R_L + r_C + Dr_C)}. \tag{4.73}$$

The magnitude of the duty cycle-to-inductor current transfer function is

$$|T_{pi}| = T_{pio} \sqrt{\frac{1 + \left(\frac{\omega}{\omega_{zi}}\right)^2}{\left[1 - \left(\frac{\omega}{\omega_o}\right)^2\right]^2 + 4\xi^2 \left(\frac{\omega}{\omega_o}\right)^2}}, \tag{4.74}$$

and the phase of the transfer function is

$$\phi_{Tpi} = -180^\circ + \tan^{-1}\left(\frac{\omega}{\omega_{zi}}\right) - \tan^{-1}\left[\frac{2\xi\left(\frac{\omega}{\omega_o}\right)}{1 - \left(\frac{\omega}{\omega_o}\right)^2}\right]. \tag{4.75}$$

For the design specification given in Section 4.1.3, the gain at DC and low frequencies is nearly $T_{pio} = 16.52$ A $= 24.36$ dBA. The natural corner frequency of the second-order low-pass filter formed by the inductor, capacitor, and load resistor is $f_o = 3.2$ kHz. Figure 4.14 shows the theoretically obtained magnitude and phase plots of the duty cycle-to-inductor current transfer function. The transfer function exhibits a real zero in the LHP and a pair of LHP complex-conjugate poles. The zero

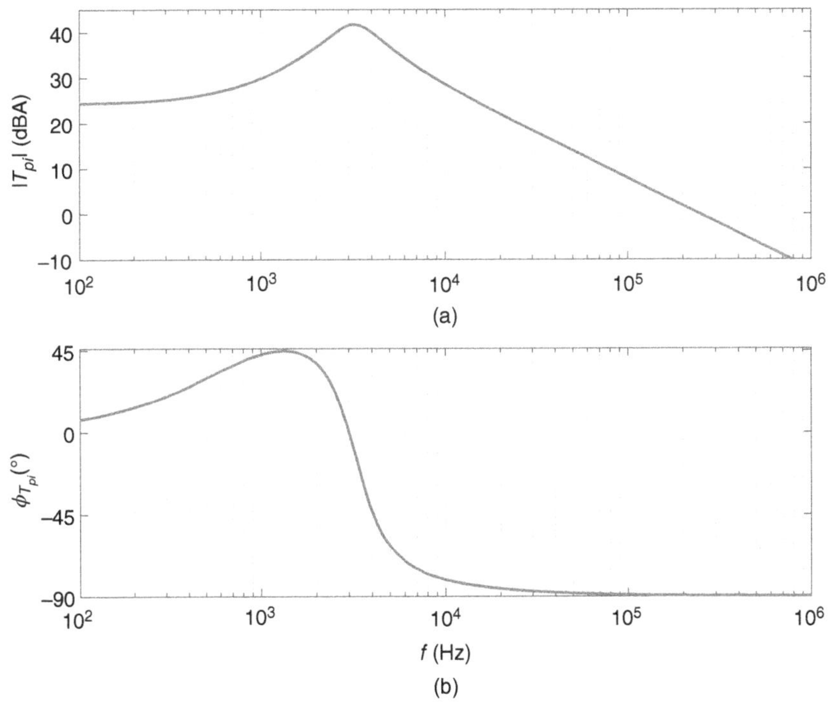

Figure 4.14 Theoretically obtained plots of duty cycle-to-inductor current transfer function T_{pi}. (a) Magnitude. (b) Phase.

frequency $f_{zi} = \omega_{zi}/2\pi = 697$ Hz, and complex-conjugate pole frequency or the natural frequency $f_o = \omega_{zp}/2\pi = 20.333/2\pi = 3.2$ kHz. The frequency of the LHP zero f_{zi} is inversely proportional to the filter capacitance and the load resistance. Beyond f_o, the gain decreases at the rate of 20 dB/dec. The phase starts at zero and rises to 45° due to f_{zi}. This means that at low frequencies, the duty cycle signal leads the inductor current by a particular phase angle. The bandwidth of T_{pi} is larger than that of T_p. Therefore, the inner-current loop presents a faster response in inductor current for change in duty cycle. The low-frequency zero f_{zi} depends on the load resistance R_L and the parameters of the filter capacitance, namely, C and r_C. As the value of R_L is decreased at a fixed C, f_{zi} moves toward f_o.

Figure 4.15 shows the trajectories of the power-stage poles p_1, p_2 and the LHP zero z_i at different values of the load resistance varying from 2.5 Ω to 10 Ω. It is observed that the real component of the complex-conjugate poles moves toward the origin with the increase in the load resistance. The LHP zero also moves toward the origin with the increase in the load resistance. Figure 4.16 shows the Bode magnitude and phase plots of the T_{pi} transfer function at selected values of the load resistance. The DC gain decreases with the increase in the load resistance. As the load resistance increases, the frequency of the LHP zero decreases. As a result, a large phase boost is observed at lower frequencies for higher values of R_L. The phase plot reaches −90° for all values of the load resistance. The location of the LHP zero z_i is independent of the duty cycle.

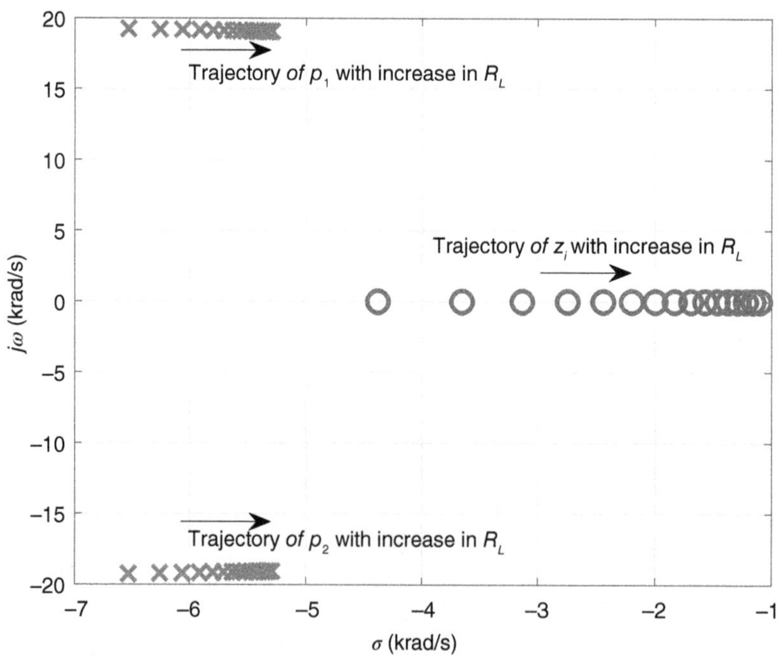

Figure 4.15 Trajectory of the poles and zeros of T_{pi} transfer function as functions of the load resistance.

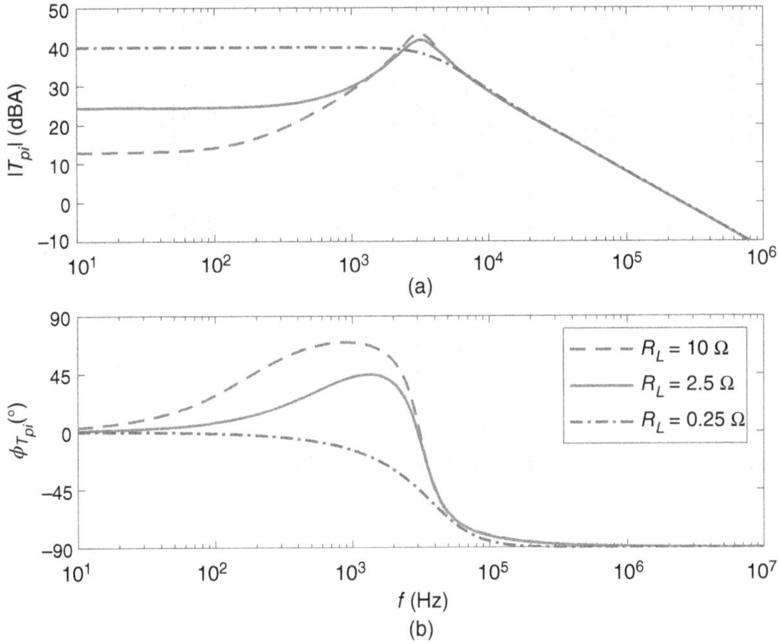

Figure 4.16 Theoretically obtained plots of duty cycle-to-inductor current transfer function T_{pi} at selected load resistances. (a) Magnitude. (b) Phase.

Figure 4.17 shows the trajectories of the power-stage poles p_1, p_2 and the LHP zero z_i at different values of the duty cycle varying from 0.2 to 0.92. The imaginary components of the poles p_1 and p_2 approach the real axis with the increase in duty cycle. Beyond critical duty cycle D_{Cr}, one of the poles moves to the origin toward the location of the zero z_i, and the other pole moves toward $-\infty$. The location of the LHP zero z_i is independent of the duty cycle. Figure 4.13 shows the Bode magnitude and phase plot of the T_{pi} transfer function at selected values of the duty cycle. As the duty cycle increases, the DC gain T_{pio} also increases and the corner frequency f_o decreases. The phase starts at 0° and saturates at −90° at high frequencies for all values of duty cycle. As the duty cycle increases, the phase boost decreases. Since the phase at high frequencies is −90°, the phase margin of the inner-current loop is nearly fixed for all duty cycle values. This offers two advantages: (a) a wide range of duty cycle can be achieved for current control, and (b) simpler controller techniques can be used.

4.3.3 Input Voltage-to-Output Voltage Transfer Function M_v

The input voltage-to-output voltage transfer function is obtained by setting $d = 0$ and $i_o = 0$ in Figure 4.6. Setting $d = 0$ in the model and applying KVL results in

$$v_o = -Dv_i + Dv_o + i_l Z_1, \tag{4.76}$$

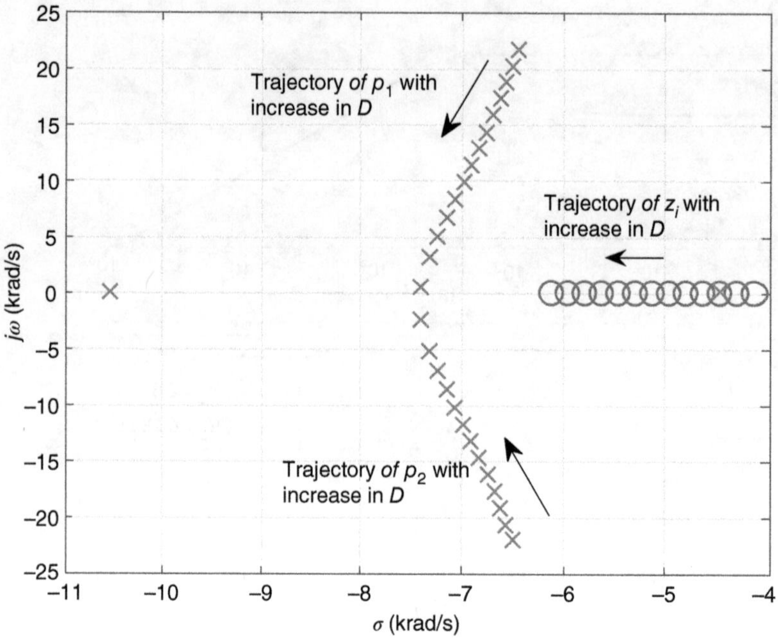

Figure 4.17 Trajectory of the poles and zeros of T_{pi} transfer function as functions of the duty cycle.

to yield

$$v_o(1-D) = -Dv_i + i_l Z_1. \tag{4.77}$$

Applying KCL at the inductor node provides

$$i_l = i_s - i_{Z_2} = Di_l - i_{Z_2} = Di_l - \frac{v_o}{Z_2}. \tag{4.78}$$

Rearranging (4.78), we get

$$i_l = -\frac{v_o}{(1-D)Z_2}. \tag{4.79}$$

Substituting (4.79) into (4.77),

$$v_o(1-D) = -Dv_i - \frac{Z_1}{(1-D)Z_2}v_o. \tag{4.80}$$

Rearranging further gives

$$v_o\left[(1-D) + \frac{Z_1}{(1-D)Z_2}\right] = -Dv_i. \tag{4.81}$$

Hence, the input voltage-to-output voltage transfer function M_v is

$$M_v(s) = \left.\frac{v_o(s)}{v_i(s)}\right|_{d=i_o=0} = -\frac{D}{1-D}\frac{1}{1+\frac{Z_1}{Z_2(1-D)^2}}. \tag{4.82}$$

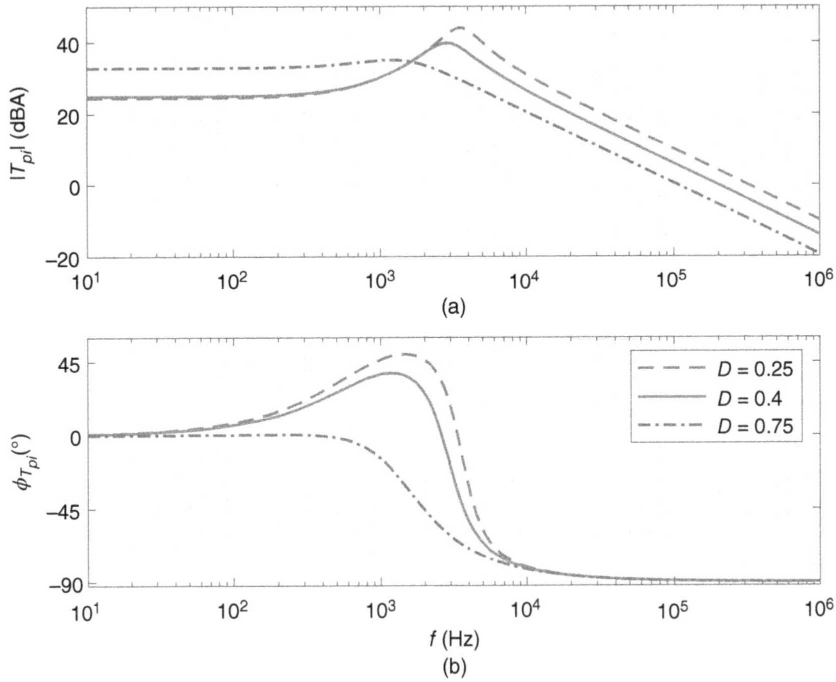

Figure 4.18 Theoretically obtained plots of duty cycle-to-inductor current transfer function T_{pi} at selected duty ratios. (a) Magnitude. (b) Phase.

Substituting (4.35) and (4.36) into (4.82) gives the input voltage-to-output voltage transfer function in the s-domain as

$$M_v(s) = \left.\frac{v_o(s)}{v_i(s)}\right|_{d=i_o=0} = M_{vx}\frac{s+\omega_{zn}}{s^2+2\xi\omega_o s+\omega_o^2} = -M_{vo}\frac{1+\dfrac{s}{\omega_{zn}}}{\left(\dfrac{s}{\omega_o}\right)^2+\dfrac{2\xi s}{\omega_o}+1}, \tag{4.83}$$

where the DC gain M_{vo} is

$$M_{vo} = M_v(0) = -\frac{R_L(1-D)D}{R_L(1-D)^2+r} \approx \frac{D}{1-D} \tag{4.84}$$

and the high-frequency gain M_{vx} is

$$M_{vx} = M_v(\infty) = \frac{R_L r_C(1-D)D}{L(R_L+r_C)}. \tag{4.85}$$

The angular frequency of the LHP zero ω_{zn} is given in (4.56). Figure 4.19 shows the theoretically obtained magnitude and phase plots of the input voltage-to-output voltage transfer function. The gain at DC is nearly equal to the duty cycle $M_{vo} = 0.44$ V/V = −7.13 dB. It is mainly dependent on duty cycle. The gain at DC is below 0 dB for the buck mode and is above 0 dB for the boost mode. In the buck-boost converter, any change in the input voltage causes a change in the output

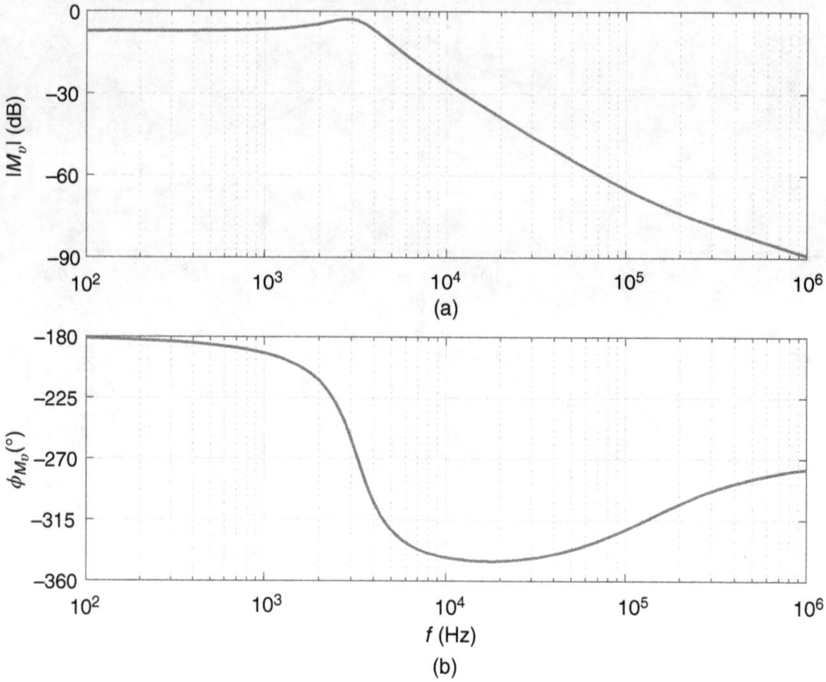

Figure 4.19 Theoretically obtained plots of input-to-output voltage transfer function M_v. (a) Magnitude. (b) Phase.

voltage during the switch ON interval. In addition, the changes in the input and output voltages take place in opposite directions, resulting in −180° phase shift at DC and low frequencies. The phase starts at −180° and decreases toward −360°. However, the high-frequency LHP zero at ω_z increases the phase to −270°.

4.3.4 Input Voltage-to-Inductor Current Transfer Function M_{vi}

The input voltage-to-inductor current transfer function is obtained by setting $d = 0$ and $i_o = 0$ in Figure 4.6. Setting $d = 0$ in the model and applying KVL results in

$$v_o = -Dv_i + Dv_o + i_l Z_1, \tag{4.86}$$

to yield

$$v_o = \frac{-Dv_i + i_l Z_1}{1 - D}. \tag{4.87}$$

Applying KCL at the inductor node provides

$$i_l = i_s - i_{Z_2} = Di_l - i_{Z_2} = Di_l - \frac{v_o}{Z_2}, \tag{4.88}$$

giving

$$i_l = -\frac{v_o}{Z_2(1 - D)}. \tag{4.89}$$

Substituting (4.87) into (4.89),

$$i_l = -\frac{-Dv_i + Z_1 i_l}{Z_2(1-D)^2}, \tag{4.90}$$

and rearranging, we get

$$i_l\left[1 + \frac{Z_1}{Z_2(1-D)^2}\right] = \frac{Dv_i}{Z_2(1-D)^2}. \tag{4.91}$$

Hence, the input voltage-to-inductor current transfer function M_{vi} is

$$M_{vi}(s) = \left.\frac{i_l(s)}{v_i(s)}\right|_{d=i_o=0} = \frac{\dfrac{D}{Z_2(1-D)^2}}{1 + \dfrac{Z_1}{Z_2(1-D)^2}} = \frac{D}{Z_1 + Z_2(1-D)^2}. \tag{4.92}$$

Substituting (4.35) and (4.36) for the expressions for impedances into (4.92) gives the input voltage-to-inductor current transfer function in the s-domain as

$$M_{vi}(s) = \left.\frac{i_l(s)}{v_i(s)}\right|_{d=i_o=0} = M_{vix}\frac{s + \omega_{zvi}}{s^2 + 2\xi\omega_o s + \omega_o^2} = M_{vio}\frac{1 + \dfrac{s}{\omega_{zvi}}}{\left(\dfrac{s}{\omega_o}\right)^2 + \dfrac{2\xi s}{\omega_o} + 1}, \tag{4.93}$$

where

$$M_{vix} = \frac{D}{L}, \tag{4.94}$$

the DC gain M_{vio} is

$$M_{vio} = M_{vi}(0) = \frac{D}{r + R_L(1-D)^2}, \tag{4.95}$$

and the angular frequency of the LHP zero is

$$\omega_{zvi} = \frac{1}{C(R_L + r_C)}. \tag{4.96}$$

Figure 4.20 shows the theoretically obtained magnitude and phase plots of the input voltage-to-inductor current transfer function. The gain at DC is $M_{vio} = 0.26\ A/V = -11.7\ dBA/V$. The phase starts at 0° and saturates at −90° at high frequencies. Through the M_{vi} current-loop relevant transfer function, one can observe that the current loop offers a lower audio-susceptibility at DC than that provided by the voltage-loop transfer function M_v. All the other frequency-domain properties remain identical to T_{pi}.

4.3.5 Reverse Current Gain A_i

The small-signal model to derive the output-to-inductor current transfer function is shown in Figure 4.21. This model is obtained by setting $v_i = 0$ and $d = 0$ in Figure 4.6.

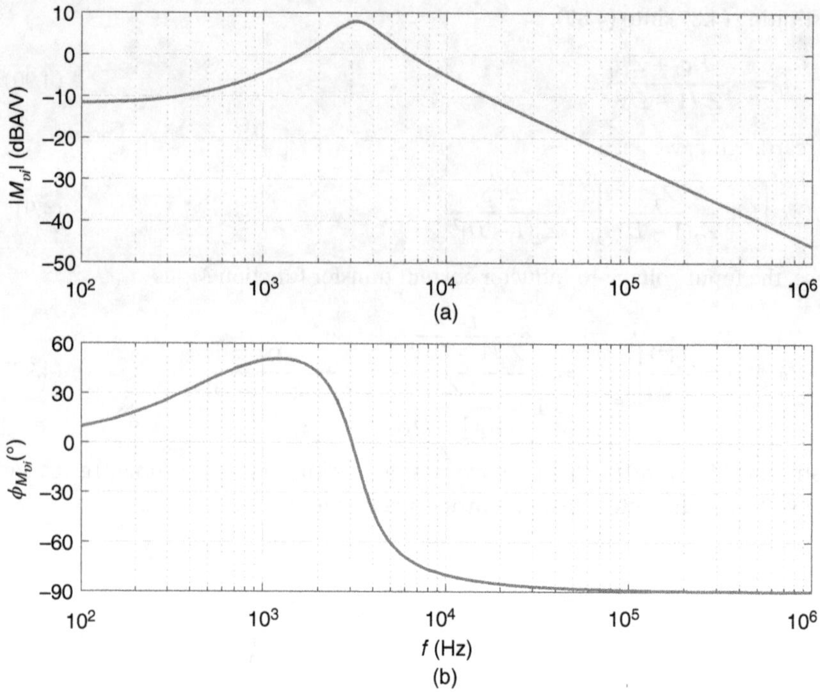

Figure 4.20 Theoretically obtained plots of input voltage-to-inductor current transfer function M_{vi}. (a) Magnitude. (b) Phase.

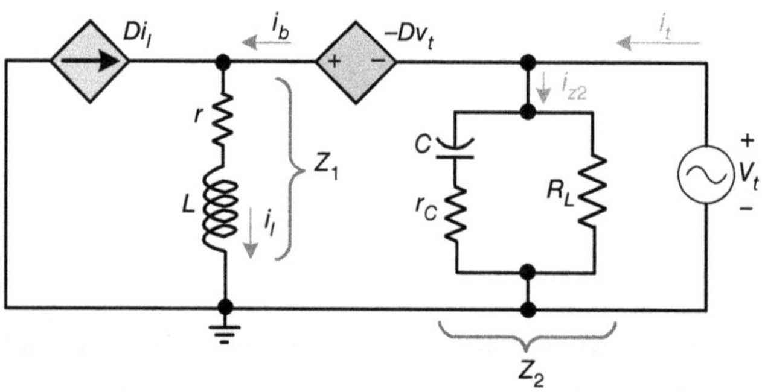

Figure 4.21 Small-signal model of the PWM buck-boost DC–DC converter in CCM to derive the output current-to-inductor current transfer function A_i.

An independent voltage source v_t is applied at the output node that forces a current i_t. Applying Kirchhoff's voltage law,

$$v_t = -(-Dv_t) + i_l Z_1 = \frac{i_l Z_1}{1 - D}. \tag{4.97}$$

Applying Kirchoff's current law,

$$i_l = Di_l + i_b. \tag{4.98}$$

Hence,

$$i_b = i_l(1 - D). \tag{4.99}$$

The current i_t is given as

$$i_t = -i_o = i_{Z_2} + i_b = \frac{v_t}{Z_2} + i_l(1 - D). \tag{4.100}$$

Substituting (4.97) into (4.100),

$$-i_o = \frac{i_l Z_1}{Z_2(1 - D)} + i_l(1 - D) = i_l\left[\frac{Z_1}{Z_2(1 - D)} + (1 - D)\right]. \tag{4.101}$$

Hence, the output current-to-inductor current, i.e. reverse current transfer function A_i, is

$$A_i(s) = \left.\frac{i_l(s)}{i_o(s)}\right|_{d=v_i=0} = -\frac{1}{1 - D}\frac{1}{1 + \dfrac{Z_1}{(1 - D)^2 Z_2}}. \tag{4.102}$$

Substituting (4.35) and (4.36) into (4.102) gives the output current-to-inductor current transfer function in the s-domain as

$$A_i(s) = \left.\frac{i_l(s)}{i_o(s)}\right|_{d=v_i=0} = A_{ix}\frac{s + \omega_{zn}}{s^2 + 2\xi\omega_o s + \omega_o^2} = A_{io}\frac{1 + \dfrac{s}{\omega_{zn}}}{\left(\dfrac{s}{\omega_o}\right)^2 + \dfrac{2\xi s}{\omega_o} + 1}, \tag{4.103}$$

where

$$A_{ix} = \frac{(1 - D)R_L r_C}{L(R_L + r_C)}, \tag{4.104}$$

the DC gain A_{io} is

$$A_{io} = A_i(0) = \frac{(1 - D)R_L}{[r + R_L(1 - D)^2]}, \tag{4.105}$$

and the angular frequency of the LHP zero ω_{zn} is given in (4.56). Figure 4.22 shows the theoretically obtained magnitude and phase plots of the output current-to-inductor current transfer function. The gain at DC is $A_{io} = 1.36$ A/A = 2.67 dB, i.e. the reverse current gain is unity at DC and low frequencies up to nearly f_o. Beyond f_o, the impedance offered by the capacitor branch is lower than the inductor, and the small-signal output current begins to flow into the capacitor. Therefore, the gain reduces by 20 dB per decade.

4.3.6 Open-Loop Input Impedance Z_i

The open-loop input impedance is obtained by setting $d = 0$ and $i_o = 0$ in Figure 4.6. Setting $d = 0$ in the model and applying KVL result in

$$v_o = -D(v_i - v_o) + i_l Z_1 = -\frac{Dv_i}{1 - D} + \frac{i_l Z_1}{1 - D}. \tag{4.106}$$

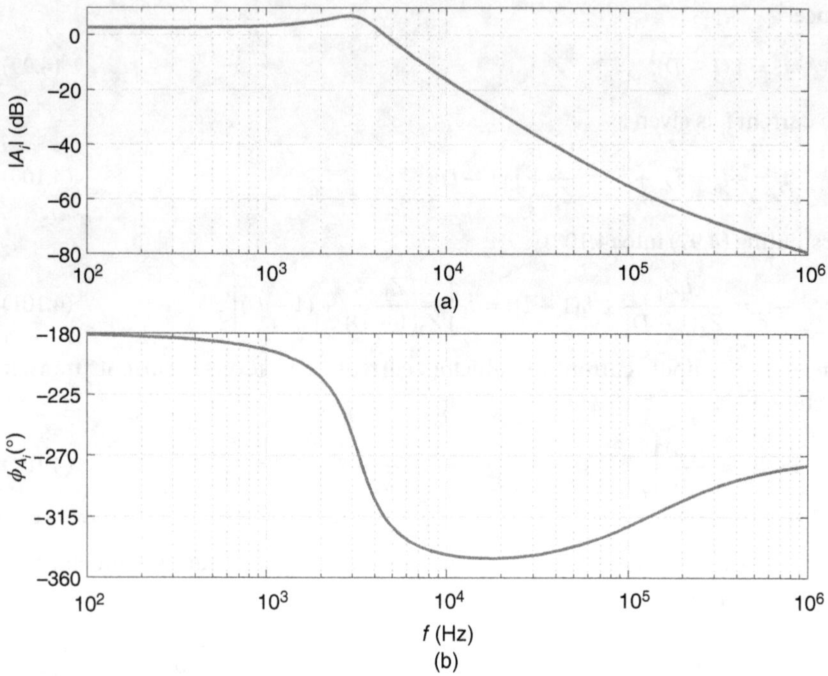

Figure 4.22 Theoretically obtained plots of output current-to-inductor current transfer function A_l. (a) Magnitude. (b) Phase.

Applying KCL at the inductor node produces

$$i_l = Di_l - i_{Z2} = Di_l - \frac{v_o}{Z_2}, \tag{4.107}$$

which yields

$$i_l = -\frac{v_o}{Z_2(1-D)}. \tag{4.108}$$

Rearranging (4.108), we get

$$v_o = -i_l Z_2(1-D). \tag{4.109}$$

Equating (4.109) and (4.106),

$$-i_l Z_2(1-D) = -\frac{Dv_i}{1-D} + \frac{i_l Z_1}{1-D}. \tag{4.110}$$

Rearranging the above equation produces

$$i_l\left[(1-D)Z_2 + \frac{Z_1}{1-D}\right] = \frac{Dv_i}{1-D}. \tag{4.111}$$

Since $d = 0$, the small-signal input current is

$$i_i = Di_l, \tag{4.112}$$

yielding

$$i_l = \frac{i_i}{D}. \tag{4.113}$$

Substituting (4.113) into (4.111), we get

$$i_i Z_2 \frac{(1-D)}{D}\left[1+\frac{Z_1}{Z_2(1-D)^2}\right] = v_i \frac{D}{1-D}. \tag{4.114}$$

Hence, the open-loop input impedance is

$$Z_i(s) = \left.\frac{v_i(s)}{i_i(s)}\right|_{d=i_o=0} = \frac{(1-D)^2}{D^2} Z_2 \left[1+\frac{Z_1}{Z_2(1-D)^2}\right] = \frac{Z_1+(1-D)^2 Z_2}{D^2}. \tag{4.115}$$

Substituting (4.35) and (4.36) into (4.115) gives the input impedance in the s-domain as

$$Z_i(s) = \left.\frac{v_i(s)}{i_i(s)}\right|_{d=i_o=0} = L_i \frac{s^2+2\xi\omega_o s+\omega_o^2}{s+\omega_{zvi}} = R_{io} \frac{\left(\frac{s}{\omega_o}\right)^2 + \frac{2\xi s}{\omega_o} + 1}{1+\frac{s}{\omega_{zvi}}}, \tag{4.116}$$

where L_i is the high-frequency input inductance and R_{io} is the DC input resistance. At high frequencies,

$$Z_i(\infty) = sL_i = \frac{sL}{D^2}. \tag{4.117}$$

At DC,

$$R_{io} = Z_i(0) = \frac{r + R_L(1-D)^2}{D^2} \tag{4.118}$$

and the angular frequency of the LHP zero ω_{zvi} is given in (4.96). Figure 4.23 shows the theoretically obtained magnitude and phase plots of the input impedance. The DC input resistance is $R_{io} = 12\ \Omega$, and the high-frequency input inductance is $L_i = 96.85\ \mu$H. The input resistance R_i at DC is proportional to the duty cycle and the load resistance. With an increase in frequency, the input impedance shows capacitive nature. Beyond f_o, the inductive reactance takes over, and the input impedance is mainly dominated by the inductive reactance. In open-loop circuit, the input current is in phase with the input voltage, i.e. a step change in the input voltage in positive direction, causing a corresponding change in the input current in the same direction.

4.3.7 Open-Loop Output Impedance Z_o

The small-signal model to derive the output impedance is shown in Figure 4.21. This model is obtained by setting $v_i = 0$ and $d = 0$. A voltage v_t is applied at the output node, producing current i_t at the output node. Applying Kirchhoff's voltage law,

$$v_t = (V_I - V_O)d - D(v_i - v_t) + i_l Z_1. \tag{4.119}$$

Because $v_i = 0$ and $d = 0$, we get

$$v_t = Dv_t + i_l Z_1 = \frac{i_l Z_1}{1-D}. \tag{4.120}$$

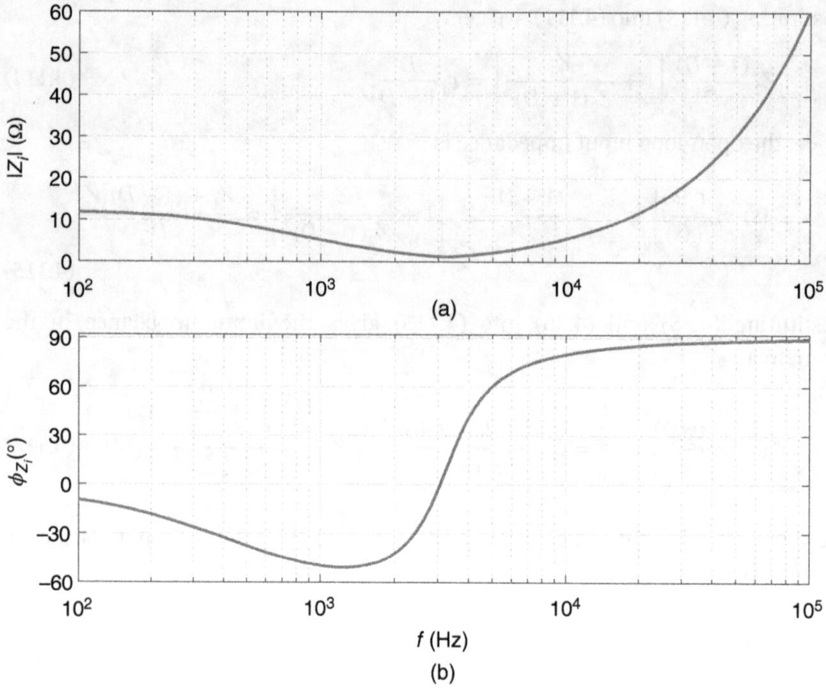

Figure 4.23 Theoretically obtained plots of input impedance Z_i. (a) Magnitude. (b) Phase.

Rearranging (4.120) produces

$$i_l = \frac{v_t(1-D)}{Z_1}. \tag{4.121}$$

Applying KCL at the inductor output node,

$$i_l = I_L d + D i_l + i_b. \tag{4.122}$$

As $d = 0$, we get

$$i_l = D i_l + i_b. \tag{4.123}$$

Hence,

$$i_b = i_l(1-D) = \frac{v_t(1-D)^2}{Z_1}. \tag{4.124}$$

Applying KCL at the output node yields the small-signal current i_t as

$$i_t = i_{Z_2} + i_b = \frac{v_t}{Z_2} + i_b. \tag{4.125}$$

Rearranging (4.125) gives

$$i_b = i_t - \frac{v_t}{Z_2}. \tag{4.126}$$

Equating (4.124) and (4.126)

$$\frac{v_t(1-D)^2}{Z_1} = i_t - \frac{v_t}{Z_2} \tag{4.127}$$

and because $i_t = -i_o$, we get

$$\frac{v_t(1-D)^2}{Z_1} = -i_o - \frac{v_t}{Z_2}. \tag{4.128}$$

Further solving (4.128) produces

$$v_t\left[\frac{1}{Z_2} + \frac{(1-D)^2}{Z_1}\right] = v_t\frac{(1-D)^2}{Z_1}\left[1 + \frac{Z_1}{Z_2(1-D)^2}\right] = -i_o. \tag{4.129}$$

Hence,

$$Z_o(s) = \left.\frac{v_t(s)}{i_o(s)}\right|_{d=v_i=0} = \frac{1}{(1-D)^2}\frac{Z_1}{1+\frac{Z_1}{Z_2(1-D)^2}} = \frac{Z_1Z_2}{Z_1 + Z_2(1-D)^2}. \tag{4.130}$$

The open-loop output impedance is

$$Z_o(s) = \left.\frac{-v_o(s)}{i_o(s)}\right|_{d=v_i=0} = \frac{Z_1Z_2}{Z_1 + Z_2(1-D)^2}. \tag{4.131}$$

Substituting (4.35) and (4.36) into (4.131) gives the open-loop output impedance in the s-domain as

$$Z_o(s) = \left.\frac{-v_o(s)}{i_o(s)}\right|_{d=v_i=0} = R_{ox}\frac{(s+\omega_{zn})(s+\omega_{rl})}{s^2 + 2\xi\omega_o s + \omega_o^2} = R_{oo}\frac{\left(1+\frac{s}{\omega_{zn}}\right)\left(1+\frac{s}{\omega_{rl}}\right)}{\left(\frac{s}{\omega_o}\right)^2 + \frac{2\xi s}{\omega_o} + 1}, \tag{4.132}$$

where the high-frequency output resistance R_{ox} is

$$R_{ox} = Z_o(\infty) = \frac{R_L r_C}{R_L + r_C}, \tag{4.133}$$

the DC output resistance R_{oo} is

$$R_{oo} = Z_o(0) = \frac{rR_L}{r + R_L(1-D)^2}, \tag{4.134}$$

and the angular frequency of the LHP zero ω_{rl} is

$$\omega_{rl} = \frac{r}{L}. \tag{4.135}$$

The angular frequency of the LHP zero ω_{zn} is given in (4.56). Figure 4.24 shows the theoretically obtained magnitude and phase plots of the output impedance. The DC output resistance is $R_{oo} = 0.19\ \Omega$. The high-frequency output resistance is $R_{ox} = 10\ \text{m}\Omega$. At DC, the inductor is a short circuit, and the filter capacitor is an open circuit. The DC resistance is due to the parallel combination of the load resistance R_L and the equivalent averaged resistance r. With the increase in frequency, the capacitive reactance is high, and the reactance is mostly offered by the inductive reactance. At a frequency f_o, the inductive and capacitive reactances

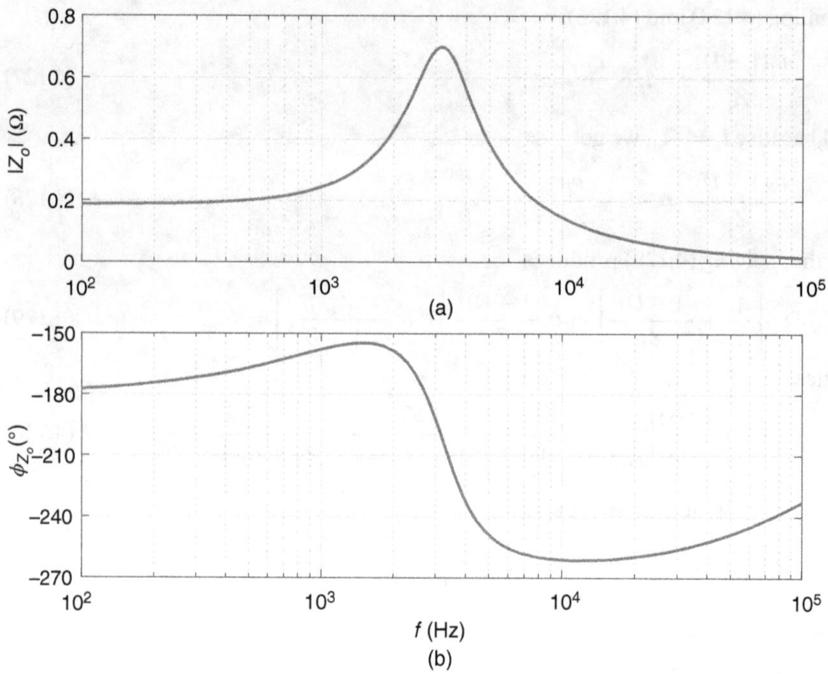

Figure 4.24 Theoretically obtained plots of output impedance Z_o. (a) Magnitude. (b) Phase.

are equal, and the converter filter network undergoes resonance. Beyond f_o, the capacitive reactance dominates and reduces the output impedance to nearly zero. A positive increase in the load current causes an increase in the output voltage in the negative direction, leading to a phase difference of −180° at DC and low frequencies. A summary of calculated values for open-loop transfer functions is given in Table 4.1.

4.4 Inner-Current Loop

Figure 4.25 shows the architecture of the inner-current loop with a low-pass filter. Figure 4.26 shows the circuit of the buck-boost DC–DC converter with inner-current loop. A sense resistor R_s is placed in the inductor branch to sense the inductor current and is a part of the feedback network. At low duty cycles, the sensed inductor current ripple is high. Therefore, the control voltage intersects with the modulator ramp voltage twice in each cycle, resulting in subharmonic instability. The low-pass filter attenuates the switching frequency components and its harmonics present in the sensed voltage $v_{RS} = R_s i_L$. Therefore, the theoretically obtained output voltage of the low-pass filter is $V_{FI} = R_s I_L$ and is governed by the DC component of the sensed inductor current. Thus, in this method, the *average* component of the inductor current is sensed, thereby reducing the problems due to the subharmonic instability and transistor misgating. The transfer functions related to the inner-current loop, which includes the filter in the feedback path, are derived and discussed in this section.

Table 4.1 Summary of calculated values for open-loop transfer functions.

Variable	Value
$\|T_{po}\|$	20.67 V = 26.31 dBV
$\|T_{pio}\|$	16.52 A = 24.36 dBA
$\|M_{vo}\|$	0.44 V/V = −7.13 dB
$\|M_{vio}\|$	0.26 A/V = −11.7 dBA/V
R_{io}	12 Ω
R_{oo}	0.19 Ω
ω_o	20.33 krad/s
f_o	3.2 kHz
ξ	0.3214
ω_{zi}	4.38 krad/s
f_{zi}	697 Hz
ω_{zvi}	3.32 krad/s
f_{zvi}	528.4 Hz
ω_{zn}	833.3 krad/s
f_{zn}	132.63 kHz
ω_{zp}	349 krad/s
f_{zp}	56 kHz
ω_{rl}	9.3 krad/s
f_{rl}	1.48 kHz

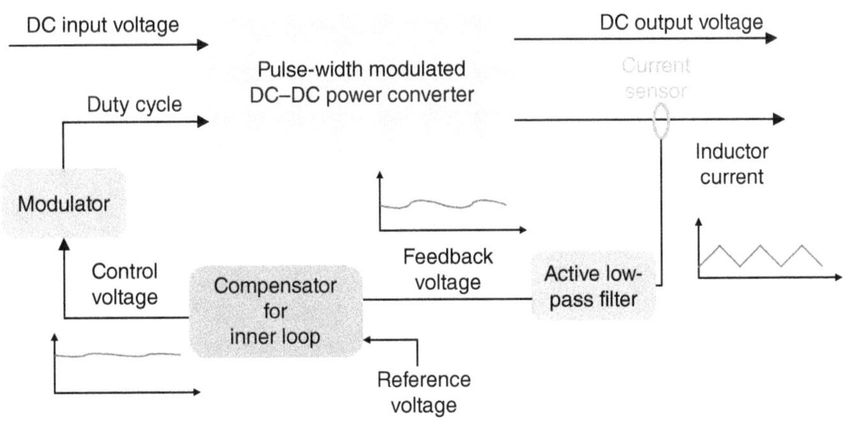

Figure 4.25 Architecture of the inner-current loop with low-pass filter in the feedback path.

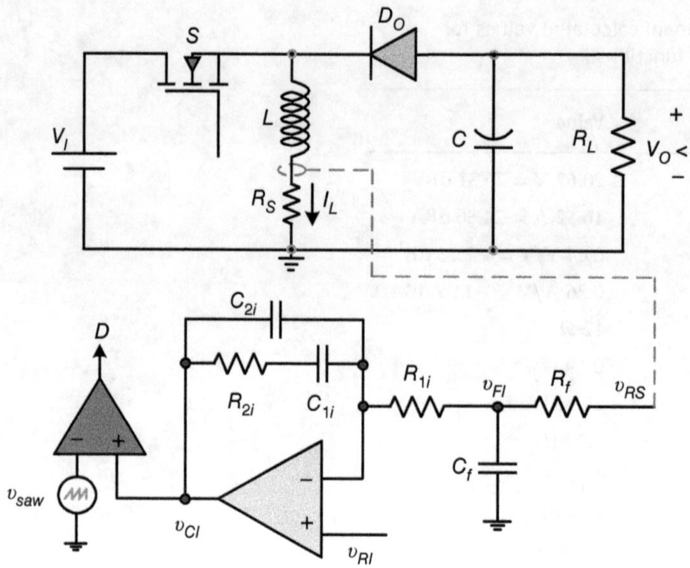

Figure 4.26 Circuit of the buck-boost DC–DC converter with inner-current loop.

4.4.1 Transfer Function of Filter T_f

The low-pass filter is composed of a resistor R_f and a capacitor C_f. The transfer function of two stages is

$$T_f(s) = \frac{v_{fi}(s)}{v_{rs}(s)} = \frac{Z_{Cf}}{R_f + Z_{Cf}} = \frac{\frac{1}{sC_f}}{R_f + \frac{1}{sC_f}} = \frac{1}{sR_fC_f + 1}$$
$$= \frac{1}{R_fC_f} \frac{1}{\frac{1}{R_fC_f} + s} = \frac{\omega_{pf}}{\omega_{pf} + s} = \frac{1}{1 + \frac{s}{\omega_{pf}}}. \tag{4.136}$$

Setting $s = j\omega$,

$$T_f(j\omega) = \frac{1}{1 + j\frac{\omega}{\omega_{pf}}}, \tag{4.137}$$

yielding

$$T_f(f) = \frac{1}{1 + j\frac{f}{f_{pf}}} = |T_f(f)|e^{j\phi(f)}, \tag{4.138}$$

where

$$|T_f(f)| = \frac{1}{\sqrt{1 + \left(\frac{f}{f_{pf}}\right)^2}} \tag{4.139}$$

and

$$\phi(f) = -\tan^{-1}\left(\frac{f}{f_{pf}}\right). \tag{4.140}$$

Using (4.139),

$$|T_f(f)|^2 = \frac{1}{1+\left(\frac{f}{f_{pf}}\right)^2}, \tag{4.141}$$

which yields

$$\left(\frac{f}{f_{pf}}\right)^2 = \frac{1}{|T_f(f)|^2} - 1 = \frac{1-|T_f(f)|^2}{|T_f(f)|^2} \tag{4.142}$$

and

$$\frac{f_{pf}}{f} = \sqrt{\frac{|T_f(f)|^2}{1-|T_f(f)|^2}}. \tag{4.143}$$

Hence,

$$f_{pf} = f\sqrt{\frac{|T_f(f)|^2}{1-|T_f(f)|^2}}. \tag{4.144}$$

The magnitude of the transfer function T_f is also given as

$$|T_f(f)| = \frac{\Delta V_{FI}}{\Delta V_{RS}}. \tag{4.145}$$

The filter frequency can be set by the amount of attenuation desired. As an example, to achieve an attenuation of 50% for $f = f_s = 200$ kHz, the filter upper cutoff frequency can be calculated using (4.144) as

$$f_{pf} = f_s\sqrt{\frac{|T_f(f)|^2}{1-|T_f(f)|^2}} = 200 \times 10^3 \times \sqrt{\frac{0.50^2}{1-0.50^2}} = 115\ \text{kHz}. \tag{4.146}$$

Assuming $R_f = 1\ \text{k}\Omega$, the value of the filter capacitor can be calculated as $C_f = 1/(2\pi f_{pf} R_f) \approx 1.38$ nF. A 1.5 nF/25 V capacitor is selected.

4.4.2 Transfer Function of Pulse-Width Modulator T_m

The control voltage-to-duty cycle transfer function of the pulse-width modulator is

$$T_m = \frac{D}{V_{CI}} = \frac{d}{v_{ci}} = \frac{1}{V_{Tm}}, \tag{4.147}$$

where V_{Tm} is the amplitude of the sawtooth waveform.

The steady-state average inductor current is $I_L = V_O/[R_L(1-D)] = 2.95$ A. At a duty cycle $D = 0.3213$ and assuming the peak sawtooth voltage of the pulse-width modulator is $V_{Tm} = 5$ V, the required steady-state control voltage is $V_{CI} = DV_{Tm} = 0.3213 \times 5 = 1.61$ V. The reference voltage to the current loop is $V_{RI} = V_{CI}$ at steady-state since the error voltage is zero, i.e. $V_{RI} = 1.61\ V$. The gain of the feedback path is controlled mainly by the sensor gain R_s such that $T_{picl} \approx 1/R_s$. For a steady-state inductor current I_L and the inner-loop reference voltage V_{RI}, which is set by the outer-voltage loop, the sense resistor can be determined using

$$R_s = \frac{V_{Rs}}{I_L} = \frac{V_{RI}}{I_L} = \frac{1.61}{2.947} = 0.5452\ \Omega. \tag{4.148}$$

Two standard 1 Ω resistors are selected and added in parallel. From Section 4.4.1, $R_f = 1$ kΩ and $C_f = 1.5$ nF.

4.4.3 Uncompensated Loop Gain T_{ki}

The natural behavior of the inner-current loop can be determined using the uncompensated loop gain as

$$T_{ki} = \frac{v_{fi}}{v_{ei}} = T_m T_{pi} R_s T_f = T_{kio} \frac{1 + \frac{s}{\omega_{zi}}}{\left(1 + \frac{s}{\omega_{pf}}\right)\left(1 + \frac{2\xi s}{\omega_o} + \frac{s^2}{\omega_o^2}\right)}, \tag{4.149}$$

where T_m, T_{pi}, and T_f are given in (4.147), (4.70), and (4.136), respectively. The DC gain T_{kio} is given by

$$T_{kio} = T_{kio}(0) = \frac{R_s}{V_{Tm}} \frac{-V_O(1 + D)}{D[r + R_L(1 - D)^2]}. \tag{4.150}$$

Figure 4.27 shows the theoretically obtained magnitude and phase plots of the uncompensated loop gain of the inner-current loop. The gain at DC is $T_{kio} = 1.65$ V/V $= 4.35$ dB. A high DC gain T_{kio} is required in view of reducing the steady-state error in the output voltage due to step changes in the input voltage or load current. Therefore, a suitable control circuit capable of providing a high gain at DC and sufficient phase boost at the crossover frequency of the loop gain becomes

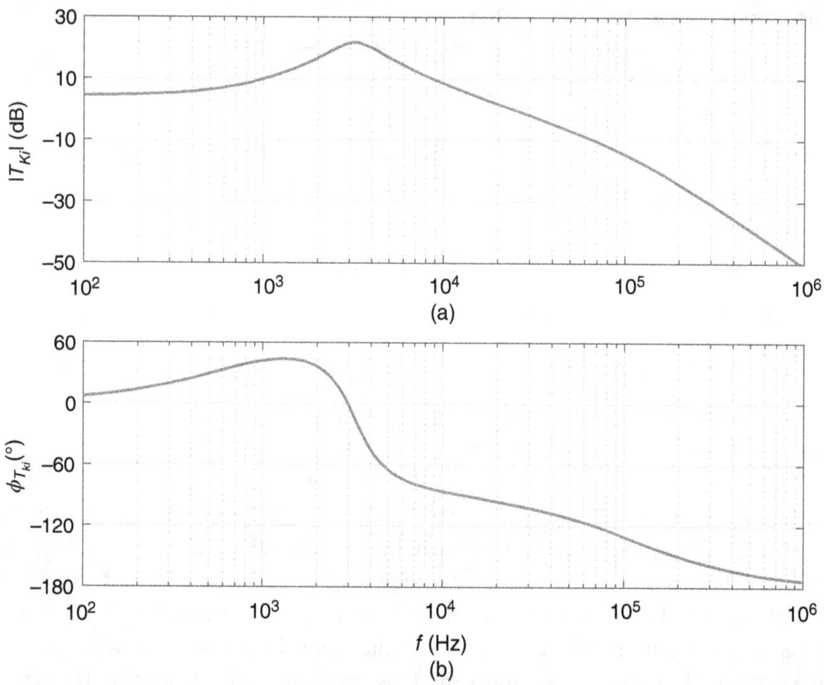

Figure 4.27 Theoretically obtained plots of uncompensated loop gain T_{ki}. (a) Magnitude. (b) Phase.

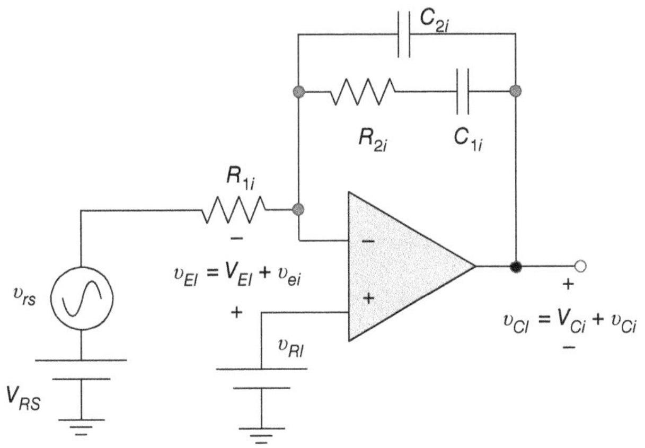

Figure 4.28 Circuit of Type-II controller used in inner-current loop.

essential. An important factor to note is that the phase of the uncompensated loop gain is zero at DC. This is because the inner-loop absorbs only the inductor branch and does not take into account the properties of the output network. Typically, the presence of a low-frequency zero in the T_{pi} transfer function enables a wider current loop bandwidth and low phase at the crossover frequency. The phase margin is in the range of 40°–90° and requires relatively simpler compensation schemes to satisfy the inner-current loop requirements. In this analysis, the Type-II integral control circuit (also known as the single-lead integral control circuit) is used.

4.4.4 Transfer Function of Compensation Circuit T_{ci}

Figure 4.28 shows the circuit of the Type-II controller used in inner-current loop. A single-lead integral control circuit introduces a pole at the origin required to boost the DC gain and a pole–zero pair to adjust crossover frequency f_c to maintain required PM. A detailed procedure to determine the transfer function of the controller to improve the DC gain (> 50 dB) and achieve an inner-current loop phase margin $PM = 60°$ is explained in [45]. The design can be performed as follows. The magnitude of the uncompensated loop gain transfer function can be determined using

$$|T_{ki}(f)| = T_{kio}\frac{\sqrt{1+\left(\frac{f}{f_{zi}}\right)^2}}{\sqrt{1+\left(\frac{f}{f_{pf}}\right)^2}\sqrt{1-\left(\frac{f}{f_o}\right)^2+4\xi^2\left(\frac{f}{f_o}\right)^2}}, \tag{4.151}$$

and the phase is

$$\phi_{Tk}(f) = -180° + \tan^{-1}\left(\frac{f}{f_{zi}}\right) - \tan^{-1}\left(\frac{f}{f_{pf}}\right) - \tan^{-1}\frac{2\xi\frac{f}{f_o}}{1-\left(\frac{f}{f_o}\right)^2}. \tag{4.152}$$

The desired crossover frequency for the loop gain is chosen as $f_c = 30$ kHz. At this frequency, the magnitude of the uncompensated loop gain transfer function is $|T_{ki}(f_c)| = 0.81$ V/V $= -1.83$ dB, and its phase is $\phi_{Tk}(f_c) = -102°$. The required phase margin is $PM = 60°$. Therefore, the phase boost required at f_c is

$$\phi_m = PM - \phi_{Tk}(f_c) - 90° = 72°, \tag{4.153}$$

giving the geometric mean between the controller corner frequencies as

$$K = \tan\left(\frac{\phi_m}{2} + 45°\right) = 6.27. \tag{4.154}$$

Assuming $R_{1i} = 1\ \text{k}\Omega/0.25\ W/2\%$, the value of C_{2i} is

$$C_{2i} = \frac{|T_k(f_c)|}{2\pi f_c K R_{1i}} = 683\ \text{pF}. \tag{4.155}$$

A 680 pF/25 V standard capacitor is selected. The expression for C_{1i} is

$$C_{1i} = C_{2i}(K^2 - 1) = 26.2\ \text{nF}. \tag{4.156}$$

A 27 nF/25 V standard capacitor is selected. The expression for R_{2i} is

$$R_{2i} = \frac{K}{2\pi f_c C_{1i}} = 1.27\ \text{k}\Omega. \tag{4.157}$$

A 1.2 $\text{k}\Omega/0.25\ W/2\%$ standard resistor is selected. The transfer function T_{ci} of the current-loop control circuit is

$$T_{ci}(s) = \frac{v_{ci}(s)}{v_{ei}(s)} = T_{cio}\frac{1 + \frac{s}{\omega_{zci}}}{s\left(1 + \frac{s}{\omega_{pci}}\right)}, \tag{4.158}$$

where

$$T_{cio} = \frac{1}{R_{1i}(C_{1i} + C_{2i})} = 36.13\ \text{kV/V}, \tag{4.159}$$

$$\omega_{zci} = \frac{1}{R_{2i}C_{1i}} = 31\ \text{krad/s}, \tag{4.160}$$

and

$$\omega_{pci} = \frac{C_{1i} + C_{2i}}{R_{2i}C_{1i}C_{2i}} = 1.26\ \text{Mrad/s}. \tag{4.161}$$

Figure 4.29 shows the theoretically obtained magnitude and phase plots of the controller transfer function used in the inner-current loop. The control transfer function is now incorporated into the critical path to determine the inner-loop gain and is discussed in Section 4.4.5.

4.4.5 Compensated Loop Gain T_i

Figure 4.30 shows the block diagram of the inner-current loop to determine the inner current-loop gain. The loop gain of the compensated inner-current loop is

$$T_i = \frac{v_{ei}}{v_{fi}} = T_{ki}T_{ci} = T_m T_{pi} R_s T_f T_{ci}, \tag{4.162}$$

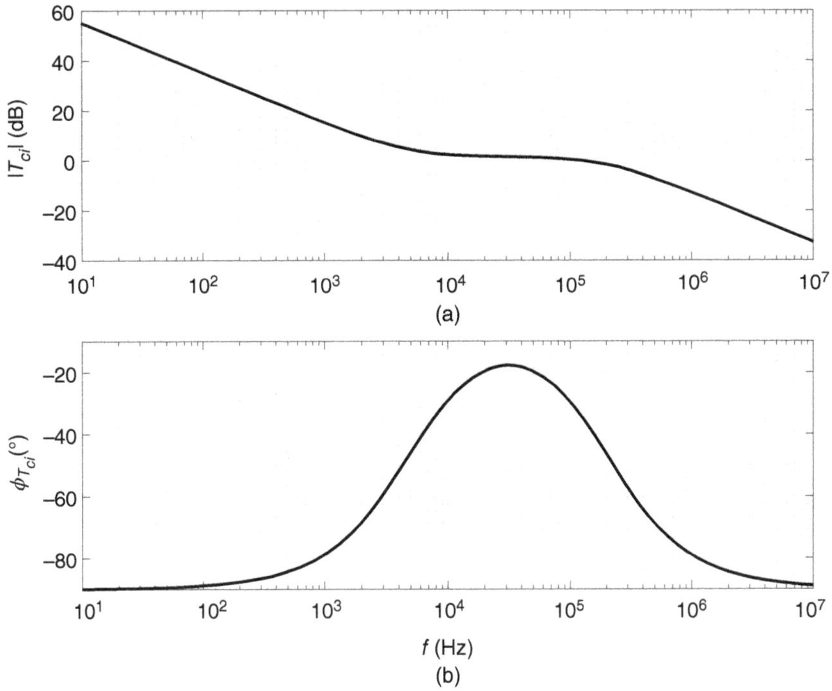

Figure 4.29 Theoretically obtained plots of controller transfer function T_{ci}. (a) Magnitude. (b) Phase.

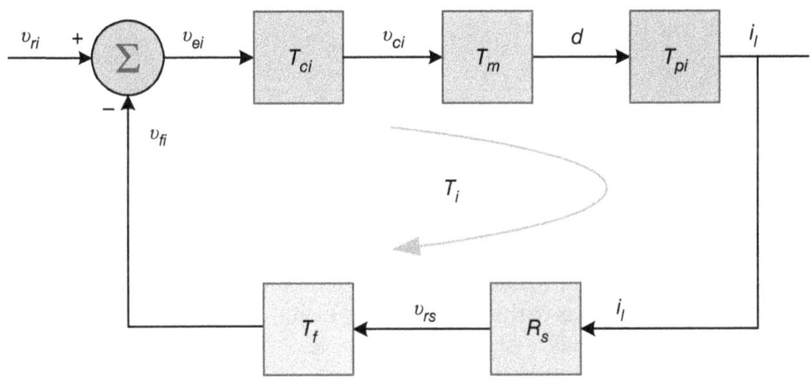

Figure 4.30 Block diagram of inner-current loop.

yielding

$$T_i = T_{io} \frac{\left(1 + \frac{s}{\omega_{zi}}\right)\left(1 + \frac{s}{\omega_{zci}}\right)}{s\left(1 + \frac{s}{\omega_{pf}}\right)\left(1 + \frac{s}{\omega_{pci}}\right)\left(1 + \frac{2\xi s}{\omega_o} + \frac{s^2}{\omega_o^2}\right)}, \tag{4.163}$$

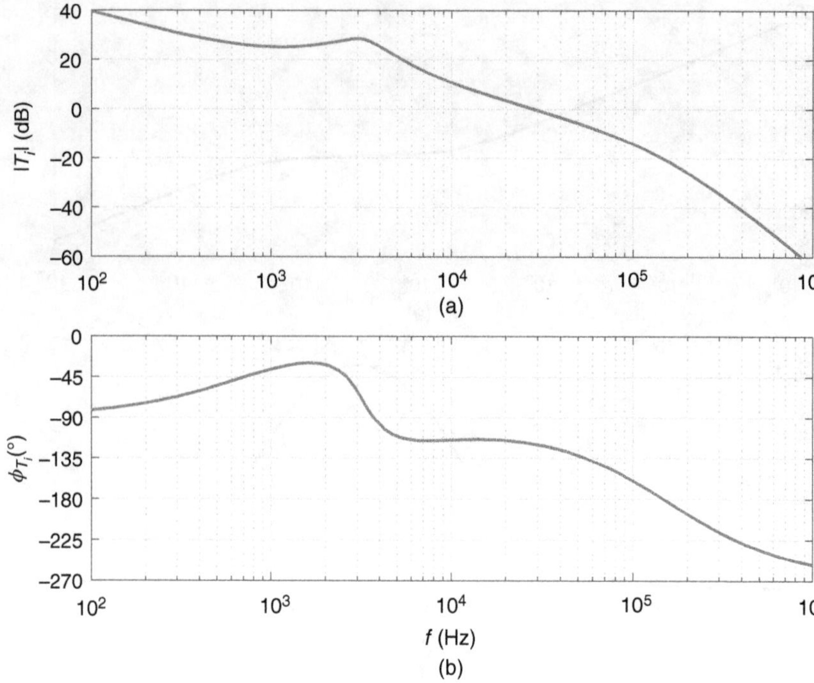

Figure 4.31 Theoretically obtained plots of compensated loop gain T_i of the inner current loop. (a) Magnitude. (b) Phase.

where T_{io} is the gain at DC given as

$$T_{io} = T_i(0) = T_{kio}T_{cio} = \frac{R_s}{V_{Tm}} - \frac{V_O(1+D)}{D(r+R_L(1-D)^2)}\frac{1}{R_{1i}(C_{1i}+C_{2i})}. \tag{4.164}$$

Figure 4.31 shows the theoretically obtained magnitude and phase plots of compensated loop gain of the inner-current loop. The gain at DC is greater than 50 dB, and the phase at DC starts at 90° due to the presence of controller pole at origin. The unity gain crossover frequency is 30 kHz, and the phase margin is $PM = 60°$. The inner loop is stable. Section 4.5 describes the various important closed-loop transfer functions of the inner-current loop.

4.5 Closed-Inner Loop Transfer Functions

The following closed-loop transfer functions relevant to the inner-loop are derived:

1. Reference voltage to inductor current transfer function T_{icl} (control).
2. Reference voltage to output voltage transfer function T_{picl} (control).
3. Input voltage-to-inductor current transfer function M_{icl} (disturbance).
4. Input voltage-to-output voltage transfer function M_{vicl} (disturbance).
5. Input voltage-to-duty cycle transfer function M_{di} (disturbance).

6. Input impedance Z_{iicl} (disturbance).
7. Output impedance Z_{oicl} (disturbance).

The block diagram required to characterize the inner current loop is as shown in Figure 4.32. The small-signal current reference v_{ri} to the inner-loop is set by the outer loop. The focus of this section is to analyze the inner-current loop only. Therefore, the outer-voltage loop has not been shown in the block diagram. The error voltage v_{ei} is formed by the difference between the current reference v_{ri} and the feedback voltage v_{fi}. The feedback voltage v_{fi} is given by

$$v_{fi} = T_f R_s i_l, \tag{4.165}$$

where T_f is the transfer function of the low-pass filter and R_s is the sense resistor. The transfer function T_f is given by

$$T_f = \frac{v_{fi}}{v_{rs}} = \frac{1}{1 + \frac{s}{\omega_{pf}}}, \tag{4.166}$$

where $\omega_{pf} = 2\pi f_{pf}$ is the filter cutoff frequency. In this analysis, the selected filter cutoff frequency is based on the allowable ripple in the sensed inductor current contributing to the concept of the average current-mode control (explained in Section 4.4.1). The gain of the low-pass filter is unity; therefore, it does not affect the DC gain of the loop gain. The transfer function T_{ci} is the voltage gain of the Type-II control circuit mentioned in Section 4.4.4. The pulse-width modulator has a linear gain $T_m = 1/V_{Tm}$, where V_{Tm} is the maximum value of the sawtooth waveform.

As the inductor current is controlled by the inner loop, the inductor current must be sensed, amplified by R_s, filtered by T_f, and compared with current reference v_{ri} to generate the small-signal error.

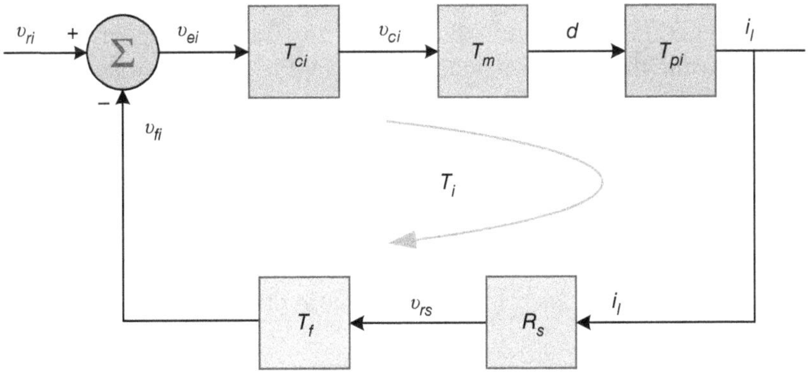

Figure 4.32 Block diagram used to derive the inner-loop control-to-inductor current transfer function T_{icl}.

4.5.1 Reference Voltage-to-Inductor Current Transfer Function T_{icl}

Using the block diagram shown in Figure 4.32, the closed-loop reference voltage-to-inductor current transfer function is

$$T_{icl}(s) = \left.\frac{i_l(s)}{v_{ri}(s)}\right|_{v_i=i_o=0}. \tag{4.167}$$

From Figure 4.32,

$$i_l = T_{ci}T_mT_{pi}v_{ei}. \tag{4.168}$$

Hence,

$$v_{ei} = \frac{i_l}{T_{ci}T_mT_{pi}}, \tag{4.169}$$

and

$$v_{fi} = T_fR_si_l. \tag{4.170}$$

From the block diagram shown in Figure 4.32, the reference voltage is

$$v_{ri} = v_{ei} + v_{fi}. \tag{4.171}$$

Substituting (4.169) and (4.170) into (4.171), we get

$$\begin{aligned} v_{ri} = v_{ei} + v_{fi} &= \frac{i_l}{T_{ci}T_mT_{pi}} + T_fR_si_l = \left(\frac{1}{T_{ci}T_mT_{pi}} + T_fR_s\right)i_l \\ &= \left(\frac{1 + T_{ci}T_mT_{pi}T_fR_s}{T_{ci}T_mT_{pi}}\right)i_l = \left(\frac{1+T_i}{T_{ci}T_mT_{pi}}\right)i_l \end{aligned} \tag{4.172}$$

to get
$$T_{icl}(s) = \left.\frac{i_l(s)}{v_{ri}(s)}\right|_{v_i=i_o=0} = \frac{T_{ci}T_mT_{pi}}{1+T_i}. \tag{4.173}$$

Figure 4.33 shows the theoretically obtained magnitude and phase plots of the inner-current loop reference voltage-to-inductor current transfer function. It can be observed that the bandwidth of the current-loop transfer function T_{icl} is high and equal to nearly 40 kHz. There is still a margin for bandwidth improvement, however at the cost of reduced DC gain. The transfer function T_{icl} does not encompass the transfer function T_p (i.e. relevant to the outer-voltage loop). Therefore, maximum speed of the converter can be achieved by the inner-current loop. This means that for any input or load disturbance, the inner-loop alone can sufficiently counteract the disturbance and provide the necessary correction in the inductor current. Ideally, the inductor current is made to follow the fixed current reference v_{ri}; therefore, this form of the current-mode control can also be referred as *current-programmed control.*

The following approximation of (4.173) holds true as the loop gain $T_i >> 1$:

$$T_{icl} = \frac{T_{ci}T_mT_{pi}}{1 + T_mT_{ci}T_{pi}T_fR_s} \approx \frac{T_{ci}T_mT_{pi}}{T_mT_{ci}T_{pi}T_fR_s} = \frac{1}{T_fR_s}. \tag{4.174}$$

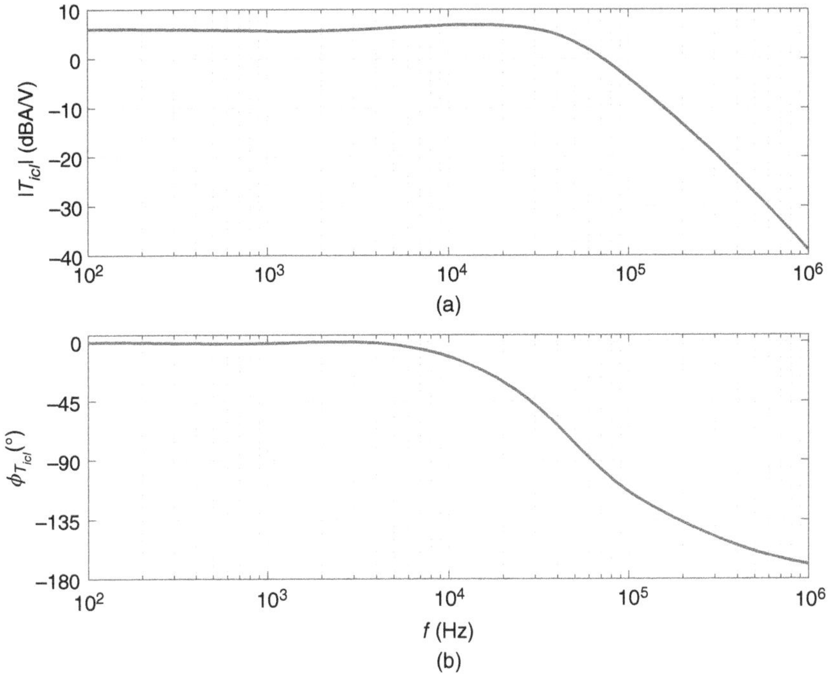

Figure 4.33 Theoretically obtained plots of inner-current loop reference voltage-to-inductor current transfer function T_{icl}. (a) Magnitude. (b) Phase.

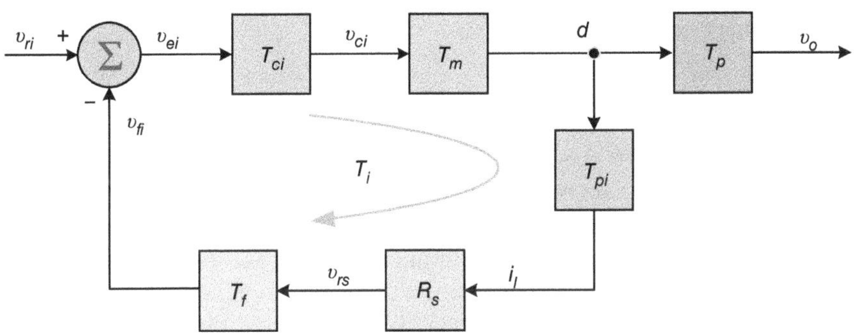

Figure 4.34 Block diagram required to derive the inner-loop control-to-output voltage transfer function T_{picl}. Source: Reproduced with permission from Dalvir Saini.

4.5.2 Reference Voltage-to-Output Voltage Transfer Function T_{picl}

Using the block diagram shown in Figure 4.34, the closed-loop reference voltage-to-output voltage transfer function is

$$T_{picl}(s) = \left.\frac{v_o(s)}{v_{ri}(s)}\right|_{v_i=i_o=0} = \frac{T_{ci}T_mT_p}{1+T_i}. \tag{4.175}$$

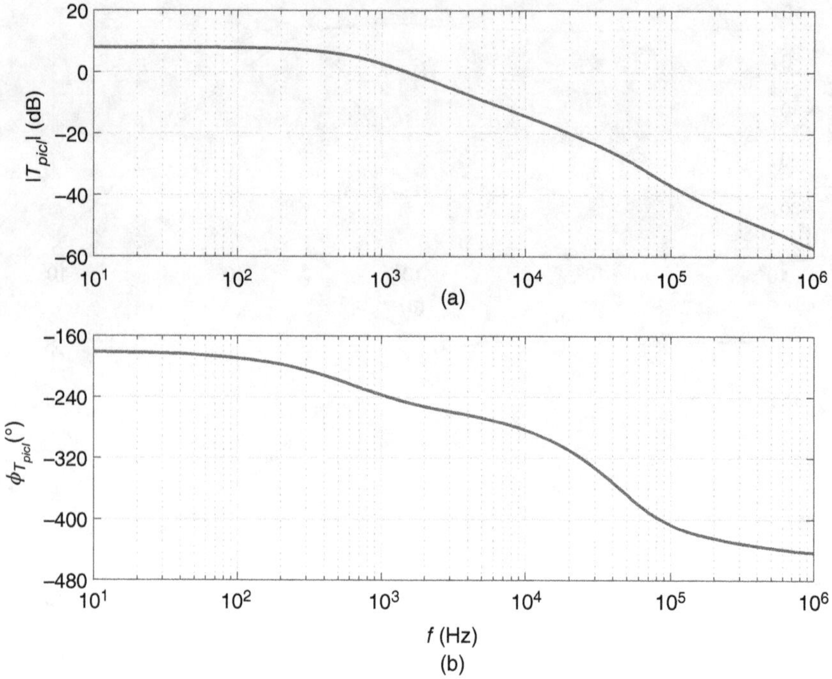

Figure 4.35 Theoretically obtained plots of reference voltage-to-output voltage transfer function T_{picl}. (a) Magnitude. (b) Phase.

At low frequencies, $T_i >> 1$. Therefore,

$$T_{picl} \approx \frac{T_{ci} T_m T_p}{T_i} = \frac{T_p}{T_f R_s}. \tag{4.176}$$

Figure 4.35 shows the theoretically obtained magnitude and phase plots of inner-current loop reference voltage-to-output voltage transfer function. The reference voltage and the output voltage are out of phase by 180°. The bandwidth of the loop from the reference to the output is limited due to the filter capacitor. A critical difference can be observed through the block diagram used to derive the T_{picl} transfer function. The output voltage is controlled by the duty cycle in the buck-boost converter, whereas the inductor current controls the output voltage in buck converters. To characterize the outer-loop transfer functions, such as T_{picl}, the critical path of the model is converted from the voltage-controlled current source (i_l/v_{ri}) to the voltage-controlled voltage source (d/v_{ri}).

4.5.3 Input Voltage-to-Inductor Current Transfer Function M_{icl}

Using the block diagram shown in Figure 4.36, the closed-loop input voltage-to-inductor current transfer function is

$$M_{icl}(s) = \left.\frac{i_l(s)}{v_i(s)}\right|_{v_{ri}=i_o=0}. \tag{4.177}$$

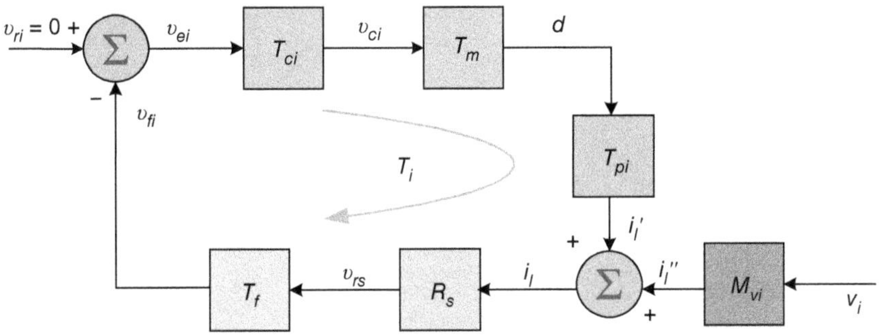

Figure 4.36 Block diagram required to derive the inner-loop input voltage-to-inductor current transfer function M_{icl}.

From the block diagram,

$$i_l = i_l' + i_l'' = (-T_i)i_l + M_{vi}v_i, \tag{4.178}$$

producing

$$i_l(1 + T_i) = M_{vi}v_i. \tag{4.179}$$

Hence,

$$M_{icl}(s) = \frac{i_l(s)}{v_i(s)} = \frac{M_{vi}}{1 + T_i}. \tag{4.180}$$

Figure 4.37 shows the theoretically obtained magnitude and phase plots of the inner-current loop input voltage-to-inductor current transfer function. As can be seen from (4.180), the M_{icl} transfer function is a representative case to show the effect of negative feedback, i.e. the negative feedback path reduces the disturbance in the input voltage to propagate through the inductor current branch by a factor of nearly T_i. However, this is possible only at low to mid frequencies and in the range of frequencies up to f_c, where $|T_i| >> 1$. Beyond this frequency range, $|T_i| << 1$ and the transfer function $M_{icl} \approx M_{vi}$. The gain at DC and low frequencies is -50 dB or lower, indicating that the inner-loop can sufficiently attenuate the input voltage disturbance. The phase starts at $-270°$ or $90°$ due to the presence the pole at the origin in the loop gain transfer function T_i.

4.5.4 Input Voltage-to-Output Voltage Transfer Function M_{vicl}

Using the block diagram shown in Figure 4.38, the closed-loop input voltage-to-output voltage transfer function is

$$M_{vicl}(s) = \left.\frac{v_o(s)}{v_i(s)}\right|_{v_{ri}=i_o=0}. \tag{4.181}$$

From the block diagram,

$$i_l = i_l' + i_l'' = T_{pi}d + M_{vi}v_i \tag{4.182}$$

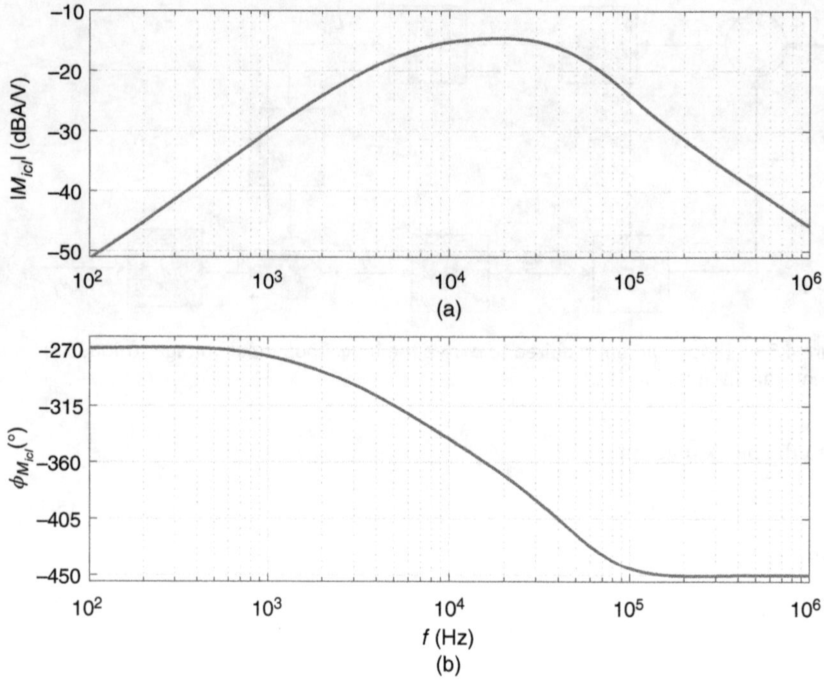

Figure 4.37 Theoretically obtained plots of input voltage-to-inductor current transfer function M_{icl}. (a) Magnitude. (b) Phase.

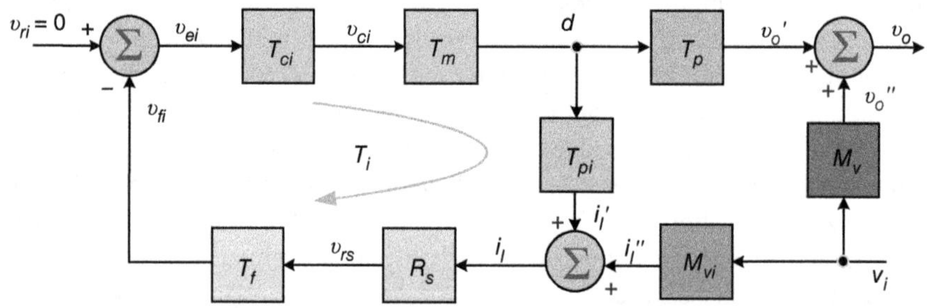

Figure 4.38 Block diagram required to derive the inner-loop input voltage-to-output voltage transfer function M_{vicl}.

and

$$\frac{-d}{T_{ix}} = T_{pi}d + M_{vi}v_i, \tag{4.183}$$

where

$$T_{ix} = \frac{T_i}{T_{pi}} = T_{ci}T_mR_sT_f. \tag{4.184}$$

Rearranging (4.183) further, we get

$$\begin{aligned} -d\left(\frac{1}{T_{ix}} + T_{pi}\right) &= M_{vi}v_i \\ -d\left(\frac{1 + T_{ix}T_{pi}}{T_{ix}}\right) &= M_{vi}v_i \\ d &= -\frac{M_{vi}T_{ix}}{1 + T_i}v_i. \end{aligned} \tag{4.185}$$

From the block diagram in Figure 4.38,

$$v_o = v_o' + v_o'' = T_p d + M_v v_i. \tag{4.186}$$

Substituting d from (4.185), we get

$$v_o = T_p\left(-\frac{M_{vi}T_{ix}}{1 + T_i}v_i\right) + M_v v_i. = \left(M_v - \frac{T_p M_{vi} T_{ix}}{1 + T_i}\right)v_i. \tag{4.187}$$

The closed-loop input-to-output transfer function is

$$M_{vicl}(s) = \frac{v_o(s)}{v_i(s)} = M_v - \frac{T_p M_{vi} T_{ix}}{1 + T_i}. \tag{4.188}$$

Figure 4.39 shows the theoretically obtained magnitude and phase plots of the inner-current loop input voltage-to-output voltage transfer function. The M_{vicl} transfer function is an important transfer function, especially for the output-voltage

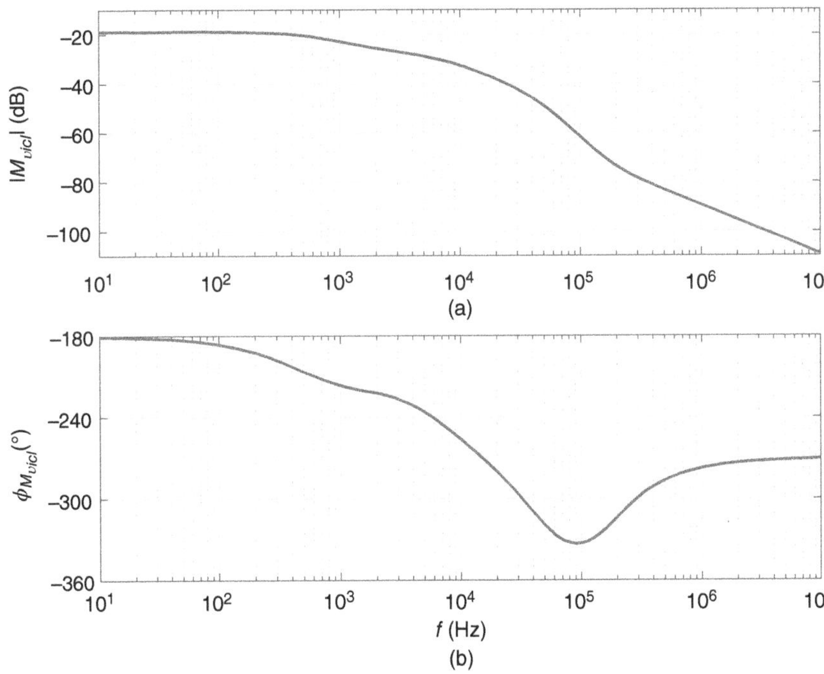

Figure 4.39 Theoretically obtained plots of input voltage-to-output voltage transfer function M_{vicl}. (a) Magnitude. (b) Phase.

loop. Provided $T_i >> 1$, the M_{vicl} transfer function can be approximated at low frequencies as

$$M_{vicl} = M_v - \frac{T_p M_{vi} T_{ix}}{T_i} = M_v - \frac{T_p M_{vi}}{T_{pi}}. \tag{4.189}$$

Unlike other disturbance transfer functions, the M_{vicl} transfer function obtained between the output and the input in the presence of the inner-current loop only depends on T_i instead of $1/T_i$.

4.5.5 Input Voltage-to-Duty Cycle Transfer Function M_{di}

Using the block diagram shown in Figure 4.38, the closed-loop input voltage-to-duty cycle transfer function is

$$M_{di}(s) = \left.\frac{d(s)}{v_i(s)}\right|_{v_{ri}=i_o=0}. \tag{4.190}$$

From (4.185),

$$d = -\frac{M_{vi} T_{ix}}{1 + T_i} v_i. \tag{4.191}$$

Hence, the closed-loop input voltage-to-duty cycle transfer function is

$$M_{di}(s) = \frac{d(s)}{v_i(s)} = -\frac{M_{vi} T_{ix}}{1 + T_i} = -\frac{M_{vi}}{T_{pi}} \frac{T_i}{1 + T_i}. \tag{4.192}$$

Figure 4.40 shows the theoretically obtained magnitude and phase plots of the inner-current loop input voltage-to-control transfer function. This transfer function determines the magnitude of change in duty cycle offered by the inner-loop and outer-loop for small-signal perturbations in the input voltage. From the Bode plot, the magnitude of the transfer function at DC is nearly $M_{dio} = 0.015$, i.e. for a step change in input voltage by 1 V, the duty cycle changes by nearly 0.015 to correct the disturbance. For example, assuming that the steady-state input voltage of the designed buck-boost converter changes from 12 V to 14 V, then the corresponding change in the duty cycle needed to maintain the output voltage V_O constant is $\Delta D = 0.033$. However, the current loop is able to provide only $\Delta D = M_{dio} \times \Delta V_I = 0.015 \times 2 = 0.03$. This means a deficit of $D = 0.003$ is required to be provided by the outer-voltage loop. In addition, for a change in the supply voltage in the positive direction, the duty cycle has to decrease to maintain the output voltage constant. Therefore, the phase difference between the input voltage and duty cycle is $-180°$.

4.5.6 Input Impedance Z_{iicl}

Using the block diagram shown in Figure 4.41, the closed-loop input impedance is

$$Z_{iicl}(s) = \left.\frac{v_i(s)}{i_i(s)}\right|_{d=i_o=0}. \tag{4.193}$$

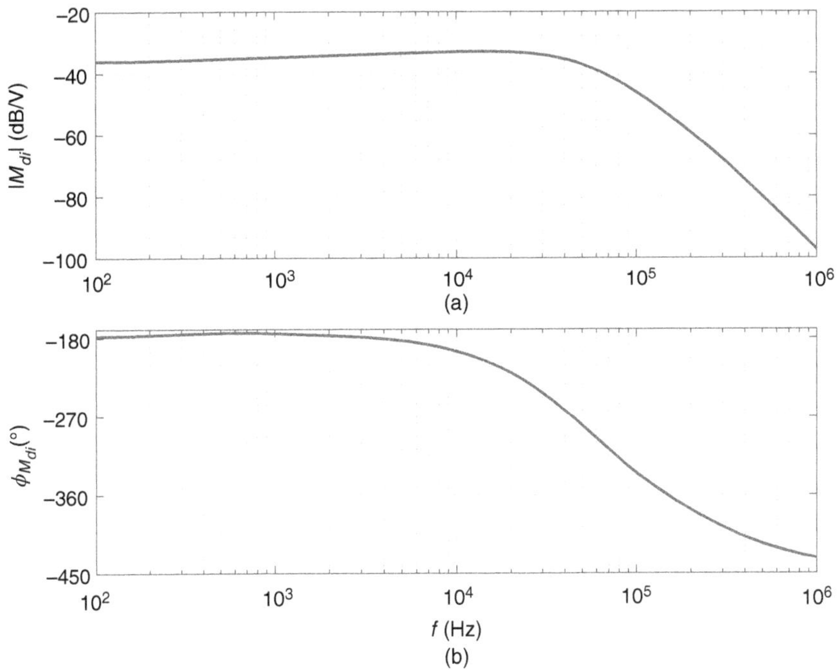

Figure 4.40 Theoretically obtained plots of input voltage-to-duty cycle transfer function M_{di}. (a) Magnitude. (b) Phase.

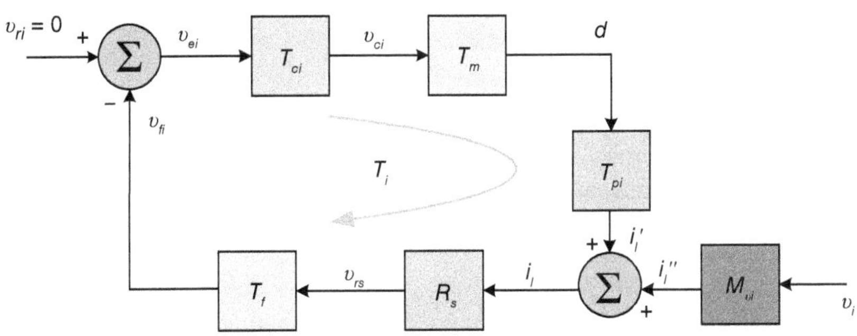

Figure 4.41 Block diagram used to derive the inner-closed loop input impeance Z_{iicl}.

From Figure 4.41,

$$i_l = i_l' + i_l'' = T_{pi}d + M_{vi}v_i \tag{4.194}$$

and solving for d yields

$$d = \frac{i_l}{T_{pi}} - \frac{M_{vi}}{T_{pi}}v_i. \tag{4.195}$$

Also,

$$i_l = T_{pi}(T_m T_{ci} v_{ei}) + M_{vi} v_i = T_{pi} T_m T_{ci}(-T_f R_s i_l) + M_{vi} v_i = -T_i i_l + M_{vi} v_i. \tag{4.196}$$

Rearranging (4.196) yields

$$i_l(1 + T_i) = M_{vi} v_i. \tag{4.197}$$

Hence,

$$i_l = \frac{M_{vi}}{1 + T_i} v_i. \tag{4.198}$$

The input current is given as

$$i_i = D i_l + I_L d. \tag{4.199}$$

Substituting d from (4.195) into (4.199), we get

$$i_i = D i_l + I_L d = D i_l + I_L \left(\frac{i_l}{T_{pi}} - \frac{M_{vi} v_i}{T_{pi}} \right) \tag{4.200}$$

$$= \left(D + \frac{I_L}{T_{pi}} \right) i_l - \frac{I_L M_{vi}}{T_{pi}} v_i. \tag{4.201}$$

Substituting (4.198) into (4.200) gives

$$i_i = \left[\left(D + \frac{I_L}{T_{pi}} \right) \frac{M_{vi}}{1 + T_i} - \frac{I_L M_{vi}}{T_{pi}} \right] v_i. \tag{4.202}$$

Hence, the closed-loop input admittance is

$$Y_{iicl}(s) = \frac{i_i(s)}{v_i(s)} = \left(D + \frac{I_L}{T_{pi}} \right) \frac{M_{vi}}{1 + T_i} - \frac{I_L M_{vi}}{T_{pi}}. \tag{4.203}$$

The closed-loop input impedance is

$$Z_{iicl}(s) = \frac{v_i(s)}{i_i(s)} = \frac{1}{Y_{iicl}(s)}. \tag{4.204}$$

Figure 4.42 shows the theoretically obtained magnitude and phase plots of the inner-current-loop input impedance.

4.5.7 Output Impedance Z_{oicl}

Using the block diagram shown in Figure 4.43, the closed-loop output impedance is

$$Z_{oicl}(s) = \left. \frac{v_o(s)}{i_o(s)} \right|_{v_{ri} = v_i = 0}. \tag{4.205}$$

From the block diagram,

$$i_l = i_l' + i_l'' = T_{pi} d + A_i i_o \tag{4.206}$$

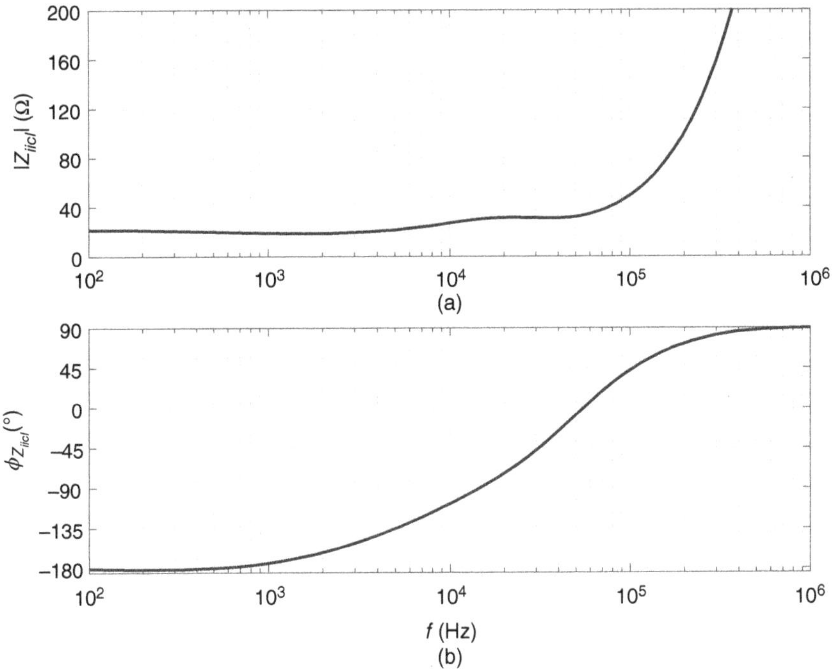

Figure 4.42 Theoretically obtained plots of closed-inner-loop input impedance Z_{iicl}. (a) Magnitude. (b) Phase.

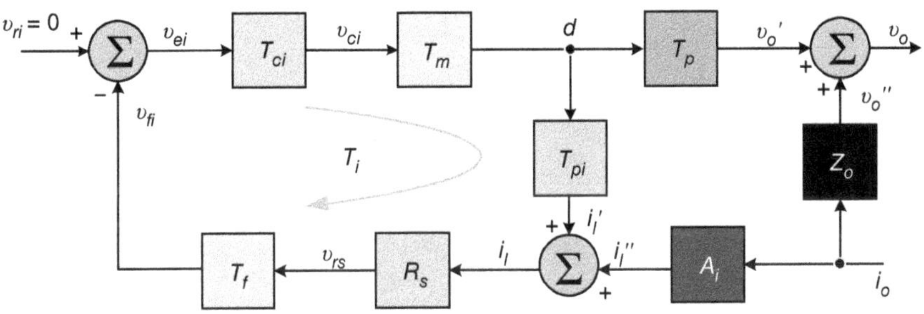

Figure 4.43 Block diagram required to derive the inner-closed-loop output impedance Z_{oicl}.

and

$$\frac{-d}{T_{ix}} = T_{pi}d + A_i i_o, \tag{4.207}$$

where

$$T_{ix} = \frac{T_i}{T_{pi}} = T_{ci}T_m R_s T_f. \tag{4.208}$$

Rearranging (4.207) produces

$$-d\left(\frac{1}{T_{ix}} + T_{pi}\right) = A_i i_o. \tag{4.209}$$

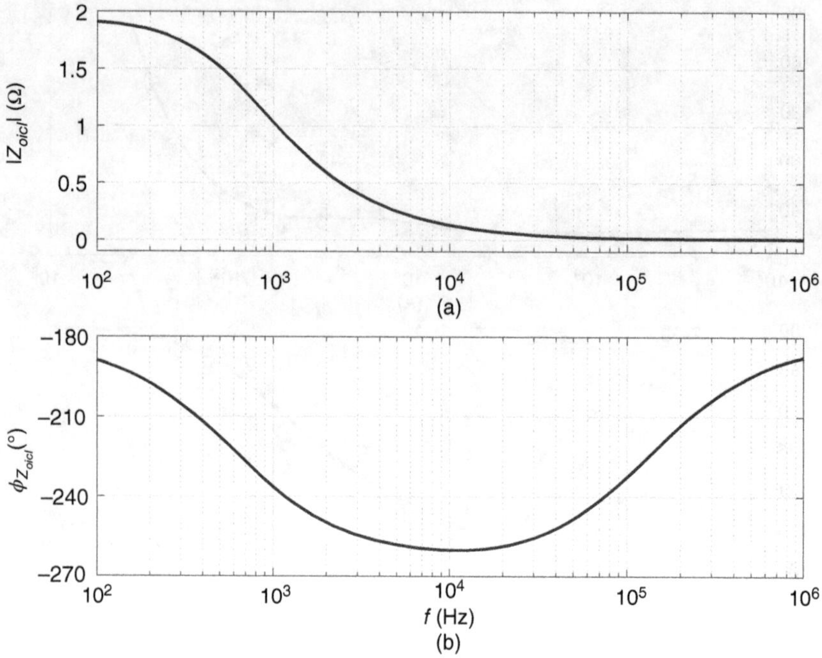

Figure 4.44 Theoretically obtained plots of closed-inner-loop output impedance Z_{oicl}. (a) Magnitude. (b) Phase.

Hence,

$$d = -\frac{A_i T_{ix}}{1 + T_i} i_o. \tag{4.210}$$

From the block diagram in Figure 4.43,

$$v_o = v_o' + v_o'' = T_p d + Z_o i_o. \tag{4.211}$$

Substituting d from (4.210) into (4.211), we get

$$v_o = T_p \left(-\frac{A_i T_{ix}}{1 + T_i} i_o \right) + Z_o i_o. = \left(Z_o - \frac{A_i T_p T_{ix}}{1 + T_i} \right) i_o. \tag{4.212}$$

The closed-loop output impedance is

$$Z_{oicl}(s) = \frac{v_o(s)}{i_o(s)} = Z_o - \frac{A_i T_p T_{ix}}{1 + T_i}. \tag{4.213}$$

Figure 4.44 shows the theoretically obtained magnitude and phase plots of the closed-inner-loop output impedance.

4.6 Outer-Voltage Loop

The outer-voltage loop, i.e. the major loop, is essential in a converter to act against the nonlinear disturbances in the load. As the load resistance changes, the load current increases or decreases causing the inductor current to change. However, to maintain constant output voltage, it is essential to introduce the outer-voltage loop with a set reference voltage V_{RV} that regulates the output voltage. In a typical two-loop system, the reference voltage to the outer-loop is a fraction of the rated output voltage that is determined by the feedback network, whose voltage gain β is determined by a network of resistors. The reference voltage must be positive. However, the output voltage is negative. Therefore, the transfer function of feedback network β must be negative to have zero phase shift in the uncompensated loop gain at DC and low frequencies. The negative feedback factor can be achieved by various means, mostly depending upon the application and the type of control (analog or digital). The block diagrams shown in this section consist of β, whose value is negative.

Figure 4.45 shows an architecture of the average current-mode controlled buck-boost converter with two-loop control. Moreover, Figure 4.46 shows the circuit schematic of the average current-mode controlled buck-boost converter with inner-current loop and outer-voltage loop. It must be noted that the circuit schematic already includes the compensation scheme required for the outer loop. Complete derivations of the transfer functions are provided in this section. The feedback gain is negative to account for the negative output voltage. The reference voltage V_{RI} for the current loop is set by the outer loop. Typically, the bandwidth of the outer loop is slow, and the inner loop is fast.

The procedure to design the outer loop is identical to that involved in designing the inner-current loop. The buck-boost power stage serves as the plant for the inner-current loop. However, in the case of outer-loop design, the power stage incorporated within the inner loop serves as the plant to be controlled. Hence, the path between the current reference v_{ri} and the duty cycle d serves as the plant for the outer-voltage loop. The transfer function between v_{ri} and d is a sixth-order

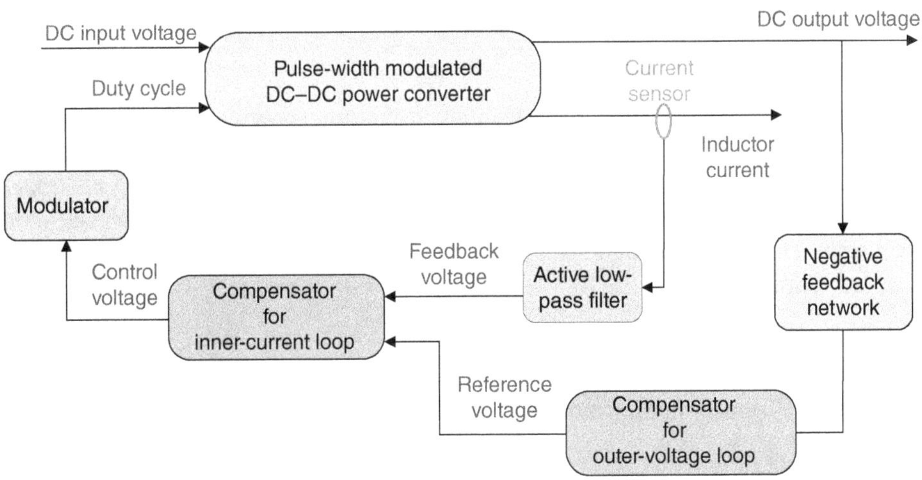

Figure 4.45 Architecture of the complete average current-mode controlled buck-boost converter with voltage-mode control.

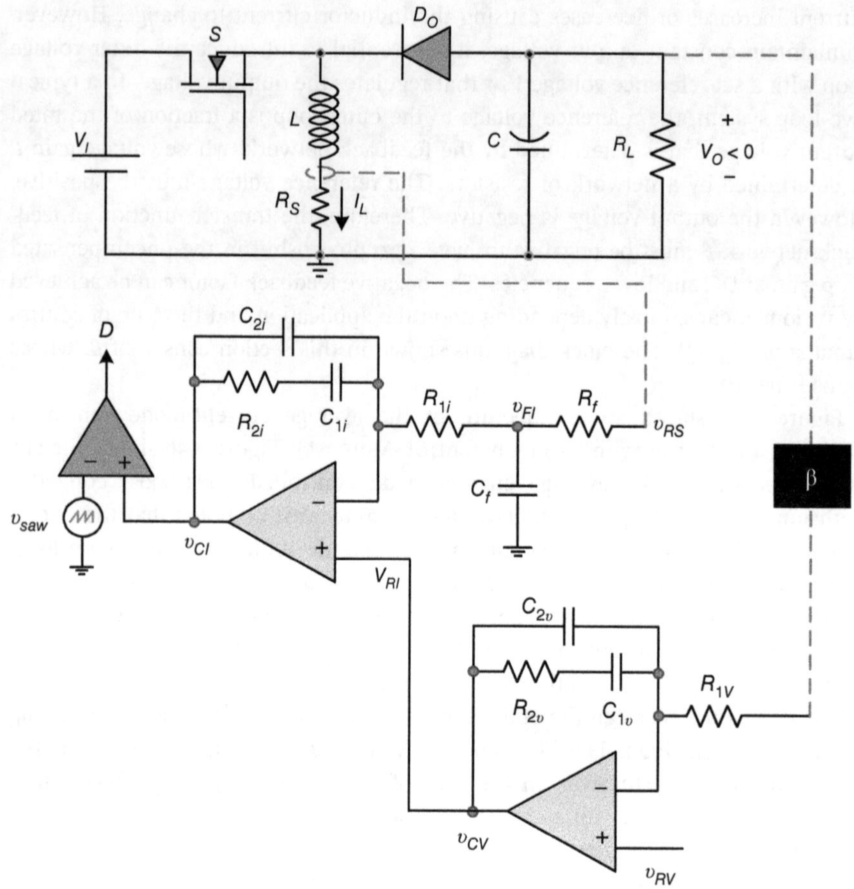

Figure 4.46 Complete circuit of the average current-mode controlled buck-boost converter with voltage-mode control.

transfer function. Therefore, it is cumbersome to express the transfer functions in individual pole–zero form. The complex-conjugate poles in T_p, the RHP zero in T_p, the low-frequency LHP zero in T_{pi}, and the inner-current loop controller poles are the ones that may exist within the crossover frequency of the voltage-loop gain. Therefore, attention must be paid in deciding the appropriate outer-loop gain crossover frequency.

4.6.1 Transfer Function of Feedback Network β

The transfer function of the feedback network is given as

$$\beta = \frac{v_{RV}}{v_O} = \frac{V_{RV}}{V_O}, \tag{4.214}$$

where V_{RV} is the reference voltage to the outer-voltage loop. The reference voltage $V_{RV} = 2.5$ V, and the transfer function of the feedback network is

$$\beta = \frac{V_{RV}}{V_O} = \frac{2.5}{-5} = -0.5. \tag{4.215}$$

4.6.2 Uncompensated Loop Gain T_{kv}

The natural behavior of the outer-voltage loop can be determined using the uncompensated loop gain as

$$T_{kv} = \frac{v_{fv}}{v_{ev}} = \frac{\beta v_o}{v_{ri}} = \beta T_{picl}. \tag{4.216}$$

Figure 4.47 shows the theoretically obtained magnitude and phase plots of the outer-voltage uncompensated loop gain T_{kv}. The gain is low at DC and is nearly $T_{kvo} = 1.25$ V/V = 1.94 dB. The crossover frequency is nearly 490 Hz. The phase margin at the crossover frequency is 145°. It is essential to provide a high gain at DC to reduce the steady-state error. The crossover frequency must be high for improved bandwidth; however, it cannot be higher than the frequency of the RHP zero present in the T_p transfer function. A significant phase boost must also be maintained at the crossover frequency to obtain a phase margin of 60°. Therefore, high-order control schemes such as Type II or Type III are preferred for this application. The

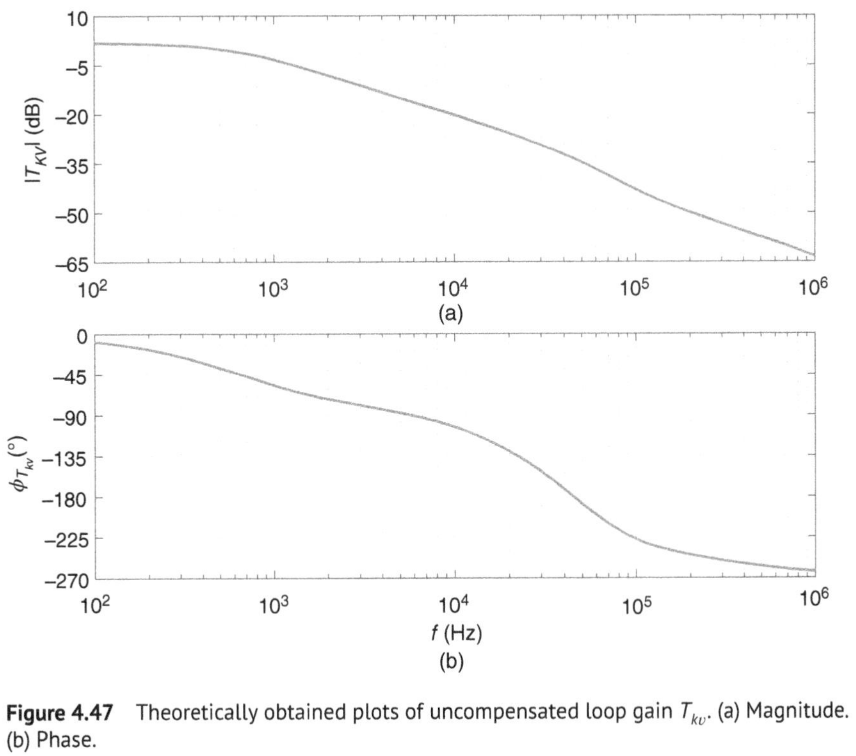

Figure 4.47 Theoretically obtained plots of uncompensated loop gain T_{kv}. (a) Magnitude. (b) Phase.

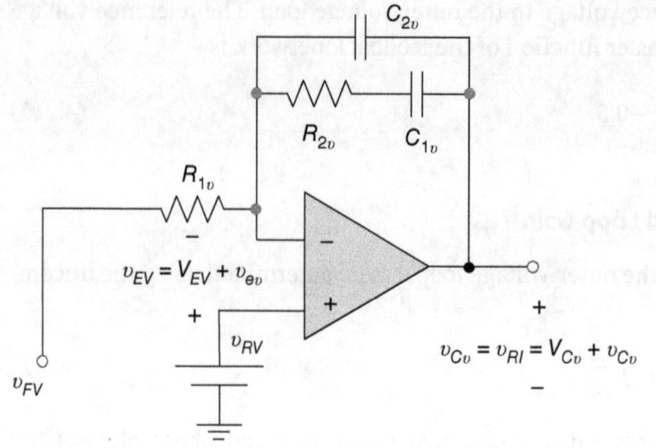

Figure 4.48 Circuit of the Type-II controller used in outer-voltage loop.

Type II, i.e. the single-lead integral control circuit, is designed in Section 4.6.3. The design was previously mentioned also for the controller of the inner-current loop. Figure 4.48 shows the control circuit for the outer-voltage loop.

4.6.3 Transfer Function of Control Circuit for Outer-Voltage Loop T_{cv}

Figure 4.48 shows the circuit of a Type-II controller used in the outer-voltage loop. The Type-II controller introduces a pole at the origin, which is required to boost the DC gain, and a pole–zero pair, which adjusts crossover frequency f_c, thus maintaining the required PM. A detailed procedure to determine the transfer function of the controller to improve the DC gain (> 20 dB) and achieve an inner-current loop phase margin $PM = 60°$ is explained in [45]. The remainder of this section describes the controller components design.

From the uncompensated loop gain transfer function T_{kv}, the crossover frequency is 495 Hz. A crossover frequency $f_c = 10$ kHz is assumed in designing the controller. At this new value of f_c, the magnitude of $T_{kv}(f_c) = 0.097$ V/V = −20.3 dB, and the phase $\phi_{Tkv}(f_c) = -104°$. The required phase margin is $PM = 60°$. Therefore, the phase boost required at f_c is

$$\phi_m = PM - \phi_{Tk}(f_c) - 90° = 74°, \tag{4.217}$$

which yields the geometric mean between the controller corner frequencies as

$$K = \tan^{-1}\left(\frac{\phi_m}{2} + 45°\right) = 7.12. \tag{4.218}$$

Assuming $R_{1v} = 1\ \text{k}\Omega/0.25\ W/2\%$, the value of C_{2v} is

$$C_{2v} = \frac{|T_k(f_c)|}{2\pi f_c K R_{1v}} = 216\ \text{pF}. \tag{4.219}$$

A 220 pF/25 V standard capacitor is selected. The expression for C_{1v} is

$$C_{1v} = C_{2v}(K^2 - 1) = 10.72\ \text{nF}. \tag{4.220}$$

A 11 nF/25 V standard capacitor is selected. The expression for R_{2v} is

$$R_{2v} = \frac{K}{2\pi f_c C_{1v}} = 10.5\ \text{k}\Omega. \tag{4.221}$$

A 10 kΩ/0.25 W/2% standard resistor is selected. The transfer function T_{cv} of the voltage-loop control circuit is

$$T_{cv} = \frac{v_{cv}(s)}{v_{ev}(s)} = T_{cvo}\frac{1+\dfrac{s}{\omega_{zcv}}}{s\left(1+\dfrac{s}{\omega_{pcv}}\right)}, \tag{4.222}$$

where

$$T_{cvo} = \frac{1}{R_{1v}(C_{1v}+C_{2v})} = 92\ \text{kV/V}, \tag{4.223}$$

$$\omega_{zcv} = \frac{1}{R_{2v}C_{1v}} = 8.8\ \text{krad/s}, \tag{4.224}$$

and

$$\omega_{pcv} = \frac{C_{1v}+C_{2v}}{R_{2v}C_{1v}C_{2v}} = 447\ \text{krad/s}. \tag{4.225}$$

Figure 4.49 shows the theoretically obtained magnitude and phase plots of the controller transfer function used in the outer-voltage loop. The poles are located at 0 Hz

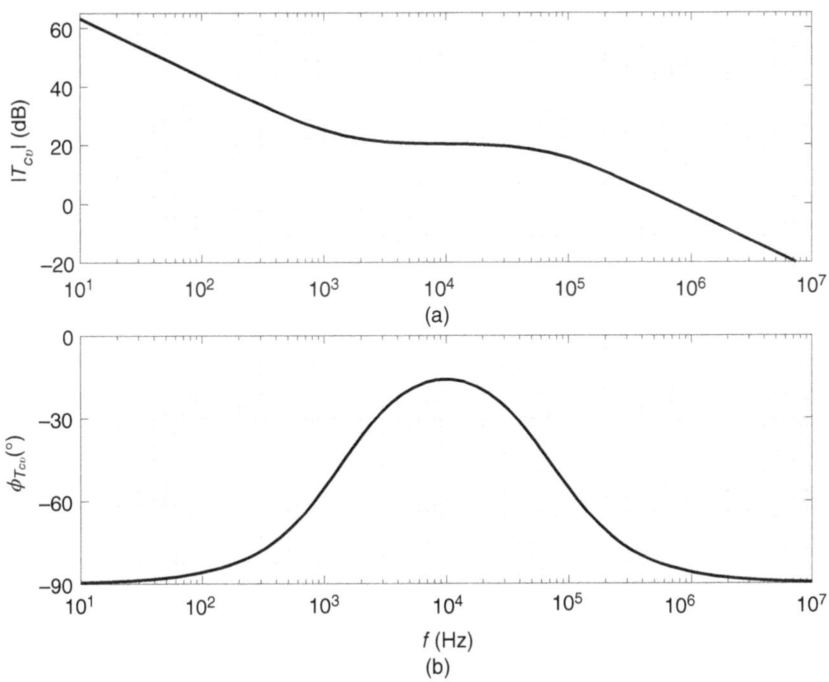

Figure 4.49 Theoretically obtained plots of controller transfer function T_{cv} of the outer-voltage loop. (a) Magnitude. (b) Phase.

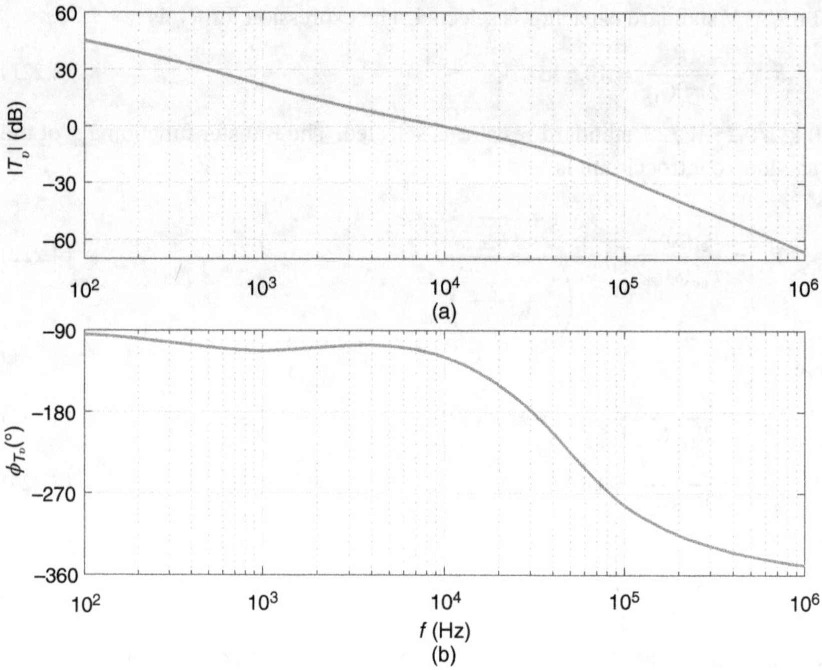

Figure 4.50 Theoretically obtained plots of loop gain T_v of the compensated outer-voltage loop. (a) Magnitude. (b) Phase.

and 71 kHz, and the zero is located at 1.4 kHz. A phase boost of nearly 74° is provided at a crossover frequency of 10 kHz.

4.6.4 Compensated Loop Gain T_v

The loop gain of the compensated outer-voltage loop is

$$T_v = \frac{v_{ev}}{v_{fv}} = T_{kv} T_{cv}. \tag{4.226}$$

Figure 4.50 shows the theoretically obtained magnitude and phase plots of the loop gain of compensated outer-voltage loop. The outer-voltage loop gain has a phase margin of 60°. The gain at DC is higher than 60 dB. A high gain at DC was already contributed by the high-gain property of the current loop plant.

4.7 Closed-Loop Transfer Functions for Outer-Voltage Loop

The following closed-loop transfer functions are derived:

1. Reference voltage-to-output voltage transfer function T_{pcl} (control).
2. Input voltage-to-output voltage transfer function M_{vcl} (disturbance).

3. Input voltage-to-duty cycle transfer function M_{dv} (disturbance).
4. Input impedance Z_{ivcl} (disturbance).
5. Output impedance Z_{ovcl} (disturbance).

4.7.1 Reference Voltage-to-Output Voltage Transfer Function T_{pcl}

Using the block diagram shown in Figure 4.51, the closed-loop reference voltage-to-output voltage transfer function is

$$T_{pcl}(s) = \frac{v_o(s)}{v_{rv}(s)} = \left.\frac{T_{cv}T_{picl}}{1+T_v}\right|_{v_i=i_o=0}. \tag{4.227}$$

Figure 4.52 shows the theoretically obtained magnitude and phase plots of the reference voltage-to-output voltage transfer function.

4.7.2 Input Voltage-to-Duty Cycle Transfer Function M_{dv}

Using the block diagram shown in Figure 4.53, the closed-loop input voltage-to-duty cycle transfer function is

$$M_{dv}(s) = \left.\frac{d(s)}{v_i(s)}\right|_{v_{rv}=i_o=0}. \tag{4.228}$$

From the block diagram,

$$v_o = v_o' + v_o'' = T_p d + M_v v_i. \tag{4.229}$$

Also,

$$v_{fv} = -\beta v_o \tag{4.230}$$

and

$$v_{cv} = v_{ri} = -\beta T_{cv} v_o. \tag{4.231}$$

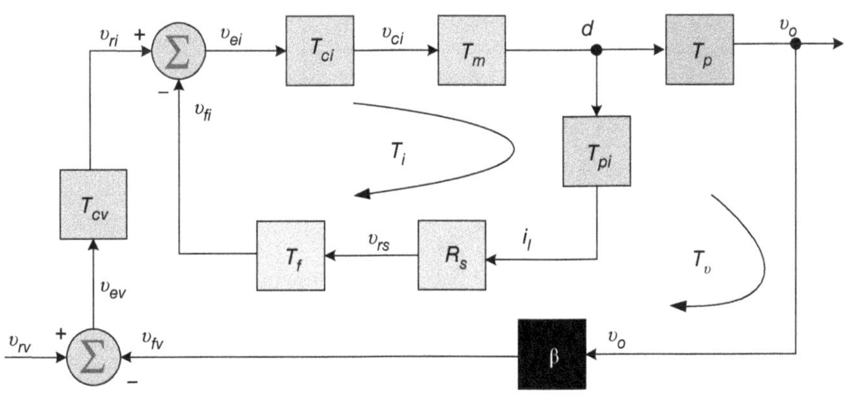

Figure 4.51 Block diagram used to derive the control-to-output voltage transfer function T_{pcl}. Reproduced with permission from Dalvir Saini.

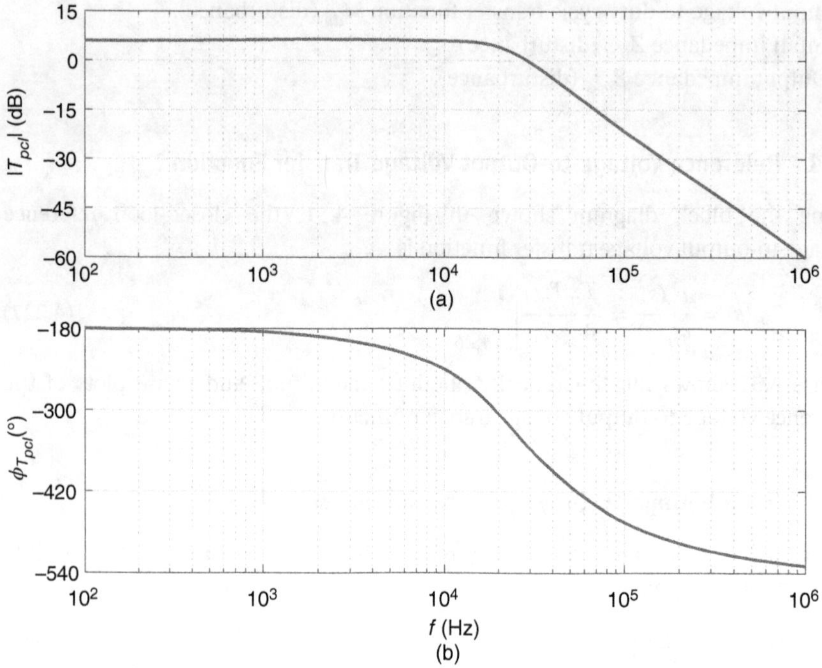

Figure 4.52 Theoretically obtained plots of the reference voltage-to-output voltage T_{pcl}. (a) Magnitude. (b) Phase.

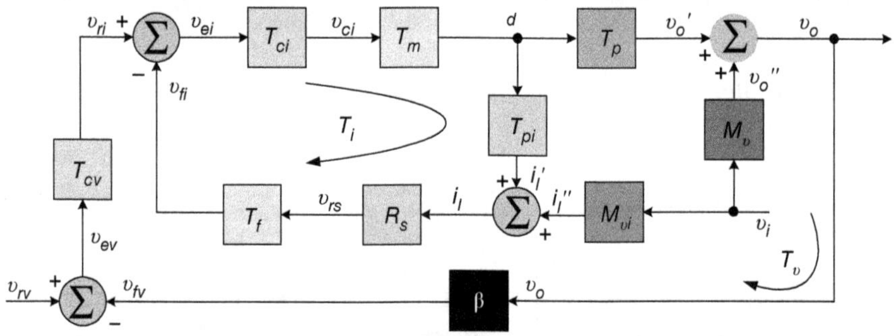

Figure 4.53 Block diagram used to derive the input voltage-to-duty cycle transfer function M_{dv}.

The error voltage v_{ei} is

$$v_{ei} = v_{ri} - v_{fi} = -T_{cv}\beta v_o - R_s T_f i_l, \tag{4.232}$$

and

$$v_{ei} = \frac{d}{T_m T_{ci}}. \tag{4.233}$$

The inductor current is

$$i_l = T_{pi} d + M_{vi} v_i. \tag{4.234}$$

Substituting (4.234) and (4.233) into (4.232) yields

$$\frac{d}{T_m T_{ci}} = -T_{cv}\beta v_o - R_s T_f (T_{pi} d + M_{vi} v_i). \tag{4.235}$$

Rearranging further, we get

$$v_o = -\frac{d}{T_{cv}\beta}\left(\frac{1}{T_m T_{ci}} + R_s T_f T_{pi}\right) - \frac{v_i}{T_{cv}\beta}\left(R_s T_f M_{vi}\right). \tag{4.236}$$

Equating (4.229) and (4.236) and rearranging, we get

$$v_i\left(-\frac{M_{vi} R_s T_f}{T_{cv}\beta} - M_v\right) = d\left(T_p + \frac{1}{T_m T_{ci} T_{cv}\beta} + \frac{R_s T_f T_{pi}}{T_{cv}\beta}\right). \tag{4.237}$$

Simplifying further gives

$$-\frac{v_i}{T_{cv}\beta}\left(M_{vi} R_s T_f + M_v T_{cv}\beta\right) = \frac{d}{T_{cv}\beta}\left(T_p T_{cv}\beta + \frac{1 + R_s T_f T_{pi} T_m T_{ci}}{T_m T_{ci}}\right) \tag{4.238}$$

and

$$-v_i\left(M_{vi} R_s T_f + M_v T_{cv}\beta\right) = d\left(T_p T_{cv}\beta + \frac{1 + R_s T_f T_{pi} T_m T_{ci}}{T_m T_{ci}}\right). \tag{4.239}$$

Hence,

$$M_{dv}(s) = \frac{d(s)}{v_i(s)} = -\frac{M_{vi} R_s T_f + \beta M_v T_{cv}}{\beta T_p T_{cv} + \dfrac{1 + T_i}{T_m T_{ci}}} = -\frac{T_{ci} T_m (M_{vi} R_s T_f + \beta M_v T_{cv})}{\beta T_p T_{cv} T_{ci} T_m + 1 + T_i}. \tag{4.240}$$

Figure 4.54 shows the theoretically obtained magnitude and phase plots of the input voltage-to-duty cycle transfer function. From the Bode plot, the magnitude of the transfer function at DC is nearly $M_{dvo} = 0.02$, i.e. for a step change in input voltage by 1 V, the duty cycle changes by nearly 0.019 to correct the disturbance. With only the inner-loop control, the change in duty cycle was $M_{dio} = 0.015$. Let us consider a case where the steady-state input voltage of the designed buck-boost converter changes from 12 V to 14 V. The corresponding change in the duty cycle needed to maintain the output voltage V_O constant would be $\Delta D = 0.033$. However, the current loop is able to provide $\Delta D = M_{dio} \times \Delta V_I = 0.015 \times 2 = 0.03$. This means that the outer loop is able to completely compensate for any remaining change in the input voltage and achieve a tightly controlled output voltage maintained at $V_O = -5$ V. In addition, for a change in the supply voltage in the positive direction, the duty cycle has to decrease to maintain the output voltage constant. Therefore, the phase difference between input voltage and duty cycle is $-180°$.

4.7.3 Input Voltage-to-Output Voltage Transfer Function M_{vcl}

Using the block diagram shown in Figure 4.55, the closed-loop input voltage-to-output voltage transfer function with outer-voltage loop is

$$M_{vcl}(s) = \left.\frac{v_o(s)}{v_i(s)}\right|_{v_{rv}=i_o=0}. \tag{4.241}$$

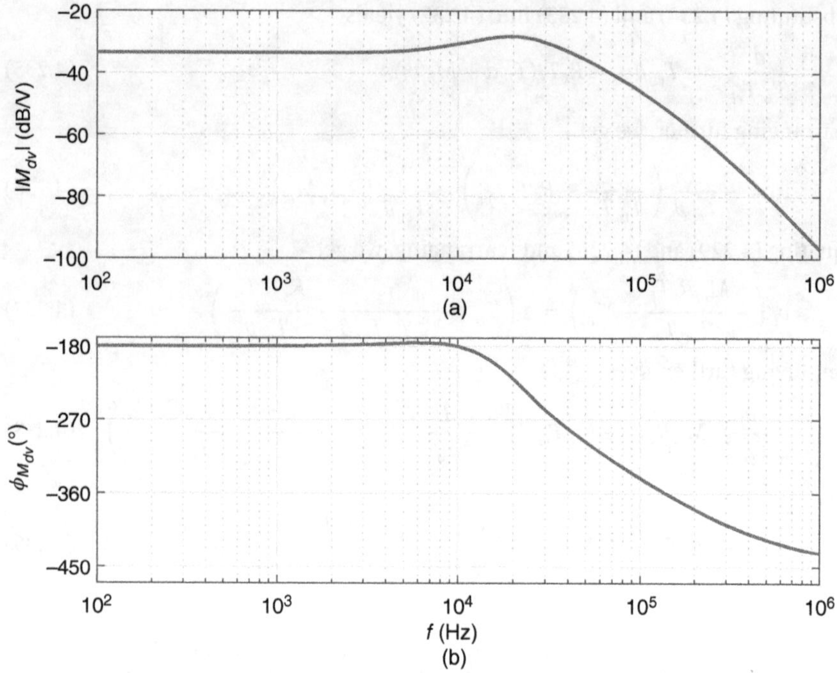

Figure 4.54 Theoretically obtained plots of the input voltage-to-duty cycle transfer function M_{dv}. (a) Magnitude. (b) Phase.

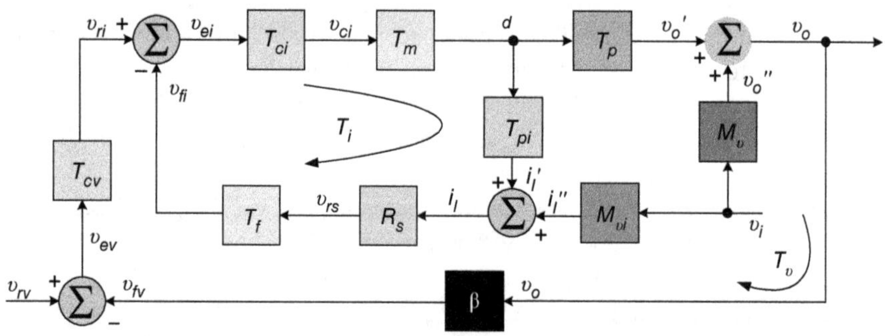

Figure 4.55 Block diagram used to derive the input voltage-to-output voltage transfer function M_{vcl}.

From (4.228),

$$d = M_{dv} v_i. \tag{4.242}$$

Substituting (4.242) into (4.229), we get

$$v_o = v_o' + v_o'' = T_p d + M_v v_i = T_p M_{dv} v_i + M_v v_i. \tag{4.243}$$

Rearrangement of (4.243) yields

$$M_{vcl}(s) = \frac{v_o(s)}{v_i(s)} = T_p M_{dv} + M_v. \tag{4.244}$$

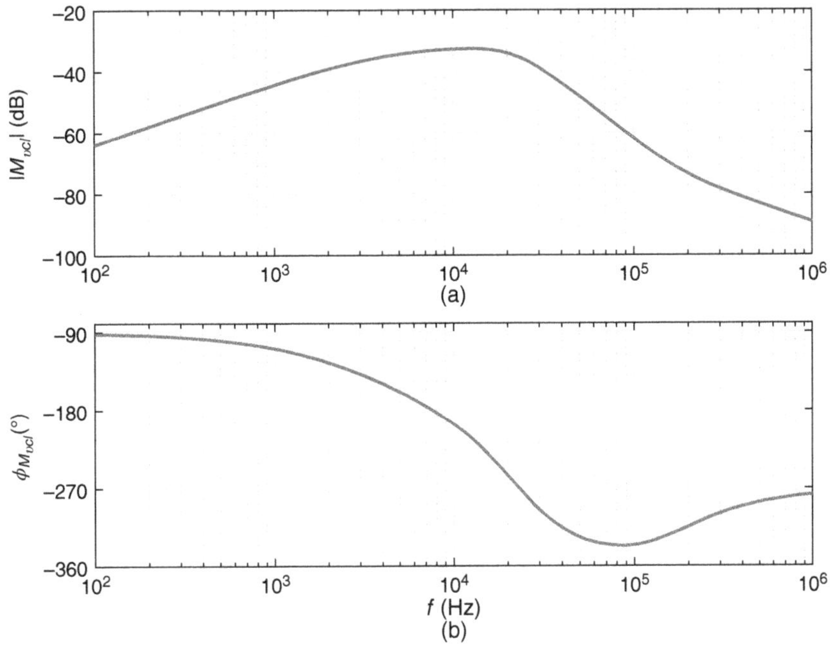

Figure 4.56 Theoretically obtained plots of the input voltage-to-output voltage transfer function M_{vcl}. (a) Magnitude. (b) Phase.

Figure 4.56 shows the theoretically obtained magnitude and phase plots of the input voltage-to-output voltage transfer function M_{vcl}. By comparing the results of M_{vicl} and M_{vcl}, it can be observed that the outer-voltage loop provides a stronger attenuation to disturbances in the input voltage than that offered by the inner-current loop alone. The attenuation is stronger at low DC and low frequencies. However, at high frequencies, approximately beyond f_c, the transfer function takes the form of M_{vicl}, where the negative feedback is almost ineffective.

4.7.4 Input Impedance Z_{ivcl}

Using the block diagram shown in Figure 4.57, the closed-loop input impedance with outer-voltage loop is

$$Z_{ivcl}(s) = \left. \frac{v_i(s)}{i_i(s)} \right|_{v_{rv}=i_o=0}. \tag{4.245}$$

From Figure 4.57,

$$v_o = v_o' + v_o'' = T_p d + M_v v_i \tag{4.246}$$

and

$$i_l = i_l' + i_l'' = T_{pi} d + M_{vi} v_i. \tag{4.247}$$

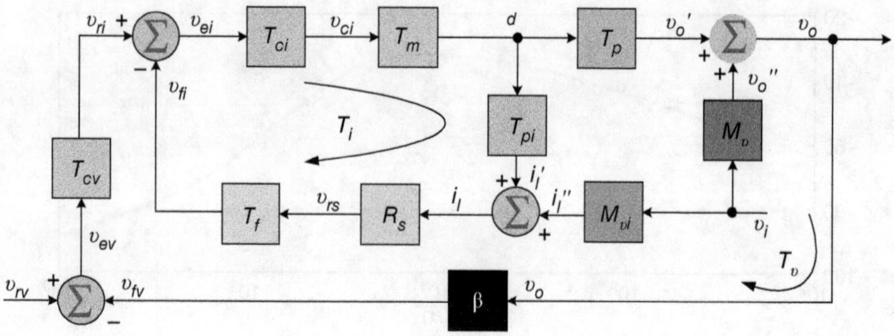

Figure 4.57 Block diagram used to derive the closed-outer-voltage-loop input impedance Z_{ivcl}.

Also,

$$v_{fv} = \beta v_o, \tag{4.248}$$

$$v_{ev} = v_{rv} - v_{fv} = 0 - \beta v_o = -\beta v_o, \tag{4.249}$$

and

$$v_{cv} = v_{ri} = T_{cv} v_{ev} = -\beta T_{cv} v_o. \tag{4.250}$$

The error voltage v_{ei} is

$$v_{ei} = \frac{d}{T_{ci} T_m}. \tag{4.251}$$

The feedback voltage v_{fi} is

$$v_{fi} = T_f R_s i_l. \tag{4.252}$$

The reference voltage for the inner-current loop is

$$v_{ri} = v_{ei} + v_{fi}. \tag{4.253}$$

Substituting (4.250), (4.251), and (4.252) into (4.253) yields

$$-\beta T_{cv} v_o = \frac{d}{T_{ci} T_m} + T_f R_s i_l. \tag{4.254}$$

Rearranging (4.254),

$$v_o = -\frac{d}{\beta T_{cv} T_{ci} T_m} - \frac{T_f R_s i_l}{\beta T_{cv}}. \tag{4.255}$$

Substituting (4.246) into (4.255),

$$T_p d + M_v v_i = -\frac{d}{\beta T_{cv} T_{ci} T_m} - \frac{T_f R_s i_l}{\beta T_{cv}}. \tag{4.256}$$

Rearranging further, we get

$$d\left(T_p + \frac{1}{\beta T_{cv} T_{ci} T_m}\right) = -\frac{T_f R_s i_l}{\beta T_{cv}} - M_v v_i. \tag{4.257}$$

Substituting (4.247) into (4.257),

$$d\left(T_p + \frac{1}{\beta T_{cv} T_{ci} T_m}\right) = -\frac{T_f R_s (T_{pi} d + M_{vi} v_i)}{\beta T_{cv}} - M_v v_i \tag{4.258}$$

$$d\left(T_p + \frac{1}{\beta T_{cv} T_{ci} T_m} + \frac{T_f R_s T_{pi}}{\beta T_{cv}}\right) = \left(-\frac{T_f R_s M_{vi}}{\beta T_{cv}} - M_v\right) v_i \tag{4.259}$$

$$d\left(\frac{\beta T_p T_{ci} T_m T_{cv} + 1 + T_f R_s T_{pi} T_{ci} T_m}{\beta T_{ci} T_m T_{cv}}\right) = \left(\frac{-T_f R_s M_{vi} - \beta T_{cv} M_v}{\beta T_{cv}}\right) v_i \tag{4.260}$$

$$d = -\frac{T_{ci} T_m (T_f R_s M_{vi} + \beta M_v T_{cv})}{\beta T_p T_{ci} T_m T_{cv} + 1 + T_f R_s T_{pi} T_{ci} T_m} v_i. \tag{4.261}$$

The input current i_i is

$$i_i = I_L d + D i_l = I_L d + D(T_{pi} d + M_{vi} v_i) = (I_L + D T_{pi}) d + D M_{vi} v_i. \tag{4.262}$$

Substituting (4.261) into (4.262) gives

$$i_i = (I_L + D T_{pi})\left[-\frac{T_{ci} T_m (T_f R_s M_{vi} + \beta M_v T_{cv})}{\beta T_p T_{ci} T_m T_{cv} + 1 + T_f R_s T_{pi} T_{ci} T_m} v_i\right] + D M_{vi} v_i \tag{4.263}$$

$$i_i = \left[-(I_L + D T_{pi})\frac{T_{ci} T_m (T_f R_s M_{vi} + \beta M_v T_{cv})}{\beta T_p T_{ci} T_m T_{cv} + 1 + T_f R_s T_{pi} T_{ci} T_m} + D M_{vi}\right] v_i. \tag{4.264}$$

Hence, the closed-loop input admittance for the outer-voltage loop is

$$Y_{ivcl}(s) = \frac{i_i(s)}{v_i(s)} = -(I_L + D T_{pi})\frac{T_{ci} T_m (T_f R_s M_{vi} + \beta M_v T_{cv})}{\beta T_p T_{ci} T_m T_{cv} + 1 + T_f R_s T_{pi} T_{ci} T_m} + D M_{vi}. \tag{4.265}$$

The closed-loop input impedance for the outer-voltage loop is

$$Z_{ivcl}(s) = \frac{v_i(s)}{i_i(s)} = \frac{1}{Y_{ivcl}(s)}. \tag{4.266}$$

Figure 4.58 shows the theoretically obtained magnitude and phase plots of the closed-loop input impedance Z_{ivcl}.

4.7.5 Output Impedance Z_{ovcl}

Using the block diagram shown in Figure 4.59, the closed-loop output impedance with the outer-voltage loop is

$$Z_{ovcl}(s) = \left.\frac{v_o(s)}{i_o(s)}\right|_{v_{rv}=v_i=0}. \tag{4.267}$$

From the Figure 4.59,

$$v_o = v_o' + v_o'' = T_p d + Z_o i_o \tag{4.268}$$

and

$$i_l = i_l' + i_l'' = A_i i_o + T_{pi} d. \tag{4.269}$$

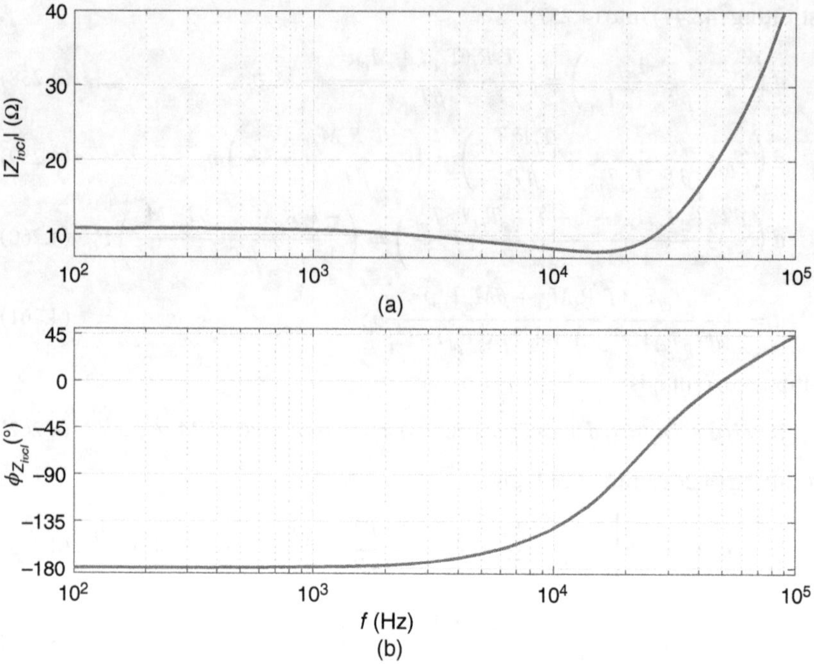

Figure 4.58 Theoretically obtained plots of the closed-outer-voltage loop input impedance Z_{ivcl}. (a) Magnitude. (b) Phase.

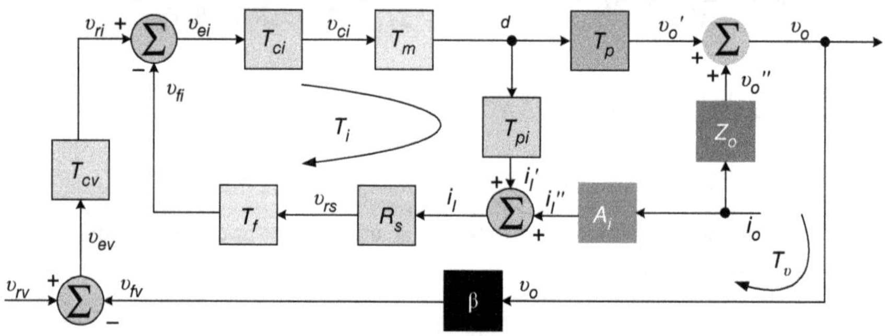

Figure 4.59 Block diagram used to derive the closed-outer-voltage-loop output impedance Z_{ovcl}.

Also,

$$v_{fv} = \beta v_o \tag{4.270}$$

and

$$v_{cv} = v_{ri} = -\beta T_{cv} v_o. \tag{4.271}$$

The error voltage v_{ei} is

$$v_{ei} = v_{ri} - v_{fi} = -T_{cv}\beta v_o - R_s T_f i_l \tag{4.272}$$

and is also expressed as

$$v_{ei} = \frac{d}{T_m T_{ci}}. \tag{4.273}$$

Substituting (4.269) and (4.273) into (4.272) yields

$$\frac{d}{T_m T_{ci}} = -T_{cv}\beta v_o - R_s T_f (A_i i_o + T_{pi} d). \tag{4.274}$$

Rearranging (4.274), we get

$$d = -\frac{(T_{cv}\beta v_o + R_s T_f A_i i_o) T_{ci} T_m}{1 + T_i}. \tag{4.275}$$

Substituting (4.275) into (4.268) yields

$$v_o = T_p \left[-\frac{(T_{cv}\beta v_o + R_s T_f A_i i_o) T_{ci} T_m}{1 + T_i} \right] + Z_o i_o. \tag{4.276}$$

Rearranging (4.276) gives

$$v_o \left(1 + \frac{\beta T_p T_{cv} T_{ci} T_m}{1 + T_i} \right) = i_o \left(Z_o - \frac{T_p R_s T_f A_i T_{ci} T_m}{1 + T_i} \right). \tag{4.277}$$

Hence, the closed-loop output impedance for the outer-voltage loop is

$$Z_{ovcl}(s) = \frac{v_o(s)}{i_o(s)} = \frac{Z_o(1 + T_i) - A_i T_p T_{ix}}{1 + T_i + T_p T_{cv} \beta T_{ci} T_m}. \tag{4.278}$$

Figure 4.60 shows the theoretically obtained magnitude and phase plots of the closed-loop output impedance Z_{ovcl}.

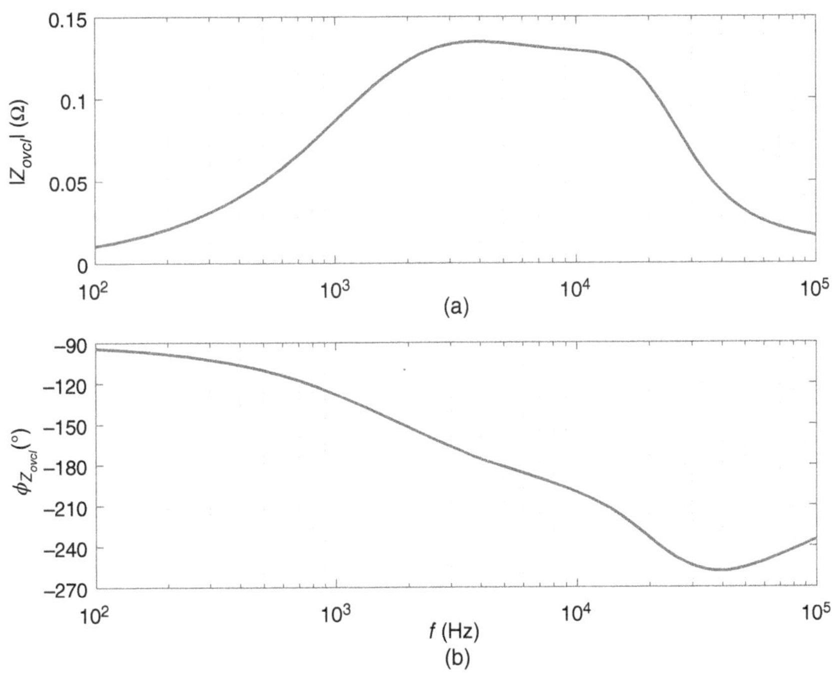

Figure 4.60 Theoretically obtained plots of the closed-outer-voltage loop output impedance Z_{ovcl}. (a) Magnitude. (b) Phase.

4.8 Comparison of Closed-Loop and Open-Loop Step Responses

4.8.1 Response of Output Voltage to Step Change in Input Voltage

Figure 4.61 compares the responses of the output voltage for unit step changes in input voltage by 1 V for (a) open loop (M_v), (b) closed-inner-current loop (M_{vicl}), and (c) closed-outer-voltage loop in the presence of the inner-current loop (M_{vcl}). In the open-loop configuration, a step change $\Delta V_I = 1$ V in the input voltage causes the output voltage to achieve a new steady-state value $V_O(\infty) = -5.44$ V. Therefore, the step change in the output voltage is $\Delta V_O = |V_O(\infty) - V_O(0)| = |-5.44 - (-5)| = 0.44$ V, yielding $M_{vo} = (\Delta V_O)/(\Delta V_I) = 0.44/1 = 0.44$ V/V, which was also observed in the magnitude response of M_v. The settling time is nearly 1 ms. For the closed-loop response determined by the current-loop transfer function M_{vicl}, the step change in input voltage causes the output voltage to rise to its new steady-state value $V_O(\infty) = -5.1$ V. The step change in the output voltage is $\Delta V_O = |V_O(\infty) - V_O(0)| = |-5.1 - (-5)| = 0.1$ V, yielding $M_{viclo} = (\Delta V_O)/(\Delta V_I) = 0.1/1 = 0.1$ V/V, which can be confirmed by the magnitude plot of M_{vicl} in Figure 4.39. However, for the closed-loop response determined by the voltage-loop transfer function M_{vcl}, the step change in the input voltage causes the output voltage to return to steady-state value $V_O(\infty) = -5$ V in 0.125 ms.

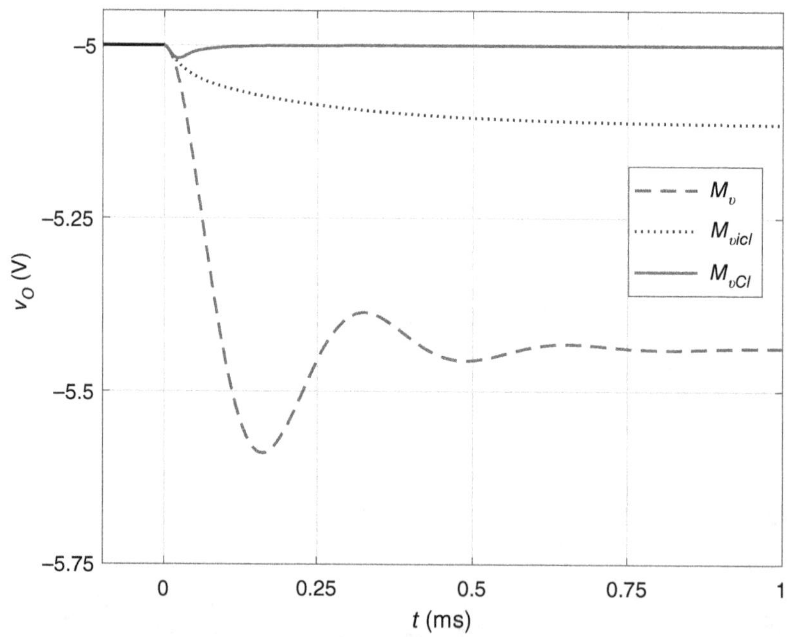

Figure 4.61 Comparison of responses in output voltage for step changes in input voltage by 1 V, obtained using M_v, M_{vicl}, and M_{vcl} transfer functions, respectively.

4.8.2 Response of Output Voltage to Step Change in Duty Cycle, Current-Loop reference Voltage, and Voltage-Loop Reference Voltage

Figure 4.62 compares the responses of the output voltage for step changes in (a) duty cycle for open-loop by 0.1 (T_p), (b) current-loop reference voltage for the closed-inner-current loop by 1 V (T_{picl}), and (c) voltage-loop reference voltage for the closed-outer-voltage loop by 1 V (T_{pcl}). In the open-loop configuration, a step change in duty cycle $\Delta D = 0.1$ causes the output voltage to achieve a new steady-state value $V_O(\infty) = -7.07$ V. Therefore, the step change in the output voltage is $\Delta V_O = |V_O(\infty) - V_O(0)| = |-7.07 - (-5)| = 2.07$ V, yielding $T_{po} = (\Delta V_O)/\Delta D = 2.07/0.1 = 20.7$, which was also observed in the magnitude plot of T_p. The step response shows a settling time of ≈ 1.25 ms. The output voltage response for the closed-inner-current loop achieves a steady-state value $V_O(\infty) = -7.5$ V in a settling time of 1.25 ms. Therefore, the step change in the output voltage is $\Delta V_O = |V_O(\infty) - V_O(0)| = |-7.5 - (-5)| = 2.5$ V, yielding $T_{piclo} = (\Delta V_O)/\Delta D = 2.5/1 = 2.5$, which was also observed in the magnitude plot of T_{picl}. The output voltage response for the closed-outer-voltage loop achieves a steady-state value $V_O(\infty) = -7$ V in a rapid settling time of nearly 0.25 ms. Therefore, the step change in the output voltage is $\Delta V_O = |V_O(\infty) - V_O(0)| = |-7 - (-5)| = 2$ V, yielding $T_{pclo} = (\Delta V_O)/\Delta D = 2/1 = 2$, which was also observed in the magnitude plot of T_{pcl}.

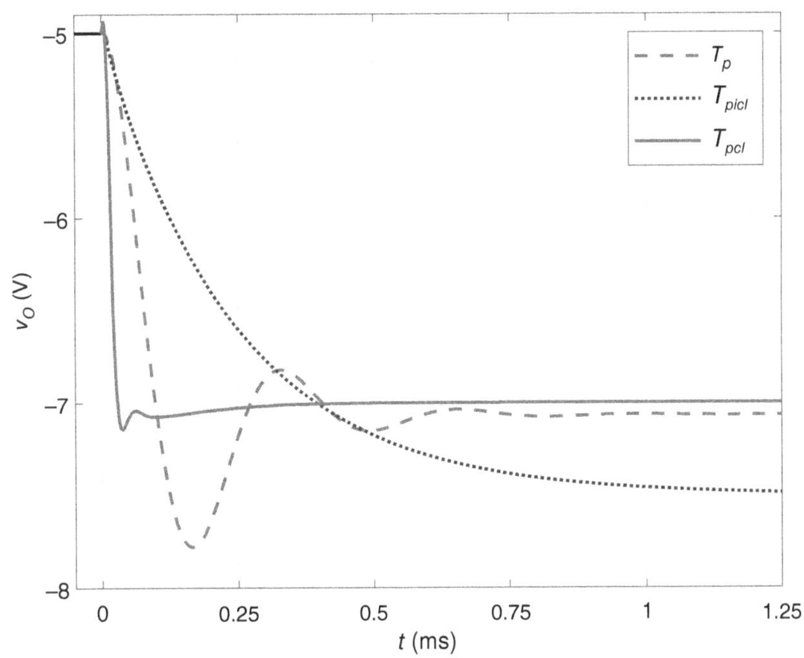

Figure 4.62 Comparison of responses in output voltage for step changes in duty cycle by 0.1, current-loop reference voltage by 1 V, and voltage-loop reference voltage by 1 V, obtained using T_p, T_{picl}, and T_{pcl} transfer functions, respectively.

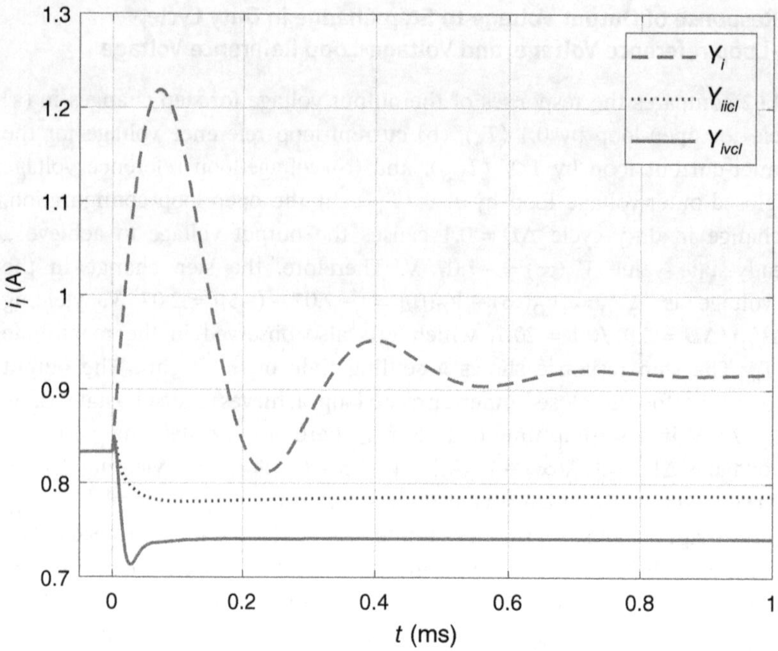

Figure 4.63 Comparison of responses in input current for step changes in the input voltage by 1 V, obtained using Y_i, Y_{iicl}, and Y_{iucl} transfer functions.

4.8.3 Response of Input Current to Step Change in Input Voltage

Figure 4.63 compares the responses of the input current for unit step change in input voltage by 1 V for (a) open-loop (Y_i), (b) closed-inner-current loop (Y_{iicl}), and (c) closed-outer-voltage loop (Y_{iucl}). In the open-loop configuration, a step change in input voltage $\Delta V_I = 1$ V causes the input current to achieve a new steady-state value $I_I(\infty) = 0.92$ A. Therefore, the step change in the input current is $\Delta I_I = I_I(\infty) - I_I(0) = 0.916 - 0.833 = 0.083$ A, yielding $Z_{io} = R_i = (\Delta I_I)/(\Delta V_I) = 1/0.083 = 12.05\ \Omega$. The step response shows a settling time of 1 ms. The input current response for the closed-inner-current loop achieves a steady-state value $I_I(\infty) = 0.79$ A in a settling time of 0.1 ms. The input current response for the closed-outer-voltage loop achieves a steady-state value $I_I(\infty) = 0.75$ A in a rapid settling time of nearly 0.1 ms. The reduction in the input current in negative direction for positive step change in output voltage is due to the negative input resistance caused at DC by the inner-current and outer-voltage loops.

4.8.4 Response of Output Voltage to Step Change in Load Current

Figure 4.64 compares the responses of the output voltage for unit step changes in output current by 1 A for (a) open-loop (Z_o), (b) closed-inner-current loop (Z_{oicl}), and (c) closed-outer-voltage loop (Z_{oucl}). In the open-loop configuration, a step change in output current $\Delta I_O = 1$ A causes the output voltage to achieve

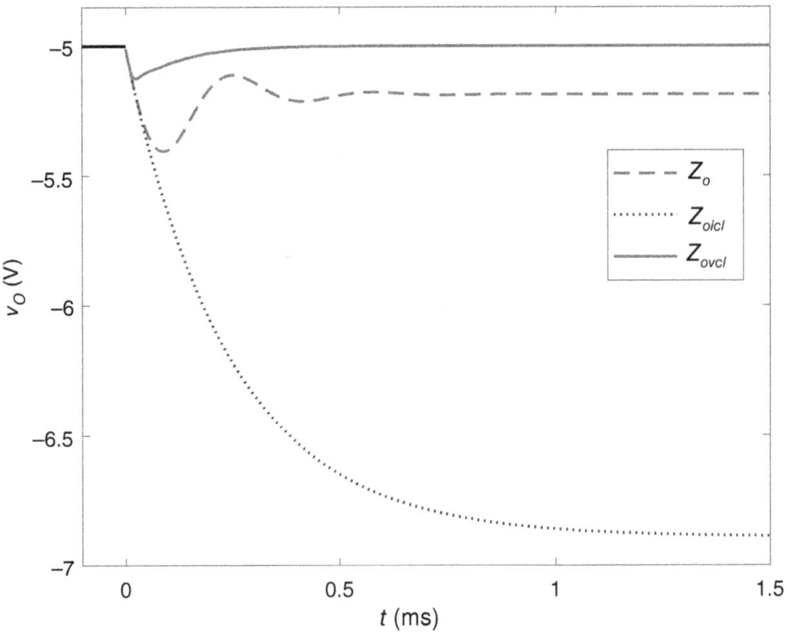

Figure 4.64 Comparison of responses in the output voltage for step changes in the output current by 1 A, obtained using Z_o, Z_{oicl}, and Z_{ovcl} transfer functions.

a new steady-state value $V_O(\infty) = -5.19$ V. Therefore, the step change in the output voltage is $\Delta V_O = |V_O(\infty) - V_O(0)| = |-5.19 - (-5)| = 0.19$ V, yielding $Z_{oo} = R_o = (\Delta V_O)/(\Delta I_O) = 0.19/1 = 0.19$ V/A. The step response shows a settling time of 1 ms. The output voltage response for the closed-inner-current loop achieves a steady-state value $V_O(\infty) = -6.89$ V in a settling time of 1.5 ms. The output voltage for the closed-outer-voltage loop returns to the steady-state value $V_O(\infty) = V_O(0) = -5$ V in a settling time of 0.5 ms.

4.9 Summary

1. The small-signal, linear model of the buck-boost converter in CCM consists of a current-controlled current source representing the transistor and a voltage-controlled voltage source representing the diode, in addition to the input AC voltage source, inductor, and filter capacitor.
2. The small-signal duty cycle-to-output voltage transfer function of a lossy buck-boost converter is a second-order transfer function constituting an LHP real zero, a RHP real zero, and a pair of complex-conjugate poles.
3. The small-signal duty cycle-to-inductor transfer function of a lossy buck-boost converter is a second-order transfer function with an LHP real zero and a pair of complex-conjugate LHP poles.
4. The small-signal input-to-output voltage transfer function of a lossy buck-boost converter is a second-order transfer function exhibiting an LHP zero and a pair of

complex-conjugate LHP poles. The location of the LHP zero and complex poles is identical to that of the duty cycle-to-output voltage transfer function.

5. The small-signal input voltage-to-inductor current transfer function of a lossy buck-boost converter is a second-order transfer exhibiting an LHP zero and a pair of complex-conjugate LHP poles.
6. The input impedance of a lossy buck-boost converter exhibits a pair of LHP complex-conjugate zeros and an LHP real zero. The input resistance is a function of the load resistance and the duty cycle. The location of the complex zero is identical to the complex pole produced by the duty cycle-to-output voltage transfer function.
7. The output impedance of a lossy buck-boost converter has two LHP real zeros and an LHP complex-conjugate poles. The output resistance is a function of the load resistance. The location of the complex pole is identical to that produced by the duty cycle-to-output voltage transfer function.
8. The current-mode control of a buck-boost converter can be achieved using a single-lead integral controller and can effectively reduce the DC steady-state error and improve the closed-loop bandwidth. This is a major advantage of the voltage-mode control, where the presence of RHP zero in the duty cycle-to-output voltage transfer function demands double-lead or higher forms of control schemes for effective compensation.

5

Average Current-Mode Control of Flyback DC–DC Converter

This chapter explains the average current-mode control of non-inverting pulse-width modulated (PWM) flyback DC–DC converter operated in continuous-conduction mode (CCM). The circuit averaging technique is presented, which is based on the principle of conservation of energy. The small-signal model of the flyback converter is derived taking into account the relevant losses. The open-loop and closed-loop transfer functions and transient responses are derived and analyzed. A design example is utilized throughout the chapter to understand open-loop and closed-loop converter responses. The objectives of this chapter are as follows:

1. To describe DC characteristics of the flyback converter.
2. To derive and analyze the small-signal linear model of the flyback converter.
3. To derive and analyze the open-loop power-stage transfer functions of the flyback converter.
4. To derive and analyze the open-loop input and output impedances of the flyback converter.
5. To design and analyze a controller circuit for the flyback converter.
6. To demonstrate the behavior of the average current-mode controlled flyback converter.
7. To derive and analyze the closed-inner-current-loop transfer functions of the flyback converter.
8. To derive and analyze the closed-outer-voltage-loop transfer functions of the flyback converter.

As the outcome of this chapter, the reader will gain sufficient knowledge in analysis, design, and characterization of flyback DC–DC converter employing average current-mode control. The subsequent analysis in this chapter is based on the following assumptions:

1. The diode capacitance and the lead inductance are zero, and thereby the switching losses are neglected.
2. Passive components are linear, time invariant, and frequency dependent.
3. The output impedance of the supply voltage source is zero for both DC and AC components.
4. The converter operates in steady state.

Average Current-Mode Control of DC-DC Power Converters, First Edition.
Marian K. Kazimierczuk, Dalvir K. Saini, and Agasthya Ayachit.

5. The switching period is much shorter than the time constants of the reactive components designed for CCM operation.
6. The DC output voltage is constant; however, the DC supply voltage, and the load resistance are variable.

5.1 Circuit Description, DC Model, and Design

5.1.1 Circuit Description

The circuit of the non-inverting PWM flyback DC–DC converter is shown in Figure 5.1. The detailed operation of the flyback converter in CCM is provided in [45]. The circuit of the flyback converter consists of a power MOSFET S, a diode D_o, a transformer with turns ratio n that provides DC isolation and changes the voltage level between input and the output, a filter capacitor C, and a load resistor R_L. The transformer is modeled by an ideal transformer and a magnetizing inductance L_m. The coupling coefficient of the transformer is assumed to be equal to 1. The leakage inductances are neglected. If the leakages inductance is taken into account, the order of the transfer functions of the flyback converter is increased. The transformer magnetizing inductance L_m stores the magnetic energy in its magnetic field. Figure 5.2 shows the equivalent circuit of the PWM flyback DC–DC converter with the magnetizing inductance L_m. The switch is turned ON and OFF at a switching frequency f_s and a duty cycle D. The equivalent series resistances (ESR) of the inductor and capacitor are r_L and r_C, respectively. The ON-state resistance of the MOSFET and the ON-state resistance of the diode are r_{DS} and R_F, respectively. The MOSFET output capacitance is C_o, the diode forward voltage is V_F, and the converter efficiency is η. The turns ratio of the transformer is

$$n = \frac{N_1}{N_2}, \tag{5.1}$$

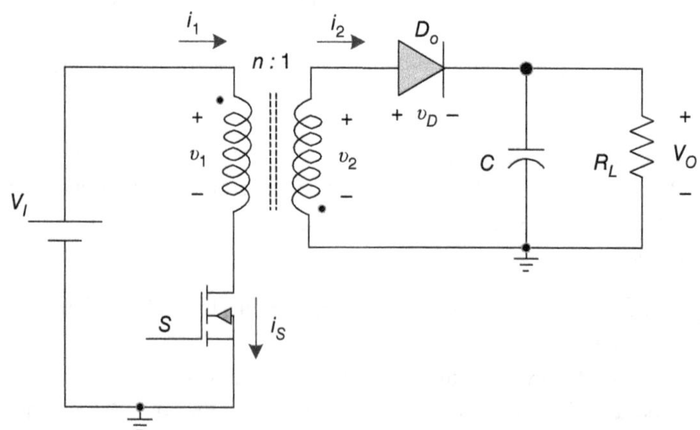

Figure 5.1 Circuit of the non-inverting PWM flyback DC–DC converter.

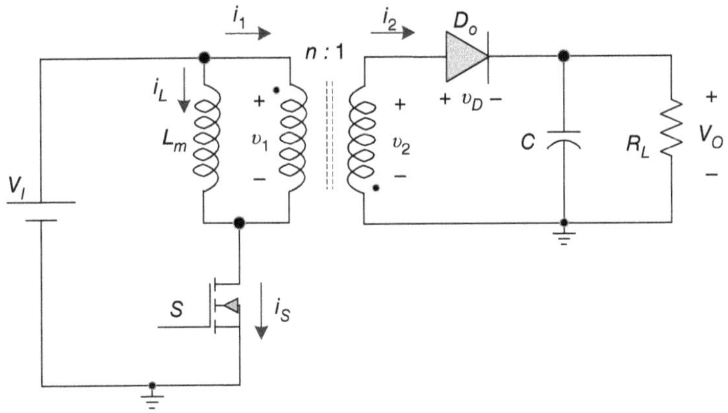

Figure 5.2 Equivalent circuit of the non-inverting PWM flyback DC–DC converter with the magnetizing inductance L_m.

where N_1 and N_2 are number of turns of the primary and secondary windings of the transformer, respectively.

5.1.2 DC Model

The DC voltage transfer function of the flyback converter for CCM is detailed in [45]. The idealized current and voltage waveforms of the switching network of a lossless flyback converter are shown in Figure 5.3. The circuit averaging technique is applied to the switching network waveforms. The switch current waveform is expressed as

$$i_S = \begin{cases} I_L & 0 < t \le DT \\ 0 & DT < t \le T. \end{cases} \tag{5.2}$$

The DC component of the switch current is defined as

$$I_S = \frac{1}{T}\int_0^{DT} i_S\,dt + \frac{1}{T}\int_{DT}^{T} i_S\,dt = \frac{1}{T}\int_0^{DT} I_L\,dt + 0 = DI_L. \tag{5.3}$$

From the circuit, the diode inverted voltage is expressed by the following relationship:

$$-v_D(t) = v_{KA}(t), \tag{5.4}$$

where v_{KA} is the voltage between diode cathode and anode. The DC component of the voltage v_{KA} is

$$V_{KA} = -V_D = V_O. \tag{5.5}$$

From Figure 5.3, the diode inverted voltage waveform can be expressed as

$$v_{KA}(t) = -v_D(t) = \begin{cases} \frac{V_I}{n} + V_O, & \text{for } 0 < t \le DT \\ 0, & \text{for } DT < t \le T. \end{cases} \tag{5.6}$$

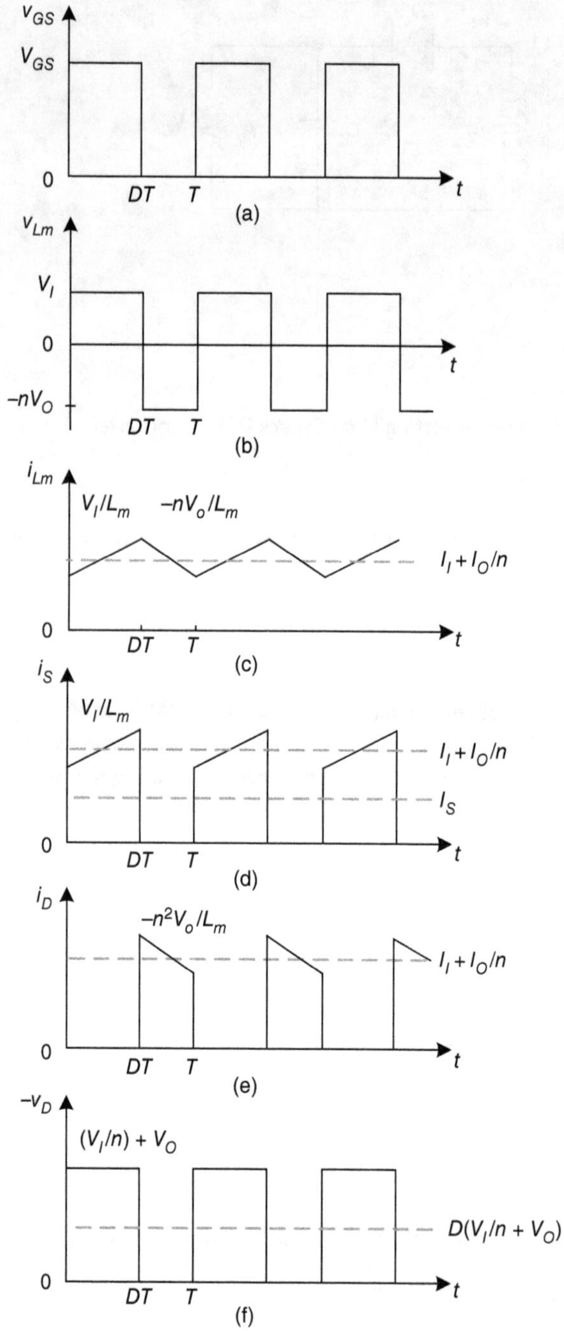

Figure 5.3 Waveforms of ideal switching network. (a) Gate-to-source voltage of switch. (b) Voltage across magnetizing inductance. (c) Current through magnetizing inductance. (d) Switch current. (e) Diode current. (f) Diode inverted voltage.

The DC component of diode inverted voltage in terms of the input and output voltage is

$$V_{KA} = -V_D = \frac{1}{T}\int_0^T v_{KA}\,dt = \frac{1}{T}\int_0^{DT} v_{KA}\,dt = \frac{1}{T}\int_0^{DT}\left(\frac{V_I}{n}+V_O\right)dt$$
$$= D\left(\frac{V_I}{n}+V_O\right) = D\left[\frac{V_I}{n}+\frac{DV_I}{n(1-D)}\right] = \frac{DV_I}{n(1-D)}. \tag{5.7}$$

As a result, the switch can be replaced by a current-controlled DC source represented by (5.3), and the diode by a voltage-controlled voltage source represented by (5.7). The DC model of the flyback converter is obtained by replacing the switch with a controlled current source and the diode by a controlled voltage source. The capacitor C is an open circuit. The magnetizing inductance L_m is a short circuit. Figure 5.4 shows the DC model of the flyback converter. Unlike the buck-boost converter, the flyback converter has a disjoint switching network, i.e. the switch S and diode D_0 do not share a common node and are decoupled by the isolated transformer windings. Furthermore, the on-state resistance r_{DS}, forward resistance R_F, and winding resistances R_{T1} and R_{T2} are lumped into an equivalent averaged resistance r placed in the magnetizing inductance L_m branch. The resistance r is obtained by applying the principle of conservation of energy [37, 45] and is derived in detail in Section 5.1.3. The resistance r is given by

$$r = Dr_{DS} + n^2(1-D)R_F + r_L. \tag{5.8}$$

Figure 5.5 shows the DC model of non-inverting flyback converter with the equivalent averaged resistance r represented in the inductor branch.

From the DC model in Figure 5.5, the input current is equal to the switch current:

$$I_I = I_S = DI_L. \tag{5.9}$$

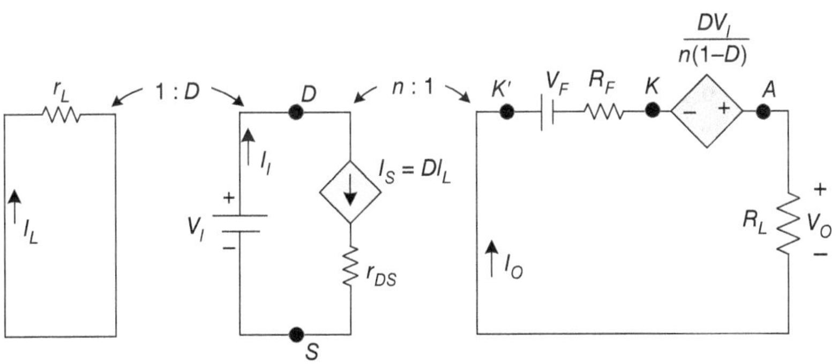

Figure 5.4 DC model of the non-inverting flyback DC–DC converter with resistances in their respective branches.

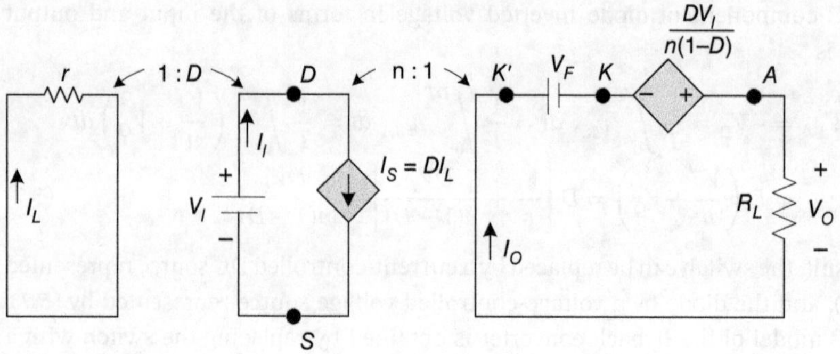

Figure 5.5 DC model of the non-inverting flyback DC–DC converter with equivalent averaged resistance *r*.

The equivalent winding resistance r_L conducts the magnetizing inductance current I_L. The input current and the magnetizing inductance current are related by the conversion ratio 1 : *D*. The input current and the output current are related by the turns ratio *n* and the duty cycle *D* as given by

$$\frac{I_I}{I_O} = \frac{DI_L}{I_O} = \frac{DI_L R_L}{V_O} = \frac{D}{n(1-D)}. \tag{5.10}$$

The DC component of the magnetizing inductance current is

$$I_L = \frac{I_I}{D} = I_I + \frac{I_O}{n} = \frac{I_O}{n(1-D)} = \frac{V_O}{n(1-D)R_L}, \tag{5.11}$$

yielding the DC input current as

$$I_I = \frac{D}{n(1-D)} I_O = \frac{D}{n(1-D)} \frac{V_O}{R_L}. \tag{5.12}$$

or the DC output current as

$$I_O = \frac{n(1-D)}{D} I_I. \tag{5.13}$$

Therefore, the dc current transfer function of the flyback converter is

$$M_{IDC} = \frac{I_O}{I_I} = \frac{n(1-D)}{D}. \tag{5.14}$$

The DC inductor current is

$$I_L = \frac{I_I}{D} = \frac{V_O}{n(1-D)R_L}. \tag{5.15}$$

For a lossless non-inverting flyback converter, $r = 0$ and $V_F = 0$.

Applying Kirchhoff's current law (KVL) to the output loop yields

$$\frac{D}{n(1-D)} V_I = V_O \tag{5.16}$$

producing the DC voltage transfer function for the lossless converter as

$$M_{VDC} = \frac{V_O}{V_I} = \frac{D}{n(1-D)}. \tag{5.17}$$

The DC voltage transfer function of the lossy flyback converter is dependent on r and V_F. The equivalent averaged resistance r_D can be placed in the diode branch and is given by

$$r_D = \frac{r_L}{n^2(1-D)^2} + \frac{Dr_{DS}}{n^2(1-D)^2} + \frac{R_F}{1-D} = \frac{r_L + Dr_{DS}}{n^2(1-D)^2} + \frac{R_F}{1-D}. \tag{5.18}$$

Detailed derivation of r_D is shown in Section 5.1.3. Subsequently, applying KVL to the output loop gives

$$v_{K'A} = r_D I_O + V_F - \frac{D}{n(1-D)} V_I + V_O = V_O \tag{5.19}$$

producing the DC voltage transfer function of a lossy non-inverting flyback converter as

$$M_{V\,DC(lossy)} = \frac{V_O}{V_I} = \frac{D}{n(1-D)} \frac{1}{1 + \frac{r_L + Dr_{DS}}{n^2(1-D)^2 R_L} + \frac{R_F}{(1-D)R_L} + \frac{V_F}{V_O}}. \tag{5.20}$$

The efficiency of the lossy flyback converter can be obtained using the DC model as

$$\eta = \frac{P_O}{P_I} = \frac{V_O I_O}{V_I I_I} = \frac{1}{1 + \frac{r_L + Dr_{DS}}{n^2(1-D)^2 R_L} + \frac{R_F}{(1-D)R_L} + \frac{V_F}{V_O}}. \tag{5.21}$$

The efficiency of the overall converter [45] is

$$\eta = \frac{P_O}{P_I} = \frac{1}{1 + \frac{r_L + Dr_{DS}}{n^2(1-D)^2 R_L} + \frac{R_F + Dr_C}{(1-D)R_L} + \frac{V_F}{V_O} + \frac{f_s C_o R_L (1 + nM_{V\,DC})^2}{M_{V\,DC}^2}}. \tag{5.22}$$

5.1.3 Derivation of Equivalent Averaged Resistance

The transformer winding resistance r_{T1} and r_{T2} can be lumped into an equivalent winding resistance r_L and placed in series with the magnetizing inductance L_m. The primary winding conducts current for the duration $0 < t \le DT$. Therefore, the primary winding resistance averaged over one switching period is

$$r_{T1(AV)} = \frac{1}{T}\int_0^T r_{T1}\,dt = \frac{1}{T}\int_0^{DT} r_{T1}\,dt = Dr_{T1}. \tag{5.23}$$

Similarly, the secondary winding conducts current for the duration $DT < t \le T$. Therefore, the secondary winding resistance averaged over one switching period is

$$r_{T2(AV)} = \frac{1}{T}\int_0^T r_{T2}\,dt = \frac{1}{T}\int_{DT}^{T} r_{T2}\,dt = (1-D)r_{T2}. \tag{5.24}$$

The secondary winding resistance is reflected to the primary using the relationship

$$n = \sqrt{\frac{r_{T2(AV)'}}{r_{T2(AV)}}}, \tag{5.25}$$

where $r'_{T2(AV)}$ is the secondary winding resistance reflected on primary side

$$r'_{T2(AV)} = n^2 r_{T2(AV)} = n^2(1-D)r_{T2}. \tag{5.26}$$

Therefore, the sum of $r_{T1(AV)}$ and $r'_{T2(AV)}$ is the ESR r_L of the transformer that produces the same heat as that individually produced by the resistances r_{T1} and r_{T2} in their respective branches. The ESR r_L is placed in series with L_m and given by

$$r_L = r_{T1(AV)} + r'_{T2(AV)} = Dr_{T1} + n^2(1-D)r_{T2}. \tag{5.27}$$

The inductor current is related to the switch current and the diode current by the relationship

$$I_L = \frac{I_S}{D} = \frac{I_D}{n(1-D)}. \tag{5.28}$$

The conduction power loss in the inductor ESR r_L is

$$P_{rL} = r_L I_L^2 = r_{L(AV)(L)} I_L^2 = \frac{r_L I_S^2}{D^2}. \tag{5.29}$$

The equivalent conduction power loss in r_L when the resistance is placed in the switch branch is

$$P_{rL} = r_{L(AV)(L)} I_L^2 = \frac{r_{L(AV)} I_S^2}{D^2} = r_{L(AV)(S)} I_S^2. \tag{5.30}$$

The equivalent conduction power loss in r_L when the resistance is placed in the diode branch is

$$P_{rL} = r_{L(AV)(L)} I_L^2 = \frac{r_L I_D^2}{n^2(1-D)^2} = r_{L(AV)(D)} I_D^2. \tag{5.31}$$

Using (5.29), (5.30), and (5.31), the ESR r_L in the inductor branch, switch branch, and the diode branch are represented, respectively, as

$$r_{L(AV)(L)} = r_L, \tag{5.32}$$

$$r_{L(AV)(S)} = \frac{r_L}{D}, \tag{5.33}$$

and

$$r_{L(AV)(D)} = \frac{r_L}{n^2(1-D)^2}. \tag{5.34}$$

The conduction power loss in the MOSFET on-state resistance r_{DS} is

$$P_{rDS} = \frac{r_{DS} I_S^2}{D} = r_{DS(AV)(S)} I_S^2. \tag{5.35}$$

The equivalent conduction power loss in r_{DS} when the resistance is placed in the inductor branch is

$$P_{rDS} = r_{DS(AV)(S)} I_S^2 = Dr_{DS} I_L^2 = r_{DS(AV)(L)} I_L^2. \tag{5.36}$$

The equivalent conduction power loss in r_{DS} when the resistance is placed in the diode branch is

$$P_{rDS} = r_{DS(AV)(S)} I_S^2 = \frac{Dr_{DS} I_D^2}{n^2(1-D)^2} = r_{DS(AV)(D)} I_D^2. \tag{5.37}$$

From (5.35), (5.36), and (5.37), the on-state resistances r_{DS} in the switch branch, inductor branch, and the diode branch are represented, respectively, as

$$r_{DS(AV)(S)} = \frac{r_{DS}}{D}, \tag{5.38}$$

$$r_{DS(AV)(L)} = Dr_{DS}, \tag{5.39}$$

and

$$r_{DS(AV)(D)} = \frac{Dr_{DS}}{n^2(1-D)^2}. \tag{5.40}$$

The conduction power loss in the diode forward resistance is

$$P_{RF} = \frac{R_F I_D^2}{1-D} = R_{F(AV)(D)} I_D^2. \tag{5.41}$$

The equivalent conduction power loss in R_F when the resistance is placed in the inductor branch is

$$P_{RF} = R_{F(AV)(D)} I_D^2 = n^2(1-D)R_F I_L^2 = R_{F(AV)(L)} I_L^2. \tag{5.42}$$

The equivalent conduction power loss in R_F when the resistance is placed in the switch branch is

$$P_{RF} = R_{F(AV)(D)} I_D^2 = \frac{n^2(1-D)R_F}{D} I_S^2 = R_{F(AV)(S)} I_S^2. \tag{5.43}$$

From (5.41), (5.42), and (5.43), the forward resistances R_F in the diode branch, inductor branch, and switch branch are represented, respectively, as

$$R_{F(AV)(D)} = \frac{R_F}{1-D}, \tag{5.44}$$

$$R_{F(AV)(L)} = n^2(1-D)R_F, \tag{5.45}$$

and

$$R_{F(AV)(S)} = \frac{n^2(1-D)R_F}{D}. \tag{5.46}$$

Therefore, the equivalent averaged resistance r placed in the inductor branch is the sum of ESR of inductor r_L, MOSFET ON-state resistance r_{DS}, and diode forward resistance R_F expressed as

$$r = r_{L(AV)(L)} + r_{DS(AV)(L)} + R_{F(AV)(L)}. \tag{5.47}$$

Using (5.32), (5.39), and (5.45) in (5.47), we get the equivalent averaged resistance r for the flyback converter in CCM as

$$r = r_L + Dr_{DS} + n^2(1-D)R_F. \tag{5.48}$$

Using (5.27) in (5.48) gives the equivalent averaged resistance as

$$\begin{aligned} r &= Dr_{T1} + n^2(1-D)r_{T2} + Dr_{DS} + n^2(1-D)R_F \\ &= D(r_{T1} + r_{DS}) + n^2(1-D)(r_{T2} + R_F). \end{aligned} \tag{5.49}$$

Figure 5.4 shows the DC model of the flyback converter with the resistances in their respective branches. Moreover, Figure 5.5 shows the DC model of the non-inverting flyback DC–DC converter with all each branch resistance lumped into an equivalent averaged resistance r. The equivalent averaged resistance $r_{S(AV)}$ can also be represented in the switch branch and the diode branch. The equivalent averaged resistance r_S in the switch branch is

$$r_S = r_{L(AV)(S)} + r_{DS(AV)(S)} + R_{F(AV)(S)} \tag{5.50}$$

yielding

$$r_S = \frac{r_L}{D} + \frac{r_{DS}}{D} + \frac{n^2(1-D)R_F}{D} = \frac{r_L + r_{DS} + n^2(1-D)R_F}{D}. \tag{5.51}$$

The equivalent averaged resistance r_D in the diode branch is

$$r_D = r_{L(AV)(D)} + r_{DS(AV)(D)} + R_{F(AV)(D)} \tag{5.52}$$

yielding

$$r_D = \frac{r_L}{n^2(1-D)^2} + \frac{Dr_{DS}}{n^2(1-D)^2} + \frac{R_F}{1-D} = \frac{r_L + Dr_{DS}}{n^2(1-D)^2} + \frac{R_F}{1-D}. \tag{5.53}$$

5.1.4 Design Example

The subsequent analysis is performed on the flyback converter designed for the following specifications:

1. Supply voltage $V_I = 12$ V.
2. Output voltage $V_O = 5$ V.
3. Load resistance $R_L = 20\ \Omega$.
4. Switching frequency $f_s = 100$ kHz.
5. Inductance to ensure CCM operation $L_m = 150\ \mu$H.
6. Filter capacitance $C = 150\ \mu$F.

The turns ratio is $n = 3$. The ESR of the capacitor is $r_C = 10$ mΩ. The ON-state resistance of the MOSFET and the ON-resistance of the diode are $r_{DS} = 0.5\ \Omega$ and $R_F = 0.025\ \Omega$, respectively. The required duty cycle to achieve the rated output voltage is $D = 0.58$. The primary and secondary winding resistances are $r_{T1} = 0.2\ \Omega$ and $r_{T2} = 0.1\ \Omega$, respectively. Using (5.49),

$$r = 0.58 \times (0.2 + 0.5) + 3^2 \times (1 - 0.58) \times (0.1 + 0.025) = 0.8785\ \Omega. \tag{5.54}$$

The diode forward voltage is $V_F = 0.3$ V. The DC input current is $I_I = 0.11$ A, DC output current is $I_O = 0.25$ A, DC inductor current is $I_L = 0.196$ A, DC switch current is $I_S = 0.11$ A, and diode voltage is $-V_D = -5.4$ V. The DC voltage transfer function is $M_{VDC} = 0.45$, and the dc current transfer function is $M_{IDC} = 2.2$. The converter efficiency using (5.22) is $\eta = 92\%$.

5.2 Large-Signal and Small-Signal Models of PWM Flyback Converter in CCM

The DC model is perturbed by small-signal, low-frequency, sinusoidal excitation about the DC operating point to obtain a large-signal nonlinear model.

The large-signal expressions for the switch current and the diode inverted voltage, respectively, are

$$i_S = d_T i_L \tag{5.55}$$

and

$$-v_D = v_{KA} = d_T\left(\frac{v_I}{n} + v_O\right) = d_T\left[\frac{v_I}{n} + \frac{d_T v_I}{n(1-d_T)}\right] = \frac{d_T v_I}{n(1-d_T)}. \tag{5.56}$$

The large-signal non-linear model of a flyback DC–DC converter in CCM is shown in Figure 5.6.

The large-signal quantities are expressed as the sum of DC and AC components as

$$i_S = I_S + i_s, \tag{5.57}$$

$$i_L = I_L + i_l, \tag{5.58}$$

$$v_{KA} = V_{KA} + v_{ka}, \tag{5.59}$$

$$v_I = V_I + v_i, \tag{5.60}$$

$$i_I = I_I + i_i, \tag{5.61}$$

$$v_O = V_O + v_o, \tag{5.62}$$

and

$$d_T = D + d, \tag{5.63}$$

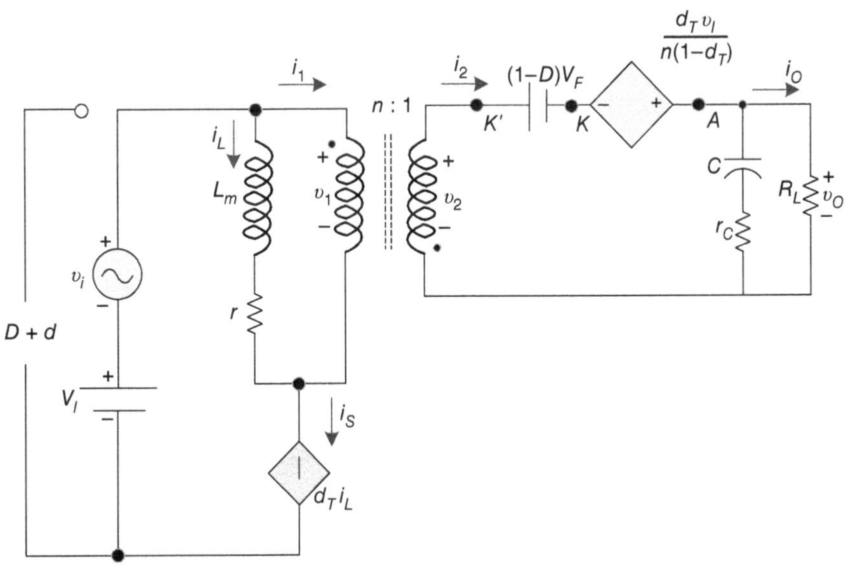

Figure 5.6 Large-signal nonlinear model of the flyback DC–DC converter in CCM.

where i_S, i_L, v_{KA}, v_I, i_I, v_O, and d_T are the large-signal components of switch current, current through magnetizing inductance, diode voltage, input voltage, input current, output voltage, and duty cycle, respectively. The quantities i_s, i_l, v_{ka}, v_i, i_i, v_o, and d are the small-signal switch current, current through magnetizing inductance, diode voltage, input voltage, input current, output voltage, and duty cycle, respectively.

Substituting (5.57), (5.58), and (5.63) into (5.55), the large-signal switch current can be expressed as

$$i_S = I_S + i_s = d_T i_L = (D+d)(I_L + i_l) = DI_L + I_L d + Di_l + i_l d. \tag{5.64}$$

Similar analysis can be performed for the diode inverted voltage. Substituting (5.59), (5.60), and (5.63) into (5.56), the large-signal diode inverted voltage can be expressed as

$$v_{KA} = \frac{(D+d)(V_I + v_i)}{n[1-(D+d)]}. \tag{5.65}$$

Let $1-(D+d) = 1-D-d = D'-d$, and rearranging (5.65) gives

$$\begin{aligned}
v_{KA}[n(D'-d)] &= (D+d)(V_I+v_i)\\
(V_{KA}+v_{ka})[n(D'-d)] &= (D+d)(V_I+v_i)\\
(V_{KA}+v_{ka})nD' - (V_{KA}+v_{ka})nd &= DV_I + Dv_i + V_I d + v_i d\\
(V_{KA}+v_{ka})nD' - V_{KA}nd - v_{ka}nd &= DV_I + Dv_i + V_I d + v_i d\\
(V_{KA}+v_{ka})nD' - v_{ka}nd &= DV_I + Dv_i + V_I d + V_{KA}nd + v_i d\\
(V_{KA}+v_{ka})nD' - v_{ka}nd &= DV_I + Dv_i + V_I d + \left[\frac{D}{n(1-D)}V_I\right]nd + v_i d.
\end{aligned} \tag{5.66}$$

Further simplifying gives

$$(V_{KA}+v_{ka})nD' - v_{ka}nd = DV_I + Dv_i + \frac{V_I d}{1-D} + v_i d. \tag{5.67}$$

The following small-signal conditions can be applied to obtain a linearized large-signal model:

$$V_I \gg v_i,$$

$$I_L \gg i_l,$$

and

$$D \gg d. \tag{5.68}$$

Applying these conditions to (5.64) and (5.67), the small-signal nonlinear products $i_l d$, $v_{ka}nd$, and $v_i d$ are eliminated, yielding a linearized large-signal model. Hence, the large-signal switch current is given as

$$i_S = DI_L + I_L d + Di_l. \tag{5.69}$$

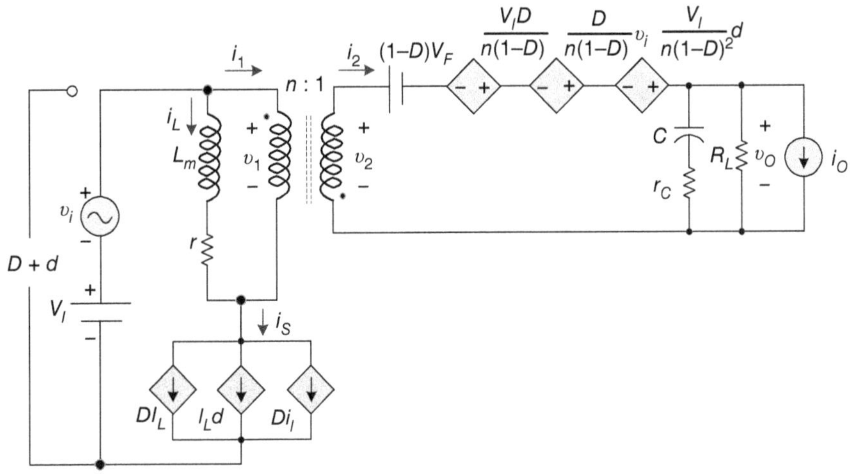

Figure 5.7 Large-signal linear model of the flyback DC–DC converter in CCM.

Similarly, the large-signal diode inverted voltage is given as

$$(V_{KA} + v_{ka})nD' = V_I D + Dv_i + \frac{V_I d}{1-D}$$
$$(V_{KA} + v_{ka}) = \frac{V_I D}{n(1-D)} + \frac{Dv_i}{n(1-D)} + \frac{V_I d}{n(1-D)^2}. \tag{5.70}$$

Using the expression for the large-signal switch current in (5.69), the switch can be represented as a parallel combination of three dependent current sources. Similarly, using the expression for the large-signal diode inverted voltage in (5.70), the diode can be represented as a series combination of three dependent voltage sources. Figure 5.7 shows the large-signal linear model of the flyback DC–DC converter.

Applying the principle of superposition to (5.69) and (5.70), the DC and AC quantities are separated. The DC terms are separated to get switch current as

$$I_S = DI_L \tag{5.71}$$

and diode inverted voltage as

$$V_{KA} = \frac{V_I D}{n(1-D)}. \tag{5.72}$$

Using (5.71) and (5.72), one obtains a DC model of the flyback converter, as shown in Figure 5.4. Extracting AC quantities from (5.69) and (5.70), we get the linear small-signal equations for switch current and diode inverted voltage as

$$i_s = I_L d + Di_l \tag{5.73}$$

and

$$v_{ka} = \frac{Dv_i}{n(1-D)} + \frac{V_I d}{n(1-D)^2}. \tag{5.74}$$

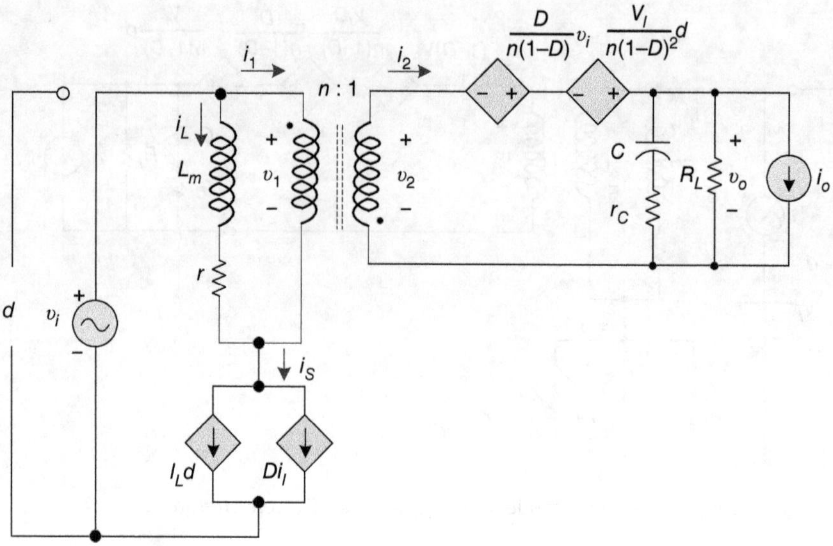

Figure 5.8 Small-signal model of the flyback DC–DC converter in CCM.

Using (5.73) and (5.74), one obtains the linear averaged small-signal model of the flyback converter, as shown in Figure 5.8. The operating point of the flyback converter is determined by V_I, D, n, I_S, and I_L.

5.3 Power-Stage Transfer Functions

The following power-stage transfer functions are discussed in this section:

1. Duty cycle-to-output voltage transfer function T_p (control).
2. Duty cycle-to-inductor current transfer function T_{pi} (control).
3. Input voltage-to-output voltage transfer function M_v (disturbance).
4. Input voltage-to-inductor current transfer function M_{vi} (disturbance).
5. Reverse current gain A_i (disturbance).
6. Input impedance Z_i (disturbance).
7. Output impedance Z_o (disturbance).

The small-signal model of the PWM flyback converter is shown in Figure 5.8. Using this model, the governing equations to derive the transfer functions are derived in this section. The impedances Z_1 and Z_2 are

$$Z_1 = r + sL_m \tag{5.75}$$

and

$$Z_2 = R_L || \left(r_C + \frac{1}{sC} \right) = \frac{R_L \left(r_C + \frac{1}{sC} \right)}{R_L + r_C + \frac{1}{sC}}. \tag{5.76}$$

The current through the parallel combination of the load resistance and the filter capacitor is

$$i_{Z_2} = \frac{v_o}{Z_2}. \tag{5.77}$$

The current through the secondary windings of the transformer is

$$i_2 = i_{Z_2} + i_o = \frac{v_o}{Z_2} + i_o. \tag{5.78}$$

The turns ratio of the transformer is

$$n = \frac{N_1}{N_2} = -\frac{v_1}{v_2} = -\frac{i_2}{i_1}, \tag{5.79}$$

where v_1 and v_2 are the voltages across the primary and secondary windings of the transformer and i_1 and i_2 are the currents through the primary and secondary windings of the transformer, respectively. The input current is

$$i_i = I_L d + D i_l. \tag{5.80}$$

The input current is related to the inductor current and primary winding current as

$$i_i = i_l + i_1 = i_l - \frac{i_2}{n} = i_l - \frac{i_{Z_2}}{n} - \frac{i_o}{n} = i_l - \frac{v_o}{nZ_2} - \frac{i_o}{n}. \tag{5.81}$$

Equating (5.80) and (5.81) yields

$$\begin{aligned}
I_L d + D i_l &= i_l - \frac{v_o}{nZ_2} - \frac{i_o}{n} \\
i_l(1-D) &= I_L d + \frac{v_o}{nZ_2} + \frac{i_o}{n} \\
i_l &= \frac{I_L d}{1-D} + \frac{v_o}{nZ_2(1-D)} + \frac{i_o}{n(1-D)}.
\end{aligned} \tag{5.82}$$

Applying Kirchhoff 's current law,

$$\begin{aligned}
v_o &= \left(\frac{V_I}{n} + V_O\right) d + D\left(\frac{v_i}{n} + v_o\right) + v_2 \\
v_o &= \left(\frac{V_I}{n} + V_O\right) d + \frac{D}{n} v_i + D v_o + v_2 \\
v_o(1-D) &= \left(\frac{V_I}{n} + V_O\right) d + \frac{D}{n} v_i - \frac{v_1}{n}.
\end{aligned} \tag{5.83}$$

The voltage across the primary winding is

$$v_1 = i_l Z_1. \tag{5.84}$$

The voltage across the primary and secondary windings of the transformer are related and expressed as

$$v_2 = -\frac{v_1}{n} = -\frac{i_l Z_1}{n}. \tag{5.85}$$

Substituting (5.85) into (5.83) gives

$$\begin{aligned} v_o(1-D) &= \left(\frac{V_I}{n} + V_O\right) d + \frac{D}{n} v_i - \frac{i_l Z_1}{n} \\ v_o &= \left[\frac{V_I}{n(1-D)} + \frac{V_O}{1-D}\right] d + \frac{Dv_i}{n(1-D)} - \frac{Z_1 i_l}{n(1-D)}. \end{aligned} \tag{5.86}$$

Substituting (5.82) into (5.86) gives

$$\begin{aligned} v_o &= \left[\frac{V_I}{n(1-D)} + \frac{V_O}{1-D}\right] d + \frac{Dv_i}{n(1-D)} \\ &\quad - \frac{Z_1}{n(1-D)} \left[\frac{I_L d}{1-D} + \frac{v_o}{nZ_2(1-D)} + \frac{i_o}{n(1-D)}\right] \\ v_o &= \left[\frac{V_I}{n(1-D)} + \frac{V_O}{1-D}\right] d + \frac{Dv_i}{n(1-D)} \\ &\quad - \frac{I_L Z_1 d}{n(1-D)^2} - \frac{Z_1 v_o}{n^2 Z_2 (1-D)^2} - \frac{Z_1 i_o}{n^2(1-D)^2}. \end{aligned} \tag{5.87}$$

From (5.15), $I_L = V_O/n(1-D)R_L$, and from (5.16), $V_I/n(1-D) = V_O/D$. Therefore,

$$\begin{aligned} \left[1 + \frac{Z_1}{n^2 Z_2 (1-D)^2}\right] v_o &= \left[\frac{V_O}{D} + \frac{V_O}{1-D} - \frac{V_O Z_1}{n^2(1-D)^3 R_L}\right] d + \frac{Dv_i}{n(1-D)} \\ &\quad - \frac{Z_1 i_o}{n^2(1-D)^2}. \end{aligned} \tag{5.88}$$

Solving further gives

$$\begin{aligned} \left[1 + \frac{Z_1}{n^2 Z_2 (1-D)^2}\right] v_o &= \frac{V_O}{D(1-D)} \left[1 - \frac{DZ_1}{n^2(1-D)^2 R_L}\right] d + \frac{Dv_i}{n(1-D)} \\ &\quad - \frac{Z_1 i_o}{n^2(1-D)^2}. \end{aligned} \tag{5.89}$$

Hence,

$$\begin{aligned} v_o &= \frac{V_O}{D(1-D)} \left[\frac{1 - \dfrac{DZ_1}{n^2(1-D)^2 R_L}}{1 + \dfrac{Z_1}{n^2 Z_2 (1-D)^2}}\right] d + \frac{D}{n(1-D)} \left[\frac{1}{1 + \dfrac{Z_1}{n^2 Z_2 (1-D)^2}}\right] v_i \\ &\quad - \left[\frac{\dfrac{Z_1}{n^2(1-D)^2}}{1 + \dfrac{Z_1}{n^2 Z_2 (1-D)^2}}\right] i_o. \end{aligned} \tag{5.90}$$

5.3.1 Duty Cycle-to-Output Voltage Transfer Function T_p

The duty cycle-to-output voltage transfer function is obtained by setting $v_i = 0$ and $i_o = 0$ in Figure 5.8. Substituting $v_i = 0$ into (5.86) gives

$$v_o = \left[\frac{V_I}{n(1-D)} + \frac{V_O}{1-D}\right] d - \frac{Z_1 i_l}{n(1-D)}. \tag{5.91}$$

Substituting (5.82) in (5.91) gives

$$v_o = \left[\frac{V_I}{n(1-D)} + \frac{V_O}{1-D}\right] d - \frac{Z_1}{n(1-D)}\left[\frac{I_L d}{1-D} + \frac{v_o}{nZ_2(1-D)}\right]$$

$$v_o\left[1 + \frac{Z_1}{n^2 Z_2(1-D)^2}\right] = \left[\frac{V_I}{n(1-D)} + \frac{V_O}{1-D} - \frac{Z_1 I_L}{n(1-D)^2}\right] d. \tag{5.92}$$

From (5.15), $I_L = V_O/n(1-D)R_L$, and from (5.16), $V_I/n(1-D) = V_O/D$. Therefore, further solving gives (5.92)

$$\left[1 + \frac{Z_1}{n^2 Z_2(1-D)^2}\right] v_o = \left[\frac{V_O}{D} + \frac{V_O}{1-D} - \frac{Z_1 V_O}{n^2(1-D)^3 R_L}\right] d$$
$$\left[1 + \frac{Z_1}{n^2 Z_2(1-D)^2}\right] v_o = \frac{V_O}{D(1-D)}\left[1 - \frac{DZ_1}{n^2(1-D)^2 R_L}\right] d. \tag{5.93}$$

Hence, the duty cycle-to-output voltage transfer function is

$$T_p(s) = \left.\frac{v_o(s)}{d(s)}\right|_{v_i=i_o=0} = \frac{V_O}{D(1-D)}\left[\frac{1 - \frac{DZ_1}{n^2(1-D)^2 R_L}}{1 + \frac{Z_1}{n^2 Z_2(1-D)^2}}\right]. \tag{5.94}$$

Substituting (5.75) and (5.76) into (5.94) gives the duty cycle-to-output voltage transfer function in the s-domain as

$$T_p(s) = \left.\frac{v_o(s)}{d(s)}\right|_{v_i=i_o=0} = T_{px}\frac{(s+\omega_{zn})(s-\omega_{zp})}{s^2 + 2\xi\omega_o s + \omega_o^2} = T_{po}\frac{\left(1 + \frac{s}{\omega_{zn}}\right)\left(1 - \frac{s}{\omega_{zp}}\right)}{\left(\frac{s}{\omega_o}\right)^2 + \frac{2\xi s}{\omega_o} + 1}, \tag{5.95}$$

where the high-frequency gain is

$$T_{px} = T_p(\infty) = -\frac{V_O}{(1-D)}\frac{r_C}{(R_L + r_C)} \tag{5.96}$$

and the DC gain is

$$T_{po} = T_p(0) = \frac{V_O}{D(1-D)}\frac{n^2(1-D)^2 R_L - rD}{n^2(1-D)^2 R_L + r}. \tag{5.97}$$

The DC gain for the ideal case $r = 0$ is

$$T_{po} = \frac{V_O}{D(1-D)}. \tag{5.98}$$

The angular corner frequency, i.e. the angular undamped natural frequency, is

$$\omega_o = \sqrt{\frac{r + n^2 R_L(1-D)^2}{L_m C(R_L + r_C)}}, \tag{5.99}$$

and the damping ratio is

$$\xi = \frac{L_m + C[r(R_L + r_C) + n^2 R_L r_C (1-D)^2]}{2\sqrt{L_m C (R_L + r_C)[n^2 R_L (1-D)^2 + r]}}. \tag{5.100}$$

The angular frequency of the left-half plane (LHP) zero is

$$\omega_{zn} = \frac{1}{r_C C}, \tag{5.101}$$

and the angular frequency of the right-half plane (RHP) zero is

$$\omega_{zp} = \frac{n^2 R_L (1-D)^2 - Dr}{DL_m} = \frac{n^2 R_L (1-D)^2 - Dr_L - D^2 r_{DS} - n^2 D(1-D) R_F}{DL_m}. \tag{5.102}$$

For the design specification given in Section 5.1.4, the gain at DC and low frequencies is nearly $T_{po} = 19.6$ V/V $= 25.8$ dB. The natural corner frequency of the second-order low-pass filter formed by the inductor, capacitor, and load resistor is $f_o = 1.37$ kHz. The damping coefficient is $\xi = 0.37$. The gain decreases by 40 dB/dec beyond f_o.

The roots of the characteristic equation of the transfer function T_p are the poles of the converter power stage. Equating the denominator in (5.95) to zero results in the poles of the transfer function

$$p_1, p_2 = \sigma \pm j\omega_d = -\xi\omega_o \pm j\omega_o\sqrt{1-\xi^2} = \omega_o(-\xi \pm j\sqrt{1-\xi^2}). \tag{5.103}$$

The poles are complex conjugate and are located on the left-half of the s-plane. The real part of the poles is dependent on the damping coefficient. At $\omega_o = 8.6$ krad/s and $\xi = 0.37$, the poles are located at

$$p_1, p_2 = -3154.4 \pm j7991.7 \text{ rad/s}. \tag{5.104}$$

The RHP zero is located at

$$z_p = \omega_{zp} = 368.6 \times 10^3 \text{ rad/s}. \tag{5.105}$$

The LHP zero is at

$$z_n = -\omega_{zn} = -666.7 \times 10^3 \text{ rad/s}. \tag{5.106}$$

Figure 5.9 shows the theoretically obtained magnitude and phase plots of duty cycle-to-output voltage transfer function. The duty cycle-to-output voltage transfer function of the flyback converter exhibits an RHP zero located at $f_{zp} = \omega_{zp}/2\pi = 58.7$ kHz. The RHP zero is a zero for magnitude but behaves like a pole for phase. The RHP zero cancels the effect of one of the complex-conjugate poles reducing the fall-off slope to −20 dB/dec. Beyond f_{zp}, the gain decreases by −20 dB/dec and is finally compensated by the LHP zero $f_{zn} = \omega_{zn}/2\pi = 106$ kHz created by the capacitor C and its ESR r_C. The gain at high frequencies is constant and equal to −45 dB.

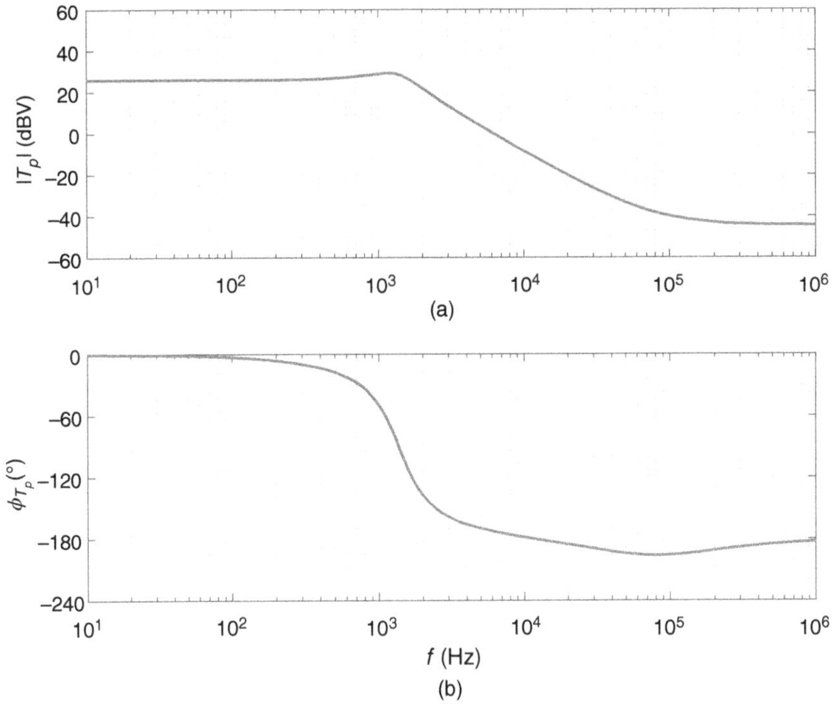

Figure 5.9 Theoretically obtained plots of duty cycle-to-output voltage transfer function T_p. (a) Magnitude. (b) Phase.

Figure 5.10 shows the locations of the power-stage poles p_1, p_2, LHP zero z_n, and RHP zero z_p at different values of turns ratio of the transformer varying from 1 to 10. The input voltage V_I, duty cycle D, and load resistance R_L are constant. The output voltage V_O, output current I_O, and inductor current I_L vary for different values of turns ratio. The location of the LHP zero z_n caused by the filter capacitor C and its ESR r_C is independent of the turns ratio. The complex-conjugate poles move away from the origin as turns ratio increases. The RHP zero z_p moves toward ∞ with the increase in turns ratio. Figure 5.11 shows the Bode magnitude and phase plots of the T_p transfer function at selected values of turns ratio of the transformer. The DC gain decreases with the increase in turns ratio. At low turns ratio values, the phase plot crosses $-180°$.

Figure 5.12 shows the locations of the power-stage poles p_1, p_2, LHP zero z_n, and RHP zero z_p at different values of the load resistance varying from 5 to 50 Ω. The input voltage V_I, output voltage V_O, turns ratio n, and the duty cycle D are constant. The output current I_O, inductor current I_L, and input current I_I vary for different values of load resistance. The relation between output current I_O, inductor current I_L, and input current I_I is given in (5.11). The location of the LHP zero z_n caused by the filter capacitor C and its ESR r_C is independent of the load resistance. It is

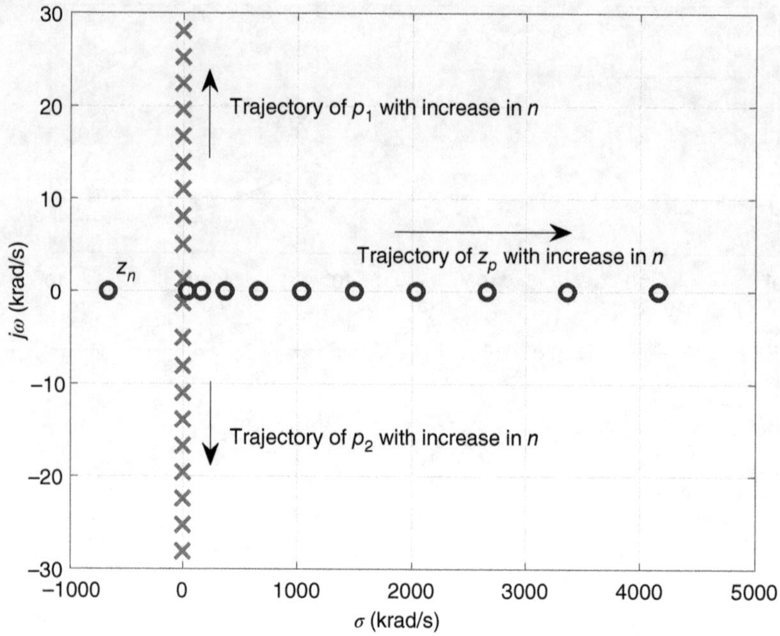

Figure 5.10 Trajectory of the poles and zeros of T_p transfer function as a function of turns ratio of the transformer. The input voltage V_I, duty cycle D, and load resistance R_L are constant. The output voltage V_O, output current I_O, and inductor current I_L vary for different values of turns ratio.

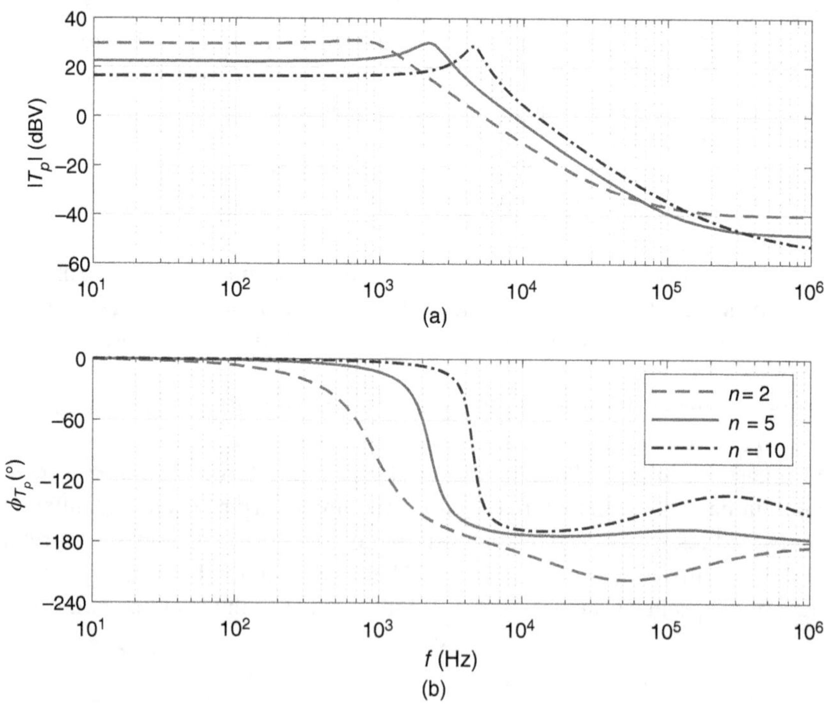

Figure 5.11 Theoretically obtained plots of duty cycle-to-output voltage transfer function T_p at selected transformer turns ratio. (a) Magnitude. (b) Phase.

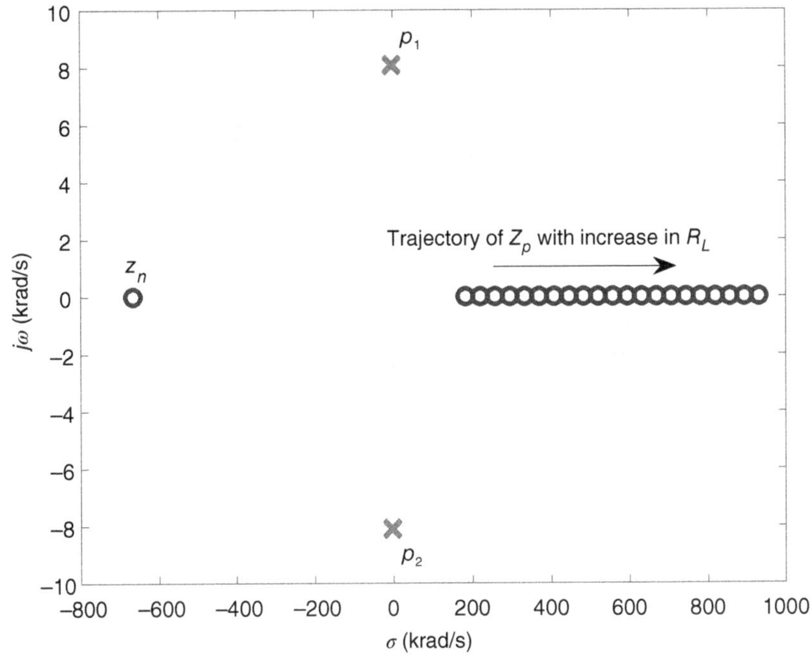

Figure 5.12 Trajectory of the poles and zeros of T_p transfer function as a function of the load resistance. The input voltage V_I, output voltage V_O, turns ratio n, and the duty cycle D are constant. The output current I_O, inductor current I_L, and input current I_I vary for different values of load resistance.

observed that the complex-conjugate poles also do not relocate with change in the load resistance. The RHP zero moves toward ∞ with the increase in the load resistance. This means that the effect of the RHP zero is not significant at high load resistance values. Figure 5.13 shows the Bode magnitude and phase plots of the T_p transfer function at selected values of the load resistance. The DC gain remains unchanged with increase in the load resistance. At low load resistance, the RHP zero is closer to the origin, and the phase plot crosses $-180°$.

Figure 5.14 shows the trajectories of the power-stage poles p_1, p_2, the LHP zero z_n, and the RHP zero z_p at different values of the duty cycle varying from 0.2 to 0.97. The input voltage V_I, turns ratio n, and load resistance R_L are constant. The output voltage V_O and output current I_O vary for different values of duty cycle. The location of the LHP zero z_n caused by the filter capacitor C and its ESR r_C is independent of the duty cycle. The imaginary components of the poles p_1 and p_2 move toward the origin with the increase in the duty cycle. The RHP zero z_p also moves toward the origin with the increases in duty cycle. At a duty cycle larger than a critical value, the poles become real with p_1 moving toward the origin and p_2 moving toward $-\infty$. An interesting phenomena occurs with the RHP zero where, as the duty cycle increases, the RHP zero moves toward the origin. At a critical duty cycle, the RHP zero is at the origin. Hence, the expression for the critical duty cycle can be derived by setting

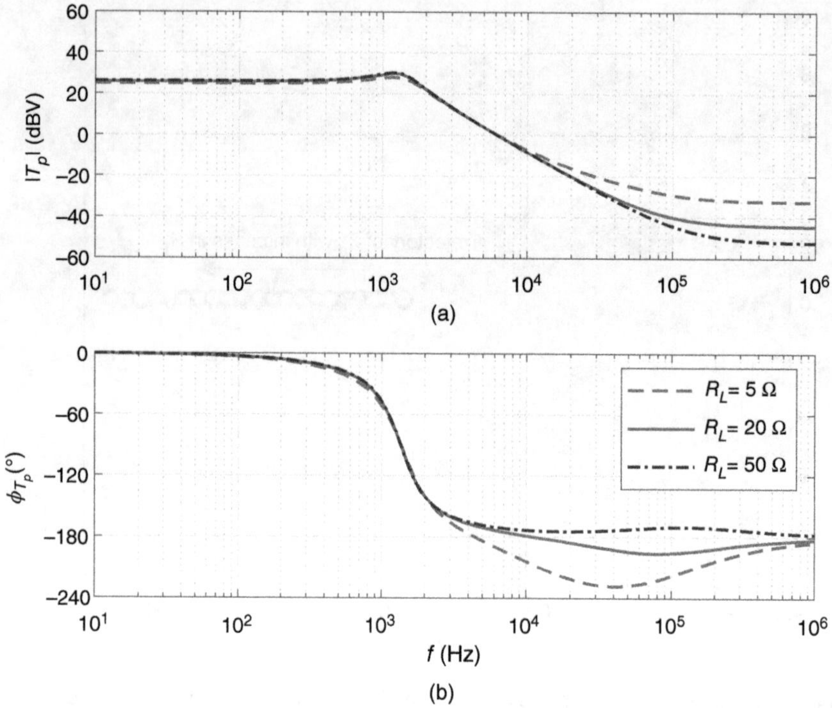

Figure 5.13 Theoretically obtained plots of duty cycle-to-output voltage transfer function T_p at selected load resistances. (a) Magnitude. (b) Phase.

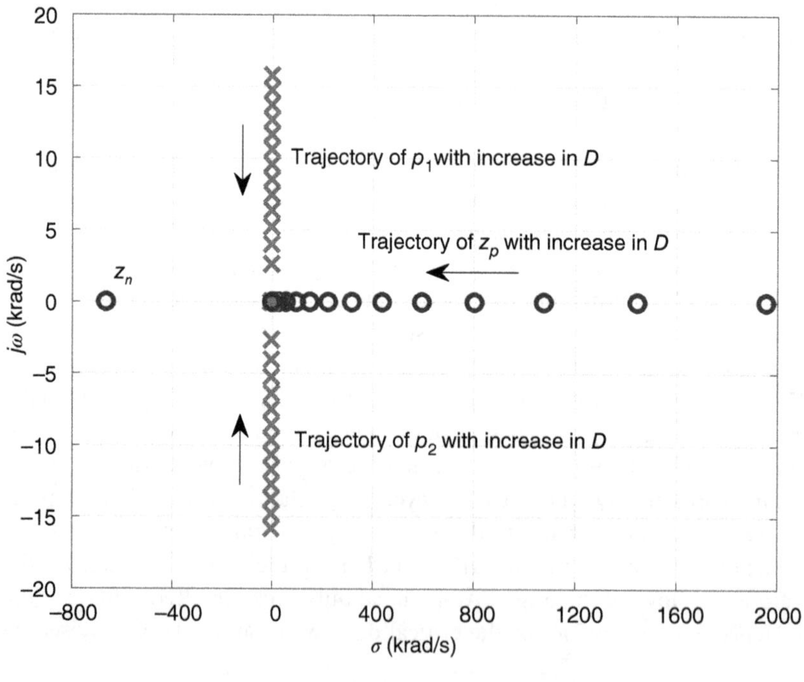

Figure 5.14 Trajectory of the poles and zeros of the T_p transfer function as a function of the duty cycle. The input voltage V_I, turns ratio n, and load resistance R_L are constant. The output voltage V_O and output current I_O vary for different values of duty cycle.

$\omega_{zp} = 0$. The critical duty cycle is expressed and calculated as

$$D_{Cr} = \frac{(2n^2R_L + r_L + n^2R_F) - \sqrt{n^2[4R_L(r_L + r_{DS}) + R_F(n^2R_F + 2r_L)] + r_L^2}}{2(n^2R_L - r_{DS} + n^2R_F)}$$
$$= 0.929. \tag{5.107}$$

Beyond the critical duty cycle D_{Cr}, by the influence of losses in the converter, the RHP zero enters the left-half of the s-plane. Figure 5.15 shows the root locus of the poles p_1 and p_2 and zero z_p of T_p closer to the origin to illustrate the behavior and the relocation beyond the critical duty cycle value. The stability of the flyback converter can also be determined using the Bode plot of the T_p transfer function analyzed at different duty cycles. Figure 5.16 shows the Bode magnitude and phase plot of the T_p transfer function at selected values of the duty cycle. As the duty cycle increases, the DC gain T_{po} increases; however, the corner frequency f_o decreases, resulting in lower bandwidth. The phase is 0° at DC and low frequencies. As duty cycle increases or decreases, the output voltage decreases or increases, respectively. With increase in frequency, the phase rolls off by −180° per decade due to the complex-conjugate poles. In addition, the presence of the RHP zero adds an additional phase of −45° per decade at f_{zp}. At high frequencies, the effect on poles (or RHP zero) is nullified by the LHP zero, and phase begins to roll up toward −180°.

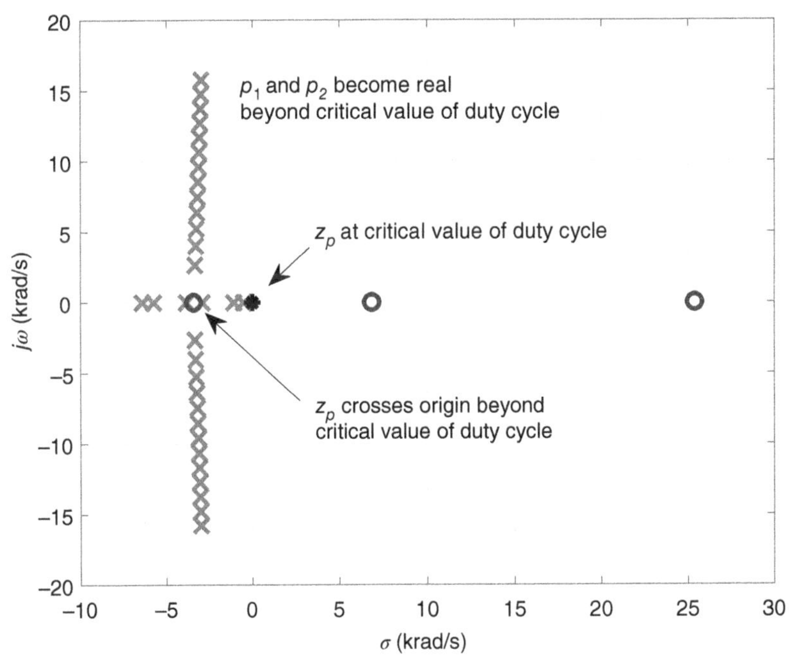

Figure 5.15 The poles and zeros of T_p transfer function beyond the critical value of the duty cycle.

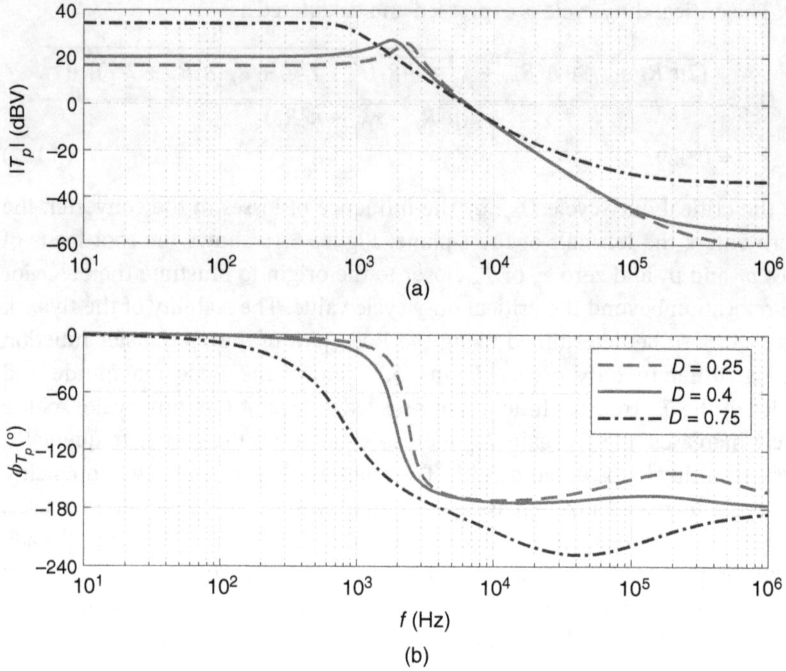

Figure 5.16 Theoretically obtained plots of duty cycle-to-output voltage transfer function T_p at selected duty ratios. (a) Magnitude. (b) Phase.

5.3.2 Duty Cycle-to-Inductor Current Transfer Function T_{pi}

The duty cycle-to-output voltage transfer function is obtained by setting $v_i = 0$ and $i_o = 0$ in Figure 5.8. By setting $v_i = 0$ in (5.86) and rearranging, we get

$$v_o = \left[\frac{V_I}{n(1-D)} + \frac{V_O}{1-D}\right] d - \frac{Z_1 i_l}{n(1-D)}. \tag{5.108}$$

Substituting (5.108) into (5.82) gives

$$\begin{aligned}
i_l &= \frac{I_L d}{1-D} + \frac{\left[\dfrac{V_I}{n(1-D)} + \dfrac{V_O}{1-D}\right] d - \dfrac{i_l Z_1}{n(1-D)}}{n Z_2 (1-D)} \\
i_l &= \frac{I_L d}{1-D} + \frac{1}{n Z_2 (1-D)} \\
&\quad \times \left\{\left[\frac{V_I}{n(1-D)} + \frac{V_O}{1-D}\right] d - \frac{Z_1 i_l}{n(1-D)}\right\} \\
i_l\left[1 + \frac{Z_1}{n^2(1-D)^2 Z_2}\right] &= \frac{I_L d}{1-D} + \frac{1}{n Z_2 (1-D)}\left[\frac{V_I d}{n(1-D)} + \frac{V_O d}{1-D}\right] \\
i_l\left[1 + \frac{Z_1}{n^2(1-D)^2 Z_2}\right] &= d\left[\frac{V_O}{n Z_2 D (1-D)^2} + \frac{I_L}{1-D}\right].
\end{aligned} \tag{5.109}$$

Hence, the duty cycle-to-inductor current transfer function is

$$T_{pi}(s) = \left.\frac{i_l(s)}{d(s)}\right|_{v_i=i_o=0} = \frac{\dfrac{V_O}{nZ_2D(1-D)^2} + \dfrac{I_L}{1-D}}{1 + \dfrac{Z_1}{n^2(1-D)^2Z_2}} = \frac{nV_O}{DR_L}\frac{R_L + DZ_2}{Z_1 + n^2(1-D)^2Z_2}. \tag{5.110}$$

Substituting (5.75) and (5.76) into (5.110) gives the duty cycle-to-inductor current transfer function in the s-domain as

$$T_{pi}(s) = T_{pix}\frac{s + \omega_{zi}}{s^2 + 2\xi\omega_o s + \omega_o^2} = T_{pio}\frac{1 + \dfrac{s}{\omega_{zi}}}{\left(\dfrac{s}{\omega_o}\right)^2 + \dfrac{2\xi s}{\omega_o} + 1}, \tag{5.111}$$

where

$$T_{pix} = \frac{nV_O}{D}\frac{r_C(1+D) + R_L}{L_m(R_L + r_C)} \tag{5.112}$$

and the DC gain is

$$T_{pio} = T_{pi}(0) = \frac{nV_O}{D}\frac{1+D}{n^2R_L(1-D)^2 + r}. \tag{5.113}$$

The DC gain for the ideal case $r = 0$ is

$$T_{pio} = \frac{V_O(1+D)}{nR_LD(1-D)^2}. \tag{5.114}$$

The angular frequency of the LHP zero is

$$\omega_{zi} = \frac{1+D}{C[r_C(1+D) + R_L]}. \tag{5.115}$$

The angular corner frequency ω_o and the damping ratio ξ were derived in (5.99) and (5.100), respectively.

For the design specification given in Section 5.1.4, the gain at DC and low frequencies is nearly $T_{pio} = 1.24$ A $= 1.87$ dBA. The natural corner frequency of the second-order low-pass filter formed by the inductor, capacitor, and load resistor is $f_o = 1.37$ kHz. Figure 5.17 shows the theoretically obtained magnitude and phase plots of the duty cycle-to-inductor current transfer function. The transfer function exhibits an LHP zero $f_{zi} = \omega_{zi}/2\pi = 84$ Hz and a pair of complex-conjugate poles at f_o. The frequency of the LHP zero f_{zi} is inversely proportional to the filter capacitance and the load resistance. Beyond f_o the gain decreases at the rate of 20 dB/dec. The phase starts at zero and rises to 45° due to f_{zi}. This means that at low frequencies, the duty cycle signal leads the inductor current. The bandwidth of T_{pi} is larger than that of T_p. Therefore, the inner-current loop presents a faster response in inductor current for changes in duty cycle.

Figure 5.18 shows the locations of the power-stage poles p_1, p_2 and LHP zero z_i at different values of turns ratio of the transformer varying from 1 to 10. The input voltage V_I, duty cycle D, and load resistance R_L are constant. The output voltage V_O,

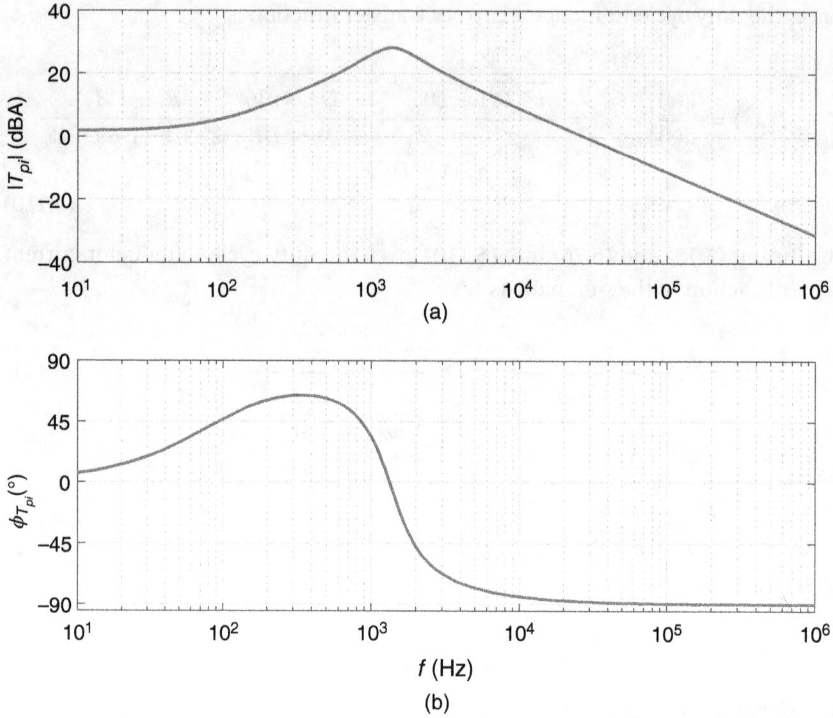

Figure 5.17 Theoretically obtained plots of duty cycle-to-inductor current transfer function T_{pi}. (a) Magnitude. (b) Phase.

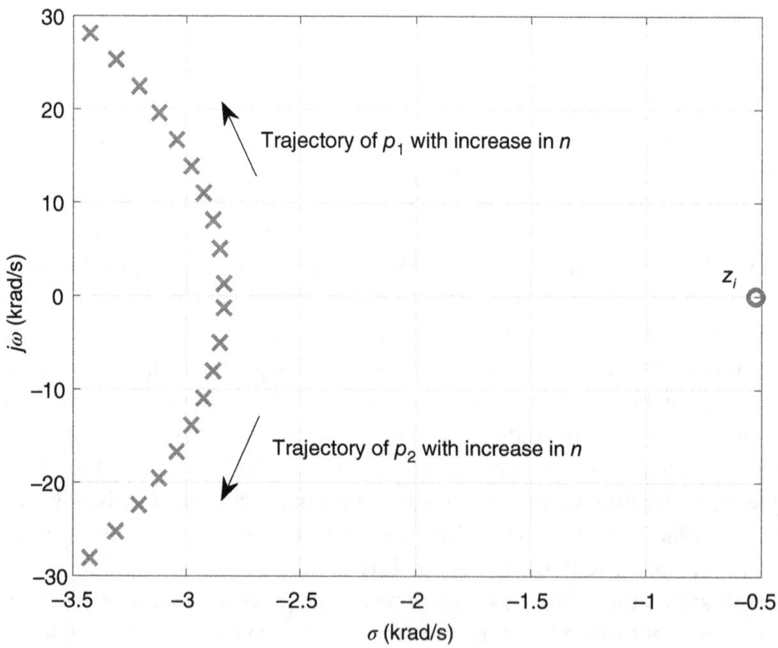

Figure 5.18 Trajectory of the poles and zeros of T_{pi} transfer function as a function of turns ratio of the transformer. The input voltage V_I, duty cycle D, and load resistance R_L are constant. The output voltage V_O, output current I_O, and inductor current I_L vary for different values of turns ratio.

output current I_O, and inductor current I_L vary for different values of turns ratio. The complex-conjugate poles move away from the origin as turns ratio increases. The complex-conjugate poles of T_{pi} transfer function are same as complex-conjugate poles of T_p transfer function. The position of LHP zero z_i remains unchanged with the increase in turns ratio. Figure 5.19 shows the Bode magnitude and phase plots of the T_p transfer function at selected values of turns ratio of the transformer. The DC gain is lower for higher value of turns ratio. As turns ratio values increase, a larger phase boost is observed at frequencies ranging between 100 Hz and 10 kHz.

Figure 5.20 shows the trajectories of the power-stage poles p_1, p_2, and the LHP zero z_i at different values of the load resistance varying from 5 Ω to 50 Ω. The input voltage V_I, output voltage V_O, turns ratio n, and the duty cycle D are constant. The output current I_O, inductor current I_L, and input current I_I vary for different values of load resistance. The relation between output current I_O, inductor current I_L, and input current I_I is given in (5.11). It is observed that the real component of the complex-conjugate poles move toward the origin with the increase in the load resistance. The LHP zero also moves toward the origin with the increase in the load resistance. Figure 5.21 shows the Bode magnitude and phase plots of the T_{pi} transfer function at selected values of the load resistance. The DC gain decreases with

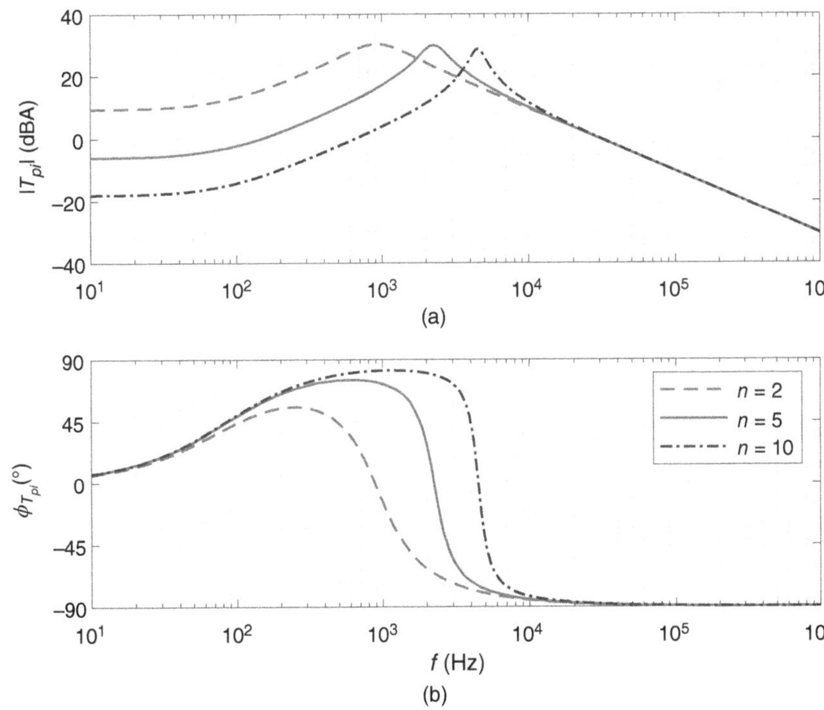

Figure 5.19 Theoretically obtained plots of duty cycle-to-inductor current transfer function T_{pi} at selected transformer turns ratio. (a) Magnitude. (b) Phase.

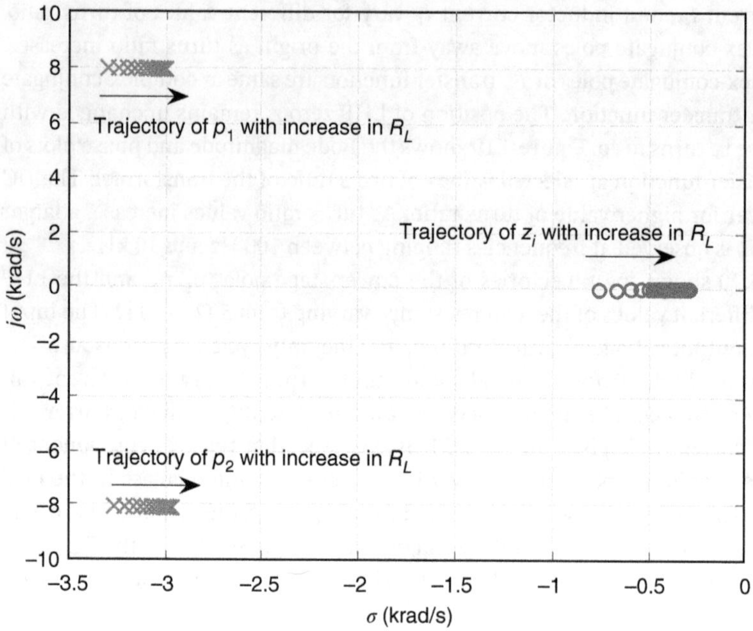

Figure 5.20 Trajectory of the poles and zeros of the T_{pi} transfer function as a function of the load resistance. The input voltage V_I, output voltage V_O, turns ratio n, and the duty cycle D are constant. The output current I_O, inductor current I_L, and input current I_I vary for different values of load resistance.

the increase in the load resistance. As the load resistance increases, the frequency of the LHP zero decreases. As a result, a large phase boost is observed at lower frequencies for higher values of R_L. The phase plot reaches −90° for all values of the load resistance.

Figure 5.22 shows the trajectories of the power-stage poles p_1, p_2, and the LHP zero z_i at different values of the duty cycle varying from 0.2 to 0.97. The input voltage V_I, turns ratio n, and load resistance R_L are constant. The output voltage V_O and output current I_O vary for different values of duty cycle. The imaginary components of the poles p_1 and p_2 move toward the origin with the increase in the duty cycle. However, the frequency of the LHP zero z_i remains fairly constant. Figure 5.23 shows the Bode magnitude and phase plot of the T_{pi} transfer function at selected values of the duty cycle. As the duty cycle increases, the DC gain T_{pio} also increases, and the corner frequency f_o decreases. The phase starts at 0° and saturates at −90° at high frequencies for all values of duty cycle. As the duty cycle increases, the phase boost decreases. Since the phase at high frequencies is −90°, the phase margin of the inner-current loop is nearly fixed for all the duty cycles. This offers two advantages: (i) a wide range of duty cycle can be achieved for current control, and (ii) simpler controller techniques can be used.

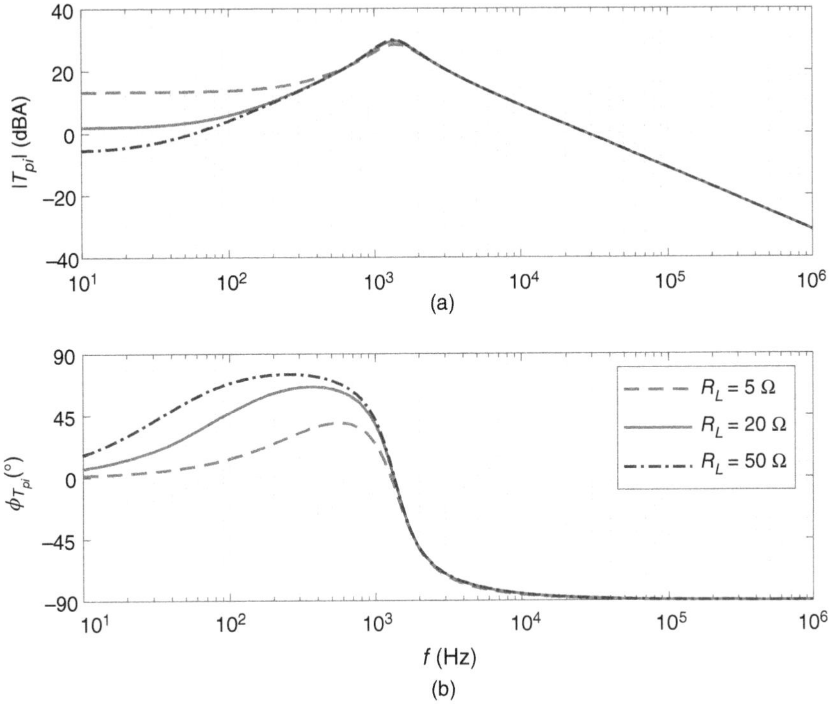

Figure 5.21 Theoretically obtained plots of duty cycle-to-inductor current transfer function T_{pi} at selected load resistances. (a) Magnitude. (b) Phase.

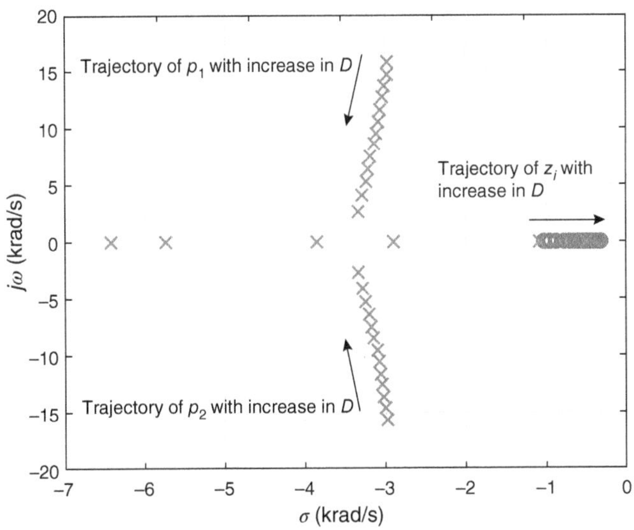

Figure 5.22 Trajectory of the poles and zeros of the T_{pi} transfer function as a function of the duty cycle. The input voltage V_I, turns ratio n, and load resistance R_L are constant. The output voltage V_O and output current I_O vary for different values of duty cycle.

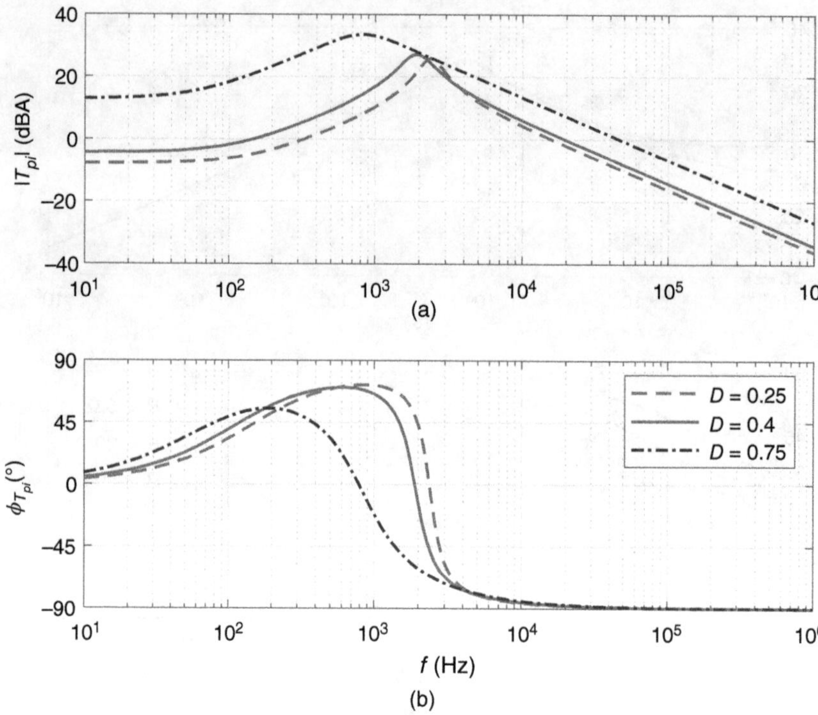

Figure 5.23 Theoretically obtained plots of duty cycle-to-inductor current transfer function T_{pi} at selected duty ratios. (a) Magnitude. (b) Phase.

5.3.3 Input Voltage-to-Output Voltage Transfer Function M_v

The input voltage-to-output voltage transfer function is obtained by setting $d = 0$ and $i_o = 0$ in Figure 5.8. By setting $d = 0$ in (5.86) and (5.90) and rearranging, we get

$$v_o = \frac{Dv_i}{n(1-D)} - \frac{i_l Z_1}{n(1-D)}. \tag{5.116}$$

Rearranging further gives

$$i_l = \frac{v_o}{nZ_2(1-D)}. \tag{5.117}$$

Substituting (5.117) into (5.116) gives

$$v_o = \frac{Dv_i}{n(1-D)} - \frac{Z_1 v_o}{n^2(1-D)^2 Z_2} \tag{5.118}$$

and

$$v_o\left[1 + \frac{Z_1}{n^2 Z_2 (1-D)^2}\right] = \frac{Dv_i}{n(1-D)}. \tag{5.119}$$

Hence, the input voltage-to-output voltage transfer function M_v is

$$M_v(s) = \left.\frac{v_o(s)}{v_i(s)}\right|_{d=i_o=0} = \frac{\dfrac{D}{n(1-D)}}{1 + \dfrac{Z_1}{n^2 Z_2 (1-D)^2}} = \frac{nD(1-D)Z_2}{Z_1 + n^2 Z_2 (1-D)^2}. \tag{5.120}$$

Substituting (5.75) and (5.76) into (5.120) gives the input voltage-to-output voltage transfer function in the s-domain as

$$M_v(s) = \left.\frac{v_o(s)}{v_i(s)}\right|_{d=i_o=0} = M_{vx}\frac{s+\omega_{zn}}{s^2+2\xi\omega_o s+\omega_o^2} = M_{vo}\frac{1+\dfrac{s}{\omega_{zn}}}{\left(\dfrac{s}{\omega_o}\right)^2+\dfrac{2\xi s}{\omega_o}+1}, \tag{5.121}$$

where the high-frequency gain is

$$M_{vx} = M_v(\infty) = \frac{nD(1-D)R_L r_C}{L_m(R_L+r_C)} \tag{5.122}$$

and the DC gain is

$$M_{vo} = M_v(0) = \frac{nD(1-D)R_L}{r+n^2R_L(1-D)^2}. \tag{5.123}$$

For the ideal case $r = 0$, it yields

$$M_{vo} = M_{VDC} = \frac{D}{n(1-D)}. \tag{5.124}$$

The angular frequency of the LHP zero ω_{zn} is given in (5.101).

Figure 5.24 shows the theoretically obtained magnitude and phase plots of the input voltage-to-output voltage transfer function. The gain at DC is nearly equal to $M_{vo} = 0.44$ V/V $= -7.11$ dB. The gain at DC is mainly dependent on the duty cycle and turns ratio. The gain at DC is below 0 dB for the buck mode and is above 0 dB for the boost mode. In the flyback converter, any change in the input voltage causes a change in the output voltage during the switch ON interval. The phase starts at zero and decreases toward −180°; however, the high-frequency LHP zero at ω_z increases the phase to −90°.

5.3.4 Input Voltage-to-Inductor Current Transfer Function M_{vi}

The input voltage-to-inductor current transfer function is obtained by setting $d = 0$ and $i_o = 0$ in Figure 5.8. By setting $d = 0$ in (5.86) and (5.90) and rearranging, we get

$$v_o = \frac{Dv_i}{n(1-D)} - \frac{i_l Z_1}{n(1-D)} \tag{5.125}$$

and

$$i_l = \frac{v_o}{nZ_2(1-D)} \tag{5.126}$$

$$v_o = nZ_2(1-D)i_l.$$

Substituting (5.126) into (5.125) gives

$$nZ_2(1-D)i_l = \frac{Dv_i}{n(1-D)} - \frac{i_l Z_1}{n(1-D)}$$

$$i_l = \frac{Dv_i}{n^2Z_2(1-D)^2} - \frac{i_l Z_1}{n^2Z_2(1-D)^2} \tag{5.127}$$

$$i_l\left[1+\frac{Z_1}{n^2Z_2(1-D)^2}\right] = \frac{Dv_i}{n^2Z_2(1-D)^2}.$$

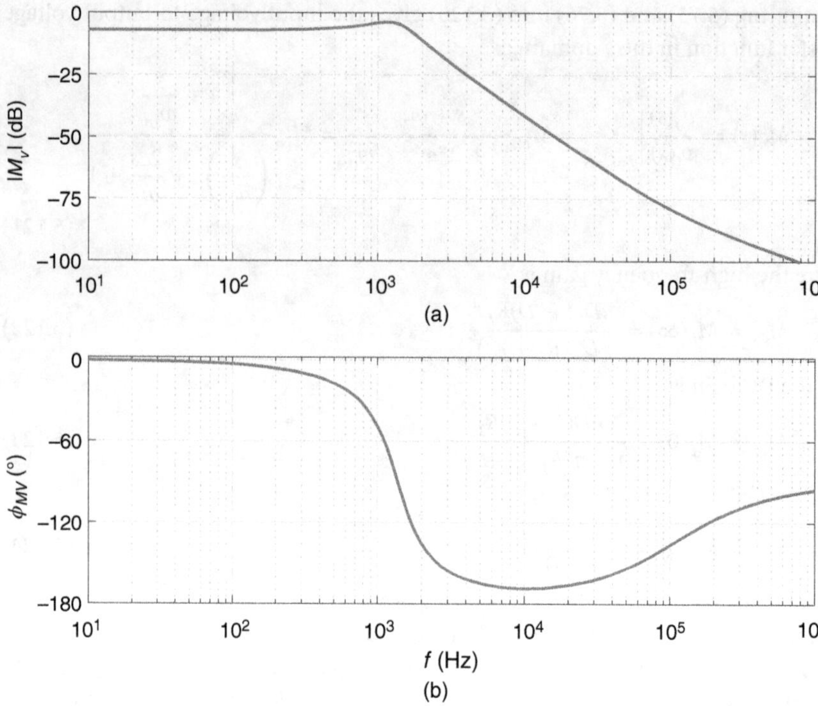

Figure 5.24 Theoretically obtained plots of input-to-output voltage transfer function M_v. (a) Magnitude. (b) Phase.

Hence, the input voltage-to-inductor current transfer function M_{vi} is

$$M_{vi}(s) = \left.\frac{i_l(s)}{v_i(s)}\right|_{d=i_o=0} = \frac{\dfrac{D}{n^2 Z_2 (1-D)^2}}{1 + \dfrac{Z_1}{n^2 Z_2 (1-D)^2}} = \frac{D}{Z_1 + n^2 Z_2 (1-D)^2}. \tag{5.128}$$

Substituting (5.75) and (5.76) into (5.128) gives the input voltage-to-inductor current transfer function in the s-domain as

$$M_{vi}(s) = M_{vix} \frac{s + \omega_{zvi}}{s^2 + 2\xi\omega_o s + \omega_o^2} = M_{vio} \frac{1 + \dfrac{s}{\omega_{zvi}}}{\left(\dfrac{s}{\omega_o}\right)^2 + \dfrac{2\xi s}{\omega_o} + 1}, \tag{5.129}$$

where M_{vix} is

$$M_{vix} = \frac{D}{L_m} \tag{5.130}$$

and the DC gain M_{vio} is

$$M_{vio} = M_{vi}(0) = \frac{D}{n^2 R_L (1-D)^2 + r}. \tag{5.131}$$

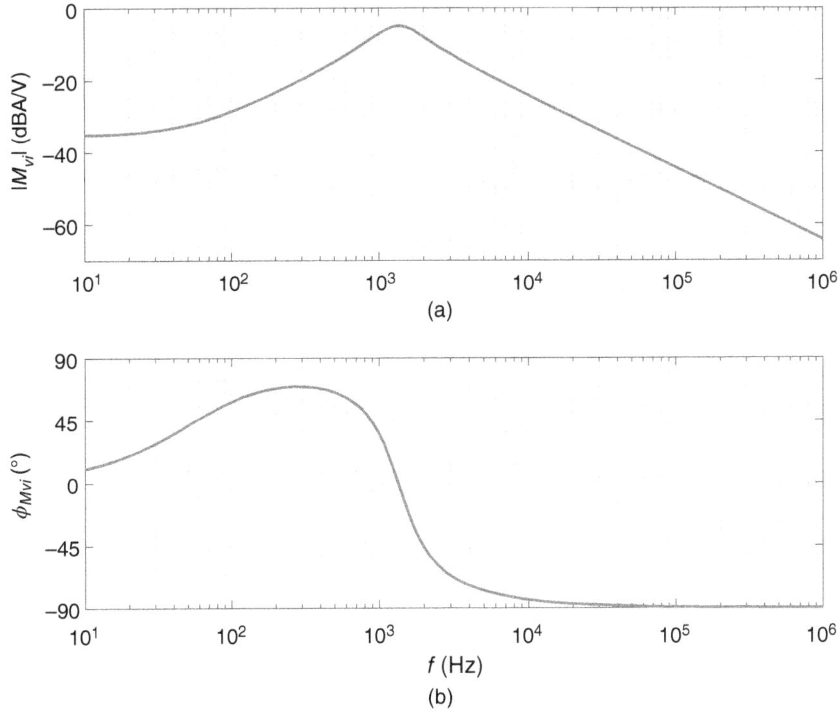

Figure 5.25 Theoretically obtained plots of input voltage-to-inductor current transfer function M_{vi}. (a) Magnitude. (b) Phase.

The DC gain M_{vio} for ideal case $r = 0$ is

$$M_{vio} = \frac{D}{n^2 R_L (1-D)^2}. \tag{5.132}$$

The angular frequency of the LHP zero is

$$\omega_{zvi} = \frac{1}{C(R_L + r_C)}. \tag{5.133}$$

The angular corner frequency and the damping ratio are given in (5.99) and (5.100), respectively. Figure 5.25 shows the theoretically obtained magnitude and phase plots of the input voltage-to-inductor current transfer function. The gain at DC is $M_{vio} = 0.02$ A/V $= -33.9$ dB. The frequency of the LHP zero is $f_{zvi} = \omega_{zvi}/(2\pi) = 53$ Hz. Through the M_{vi} current-loop relevant transfer function, one can observe that the current loop offers a lower audio susceptibility at DC than that provided by the voltage-loop transfer function M_v. All the other frequency-domain properties remain identical to T_{pi}.

5.3.5 Reverse Current Gain A_i

The small-signal model to derive the output-to-inductor current transfer function is shown in Figure 5.26. This model is obtained by setting $v_i = 0$ and $d = 0$. Substituting

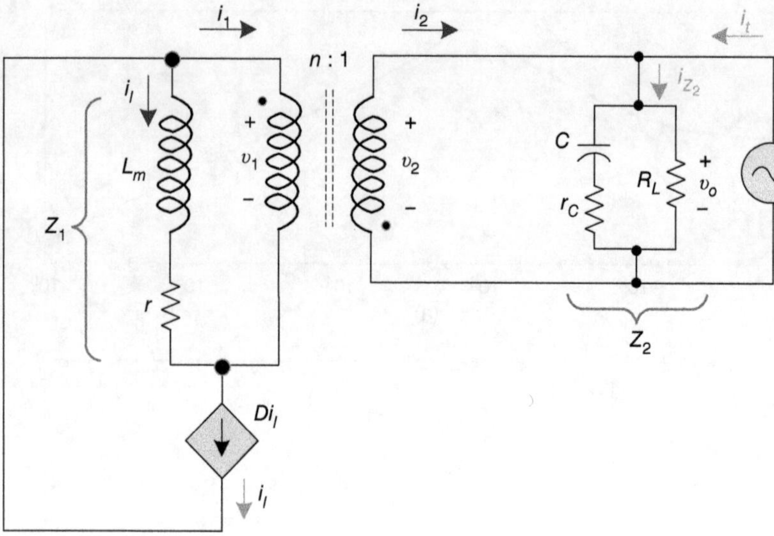

Figure 5.26 Small-signal model of the PWM flyback DC–DC converter in CCM to derive the output current-to-inductor current transfer function A_i.

$d = 0$ into (5.80) gives

$$i_i = Di_l. \tag{5.134}$$

An independent voltage source v_t is applied at the output, which forces a current i_t. By applying Kirchhoff's current law and rearranging, we get

$$i_t = i_{Z_2} - i_2 = \frac{v_o}{Z_2} + ni_1 = \frac{v_o}{Z_2} + n(i_i - i_l). \tag{5.135}$$

Substituting (5.134) into (5.135) gives

$$i_t = \frac{v_o}{Z_2} + n(Di_l - i_l) = \frac{v_o}{Z_2} - ni_l(1 - D). \tag{5.136}$$

Setting $d = 0$ and $v_i = 0$ in (5.86) gives

$$v_o = -\frac{i_l Z_1}{n(1 - D)}. \tag{5.137}$$

Substituting (5.137) into (5.136) gives

$$i_t = -\frac{i_l Z_1}{n(1 - D)Z_2} - ni_l(1 - D) = -\left[\frac{Z_1}{n(1 - D)Z_2} - n(1 - D)\right] i_l. \tag{5.138}$$

Hence, the output current-to-inductor current transfer function, i.e. reverse current transfer function A_i, is

$$A_i(s) = \left.\frac{i_l(s)}{i_t(s)}\right|_{d=v_i=0} = -\frac{n(1 - D)Z_2}{Z_1 + n^2(1 - D)^2 Z_2}. \tag{5.139}$$

Substituting (5.75) and (5.76) into (5.139) gives the output current-to-inductor current transfer function in the s-domain as

$$A_i(s) = \left.\frac{i_l(s)}{i_t(s)}\right|_{d=v_i=0} = A_{ix}\frac{s+\omega_{zn}}{s^2+2\xi\omega_o s+\omega_o^2} = A_{io}\frac{1+\dfrac{s}{\omega_{zn}}}{\left(\dfrac{s}{\omega_o}\right)^2+\dfrac{2\xi s}{\omega_o}+1}, \quad (5.140)$$

where

$$A_{ix} = -\frac{nR_L r_C(1-D)}{L_m(R_L+r_C)}, \quad (5.141)$$

and the DC gain A_{io} is

$$A_{io} = A_i(0) = \frac{-nR_L(1-D)}{n^2R_L(1-D)^2+r}. \quad (5.142)$$

The DC gain A_{io} for ideal case $r = 0$ is

$$A_{io} = \frac{-1}{n(1-D)}. \quad (5.143)$$

The angular frequency of the LHP zero ω_{zn} is given in (5.101).

Figure 5.27 shows the theoretically obtained magnitude and phase plots of the output current-to-inductor current transfer function. The gain at DC is

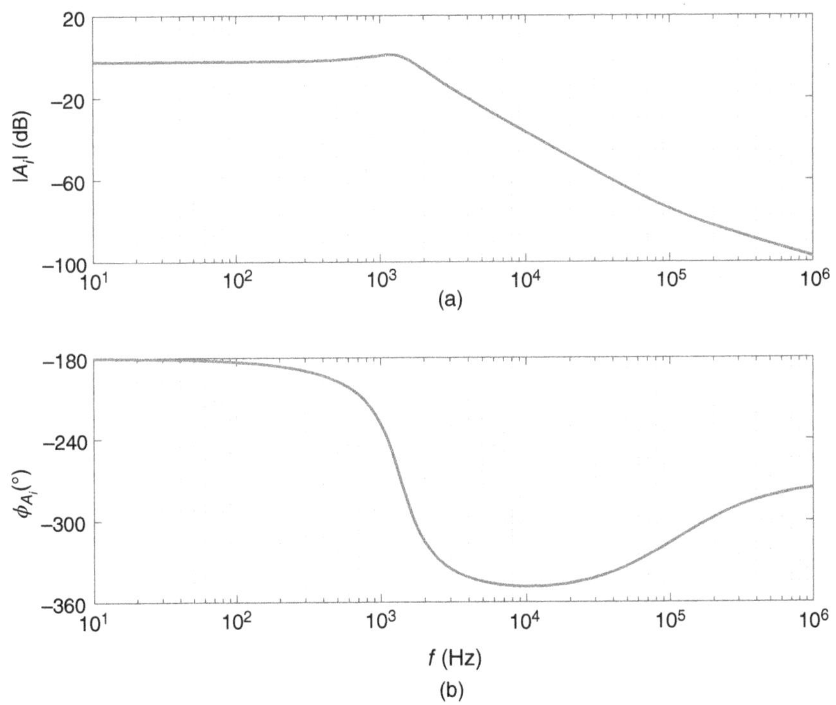

Figure 5.27 Theoretically obtained plots of output current-to-inductor current transfer function A_i. (a) Magnitude. (b) Phase.

$A_{io} = 0.77$ A/A $= -2.27$ dB, i.e. the reverse current gain is unity at DC and low-frequencies up to nearly f_o. Beyond f_o, the impedance offered by the capacitor branch is lower than the inductor, and the small-signal output current begins to flow into the capacitor. Therefore, the gain reduces by 20 dB per decade.

5.3.6 Open-Loop Input Impedance Z_i

The open-loop input impedance is obtained by setting $d = 0$ and $i_o = 0$ in (5.86) to yield

$$v_o = \frac{Dv_i}{n(1-D)} - \frac{Z_1 i_l}{n(1-D)} \tag{5.144}$$

and

$$i_l = \frac{v_o}{nZ_2(1-D)}. \tag{5.145}$$

Substituting (5.145) into (5.144) gives

$$\begin{aligned} v_o &= \frac{Dv_i}{n(1-D)} - \frac{Z_1 v_o}{n^2 Z_2 (1-D)^2} \\ v_o\left[1 + \frac{Z_1}{n^2(1-D)^2 Z_2}\right] &= \frac{Dv_i}{n(1-D)} \\ v_o &= \frac{nD(1-D)Z_2 v_i}{Z_1 + n^2 Z_2 (1-D)^2}. \end{aligned} \tag{5.146}$$

Substituting (5.145) into (5.81) gives

$$\begin{aligned} i_i &= \frac{v_o}{nZ_2(1-D)} - \frac{v_o}{nZ_2} \\ &= v_o\left[\frac{1}{nZ_2(1-D)} - \frac{1}{nZ_2}\right] \\ &= \frac{Dv_o}{nZ_2(1-D)}. \end{aligned} \tag{5.147}$$

Substituting (5.146) into (5.147) gives

$$i_i = \frac{D}{nZ_2(1-D)} \frac{nD(1-D)Z_2 v_i}{Z_1 + n^2 Z_2 (1-D)^2} = \frac{D^2}{Z_1 + n^2 Z_2 (1-D)^2} v_i. \tag{5.148}$$

The open-loop input impedance is

$$Z_i(s) = \left.\frac{v_i(s)}{i_i(s)}\right|_{d=i_o=0} = \frac{Z_1 + n^2 Z_2 (1-D)^2}{D^2}. \tag{5.149}$$

Substituting (5.75) and (5.76) into (5.149) gives the output current-to-inductor current transfer function in the s-domain as

$$Z_i(s) = \left.\frac{v_i(s)}{i_i(s)}\right|_{d=i_o=0} = L_i \frac{s^2 + 2\xi\omega_o s + \omega_o^2}{(s + \omega_{zvi})} = R_{io} \frac{\left(\frac{s}{\omega_o}\right)^2 + \frac{2\xi s}{\omega_o} + 1}{\left(1 + \frac{s}{\omega_{zvi}}\right)}, \tag{5.150}$$

where L_i is the high-frequency input inductance and R_{io} is the DC input resistance. At high frequencies,

$$Z_i(\infty) = sL_i = s\frac{L_m}{D^2}. \tag{5.151}$$

At DC,

$$R_{io} = Z_i(0) = \frac{n^2R_L(1-D)^2 + r}{D^2}. \tag{5.152}$$

For the ideal case $r = 0$, the DC input resistance is

$$R_{io} = \frac{n^2R_L(1-D)^2}{D^2}, \tag{5.153}$$

and the angular frequency of the LHP zero ω_{zvi} is given in (5.133).

Figure 5.28 shows the theoretically obtained magnitude and phase plots of the input impedance. The input resistance R_{io} at DC is proportional to the duty cycle and the load resistance. The DC input resistance is $R_{io} = 100\ \Omega$. The high-frequency inductance is $L_i = 452\ \mu$H. With increase in frequency, the input impedance shows capacitive nature. Beyond f_o, the inductive reactance takes over and the input impedance is mainly dominated by the inductive reactance. In open loop, the input current is in phase with the input voltage, i.e. a step change in the input voltage in the positive direction causes a corresponding change in the input current in the same direction.

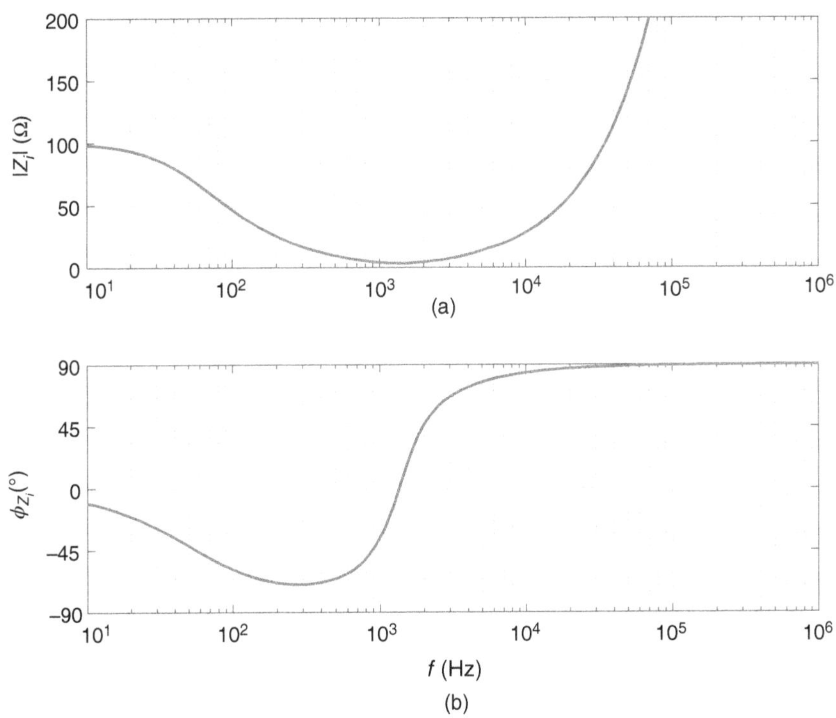

Figure 5.28 Theoretically obtained plots of input impedance Z_i. (a) Magnitude. (b) Phase.

5.3.7 Open-Loop Output Impedance Z_o

The small-signal model to derive the output impedance is shown in Figure 5.26. This model is obtained by setting $v_i = 0$ and $d = 0$. An independent voltage source v_t is applied at the output, which forces a current i_t. From (5.136) and (5.137),

$$i_t = \frac{v_o}{Z_2} - ni_l(1 - D) \tag{5.154}$$

and

$$v_o = -\frac{i_l Z_1}{n(1 - D)}$$
$$i_l = -\frac{n(1 - D)v_o}{Z_1}. \tag{5.155}$$

Substituting (5.155) into (5.154) gives

$$i_t = \frac{v_o}{Z_2} - \frac{n^2(1 - D)^2 v_o}{Z_1} = v_o\left[\frac{1}{Z_2} - \frac{n^2(1 - D)^2}{Z_1}\right]. \tag{5.156}$$

Hence, the open-loop output impedance is

$$Z_o(s) = \left.\frac{v_o(s)}{i_t(s)}\right|_{d=i_o=0} = \frac{Z_1 Z_2}{Z_1 + n^2 Z_2(1 - D)^2}. \tag{5.157}$$

Substituting (5.75) and (5.76) into (5.157) gives the open-loop output impedance in the s-domain as

$$Z_o(s) = \left.\frac{v_o(s)}{i_t(s)}\right|_{d=i_o=0} = R_{ox}\frac{(s + \omega_{zn})(s + \omega_{rl})}{s^2 + 2\xi\omega_o s + \omega_o^2} = R_{oo}\frac{\left(1 + \frac{s}{\omega_{zn}}\right)\left(1 + \frac{s}{\omega_{rl}}\right)}{\left(\frac{s}{\omega_o}\right)^2 + \frac{2\xi s}{\omega_o} + 1}, \tag{5.158}$$

where the high-frequency output resistance gain R_{ox} is

$$R_{ox} = Z_o(\infty) = \frac{R_L r_C}{R_L + r_C}, \tag{5.159}$$

and the DC output resistance R_{oo} is

$$R_{oo} = Z_o(0) = \frac{rR_L}{r + n^2 R_L(1 - D)^2}. \tag{5.160}$$

The DC gain R_{oo} for ideal case $r = 0$ is

$$R_{oo} = \frac{1}{n^2(1 - D)^2}, \tag{5.161}$$

and the angular frequency of the LHP zero ω_{rl} is

$$\omega_{rl} = \frac{r}{L_m}. \tag{5.162}$$

The angular frequency of the LHP zero ω_{zn} is given in (5.101).

Figure 5.29 shows the theoretically obtained magnitude and phase plots of the output impedance. The output impedance at DC is approximately equal to

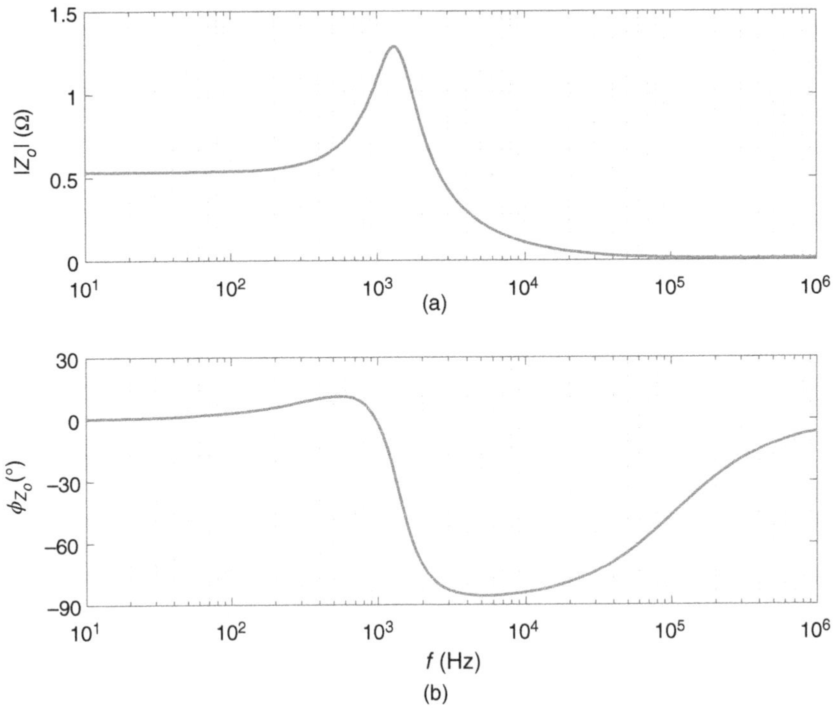

Figure 5.29 Theoretically obtained plots of output impedance Z_o. (a) Magnitude. (b) Phase.

$Z_o(0) = R_o = 0.53\ \Omega$. At DC, the inductor is a short circuit, and the filter capacitor is an open circuit. The DC resistance is due to the parallel combination of the equivalent averaged resistance r and the effective load resistance R_L reflected on to the primary side. At lower frequencies, $f < f_o$, the magnitude of impedance increases due to the inductive reactance. At a frequency f_o, the inductive and capacitive reactances are equal, and the converter filter network undergoes resonance. Beyond f_o, the capacitive reactance dominates and reduces the output impedance to nearly zero. A summary of calculated values for open-loop transfer functions is given in Table 5.1.

5.4 Inner-Current Loop

Figure 5.30 shows the architecture of the inner-current loop with a low-pass filter in the feedback path. Moreover, Figure 5.31 shows the circuit of the flyback DC–DC converter with inner-current loop. A sense resistor R_s is placed in the inductor branch to sense the inductor current and is a part of the feedback network. At low duty cycles, the sensed inductor current ripple is high. Therefore, the control voltage intersects with the modulator ramp voltage twice in each cycle, resulting in subharmonic instability. The low-pass filter attenuates the switching frequency components and its harmonics present in the sensed voltage $v_{RS} = R_s i_L$. Therefore, the theoretically obtained output voltage of the low-pass filter is $V_{FI} = R_s I_L$ and

Table 5.1 Summary of calculated values for the open-loop transfer functions.

Variable	Value
$\lvert T_{po}\rvert$	19.6 V = 25.85 dBV
$\lvert T_{pio}\rvert$	1.24 A = 1.87 dBA
$\lvert M_{vo}\rvert$	0.44 V/V = −7.13 dB
$\lvert M_{vio}\rvert$	0.02 A/V = −33.9 dBA/V
R_{io}	100 Ω
R_{oo}	0.53 Ω
ω_o	8.6 krad/s
f_o	1.4 kHz
ξ	0.37
ω_{zi}	525 rad/s
f_{zi}	84 Hz
ω_{zvi}	333 rad/s
f_{zvi}	53 Hz
ω_{zn}	667 krad/s
f_{zn}	106 kHz
ω_{zp}	369 krad/s
f_{zp}	59 kHz
ω_{rl}	5.9 krad/s
f_{rl}	939 Hz

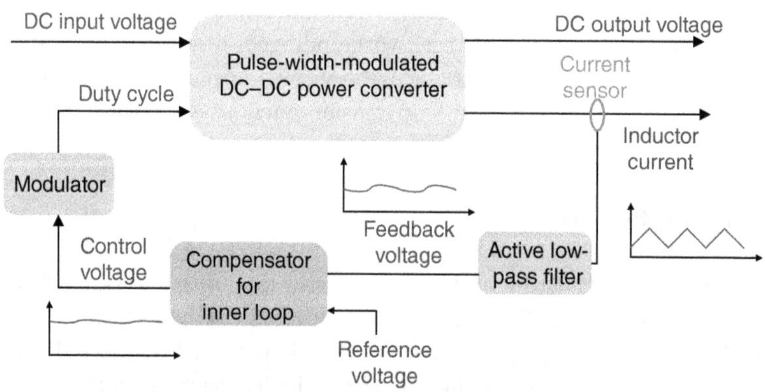

Figure 5.30 Architecture of the inner-current loop with filter block.

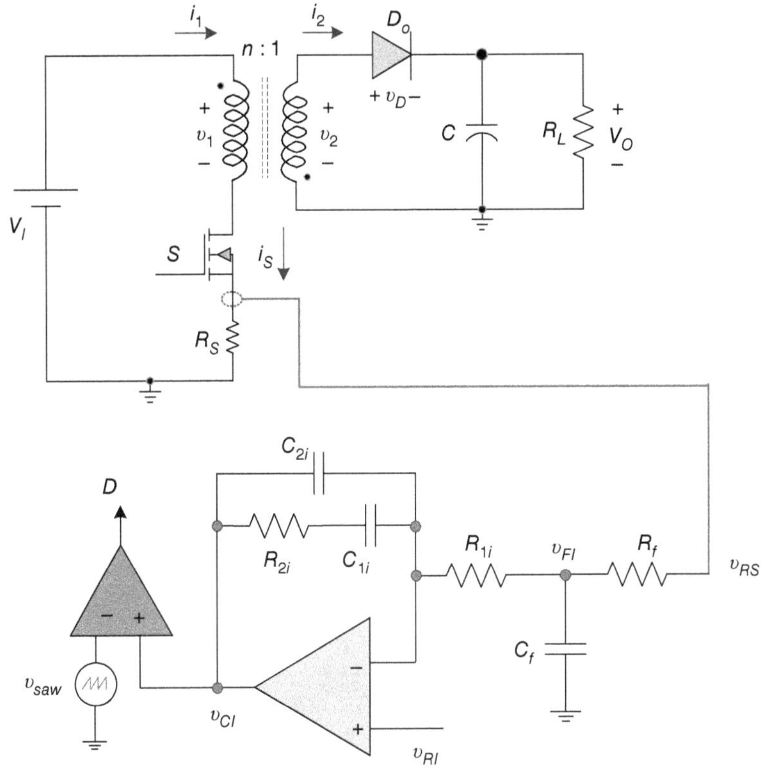

Figure 5.31 Circuit of the flyback DC–DC converter with inner-current loop.

is governed by the DC component of the sensed inductor current. Thus, in this method, the *average* component of the inductor current is sensed, thereby reducing the problems due to the subharmonic instability and transistor misgating. The transfer functions related to the inner-current loop, which includes the filter in the feedback path, are derived and discussed in this section.

5.4.1 Transfer Function of Filter and Non-inverting Amplifier T_f

The low-pass filter is composed of a resistor R_f and a capacitor C_f. The transfer function of two stages is

$$T_f(s) = \frac{\upsilon_{fi}(s)}{\upsilon_{rs}(s)} = \frac{Z_{Cf}}{R_f + Z_{Cf}} = \frac{\frac{1}{sC_f}}{R_f + \frac{1}{sC_f}} = \frac{1}{sR_fC_f + 1}$$
$$= \frac{1}{R_fC_f}\frac{1}{\frac{1}{R_fC_f} + s} = \frac{\omega_{pf}}{\omega_{pf} + s} = \frac{1}{1 + \frac{s}{\omega_{pf}}}. \tag{5.163}$$

Setting $s = j\omega$ gives

$$T_f(j\omega) = \frac{1}{1 + j\frac{\omega}{\omega_{pf}}}, \tag{5.164}$$

which yields

$$T_f(f) = \frac{1}{1 + j\frac{f}{f_{pf}}} = |T_f(f)|e^{j\phi(f)}, \tag{5.165}$$

where

$$|T_f(f)| = \frac{1}{\sqrt{1 + \left(\frac{f}{f_{pf}}\right)^2}} \tag{5.166}$$

and

$$\phi(f) = -\tan^{-1}\left(\frac{f}{f_{pf}}\right). \tag{5.167}$$

Using (5.166),

$$|T_f(f)|^2 = \frac{1}{1 + \left(\frac{f}{f_{pf}}\right)^2}, \tag{5.168}$$

which yields

$$\left(\frac{f}{f_{pf}}\right)^2 = \frac{1}{|T_f(f)|^2} - 1 = \frac{1 - |T_f(f)|^2}{|T_f(f)|^2} \tag{5.169}$$

and

$$\frac{f_{pf}}{f} = \sqrt{\frac{|T_f(f)|^2}{1 - |T_f(f)|^2}}. \tag{5.170}$$

Hence,

$$f_{pf} = f\sqrt{\frac{|T_f(f)|^2}{1 - |T_f(f)|^2}}. \tag{5.171}$$

The magnitude of the transfer function T_f is also given as

$$|T_f(f)| = \frac{\Delta V_{FI}}{\Delta V_{RS}}. \tag{5.172}$$

The filter frequency can be set by the amount of attenuation desired. As an example, to achieve an attenuation of 50% for $f = f_s = 100$ kHz, the filter upper cutoff frequency can be calculated using (5.171) as

$$f_{pf} = f_s\sqrt{\frac{|T_f(f)|^2}{1 - |T_f(f)|^2}} = 100 \times 10^3 \times \sqrt{\frac{0.50^2}{1 - 0.50^2}} = 57 \text{ kHz}. \tag{5.173}$$

Assuming $R_f = 1\ \text{k}\Omega/0.25\ \text{W}/1\%$, the value of the filter capacitor can be calculated as $C_f = 1/(2\pi f_{pf} R_f) \approx 2.8$ nF. A 4.7 nF/25 V capacitor is selected.

5.4.2 Transfer Function of Pulse-Width Modulator T_m

The control voltage-to-duty cycle transfer function of the pulse-width modulator is

$$T_m = \frac{D}{V_{CI}} = \frac{d}{v_{ci}} = \frac{1}{V_{Tm}}, \tag{5.174}$$

where V_{Tm} is the amplitude of the sawtooth waveform.

The steady-state average inductor current is $I_L = I_O/[n(1-D)] = 0.196$ A. At a duty cycle $D = 0.58$ and assuming the peak sawtooth voltage of the pulse-width modulator is $V_{Tm} = 0.5$ V, the required steady-state control voltage is $V_{CI} = DV_{Tm} = 0.58 \times 0.5 = 0.29$ V. The reference voltage to the current loop is $V_{RI} = V_{CI}$ at steady-state since the error voltage is zero, i.e. $V_{RI} = 0.29$ V. The gain of the feedback path is controlled mainly by the sensor gain R_s. For a steady-state inductor current I_L and the inner-loop reference voltage V_{RI}, which is set by the outer-voltage loop, the sense resistor can be determined using

$$R_s = \frac{V_{Rs}}{I_L} = \frac{V_{RI}}{I_L} = \frac{0.29}{0.2} = 1.45\ \Omega. \tag{5.175}$$

A standard resistor $R_s = 1.5\ \Omega/0.25$ W/1% is selected. From Section 5.4.1, $R_f =$ 1 kΩ/0.25 W/1% and $C_f = 4.7$ nF/25 V.

5.4.3 Uncompensated Loop Gain T_{ki}

The natural behavior of the inner-current loop can be determined using the uncompensated loop gain as

$$T_{ki} = \frac{v_{fi}}{v_{ei}} = T_m T_{pi} R_s T_f = T_{kio} \frac{1 + \frac{s}{\omega_{zi}}}{\left(1 + \frac{s}{\omega_{pf}}\right)\left(1 + \frac{2\xi s}{\omega_o} + \frac{s^2}{\omega_o^2}\right)}, \tag{5.176}$$

where T_m, T_{pi}, and T_f are given in (5.174), (5.111), and (5.163), respectively. The DC gain T_{kio} is given by

$$T_{kio} = T_{ki}(0) = \frac{R_s}{V_{Tm}} \frac{nV_O}{D} \frac{1+D}{n^2 R_L (1-D)^2 + r}. \tag{5.177}$$

Figure 5.32 shows the theoretically obtained magnitude and phase plots of the uncompensated loop gain of the inner-current loop. The gain at DC is $T_{kio} = 3.6$ V/V $= 11.2$ dB. A high DC gain T_{kio} is required to reduce the steady-state error in the output voltage due to step changes in the input voltage or load current. Therefore, a suitable control circuit capable of providing a high gain at DC and sufficient phase boost at the crossover frequency of the loop gain becomes essential. The phase of the uncompensated loop gain is zero at DC. This is because the inner-loop takes into account only the inductor branch and not the properties of the output network. Typically, the presence of a low-frequency zero in the T_{pi} transfer function enables a wider current loop bandwidth and low phase at the crossover frequency. The phase margin is in the range of 40°–90° and requires relatively simpler compensation schemes to satisfy the inner current loop requirements. A Type-II

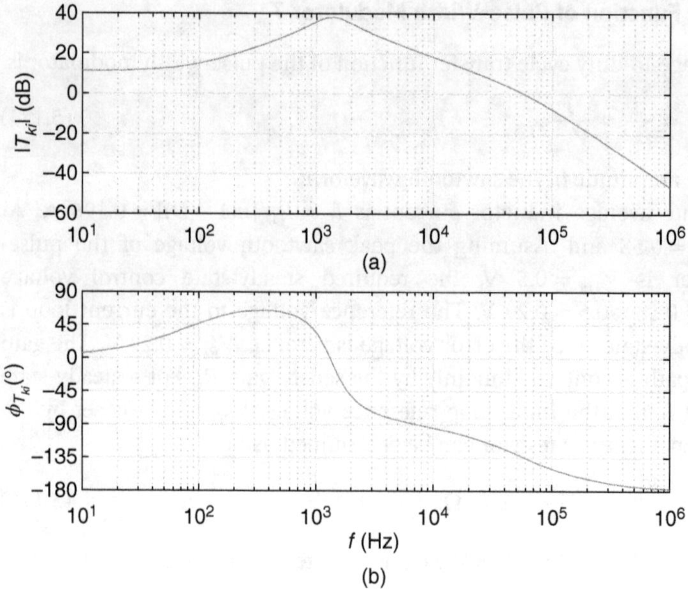

Figure 5.32 Theoretically obtained plots of uncompensated loop gain T_{ki}. (a) Magnitude. (b) Phase.

controller (also known as the single-lead integral) is used for the inner-current loop compensation.

5.4.4 Transfer Function of Compensation Circuit T_{ci}

Figure 5.33 shows the circuit of a Type-II controller. A Type-II control circuit introduces a pole at the origin, which is required to boost the DC gain and a pole–zero pair, which adjusts the crossover frequency f_c, thus maintaining the required PM.

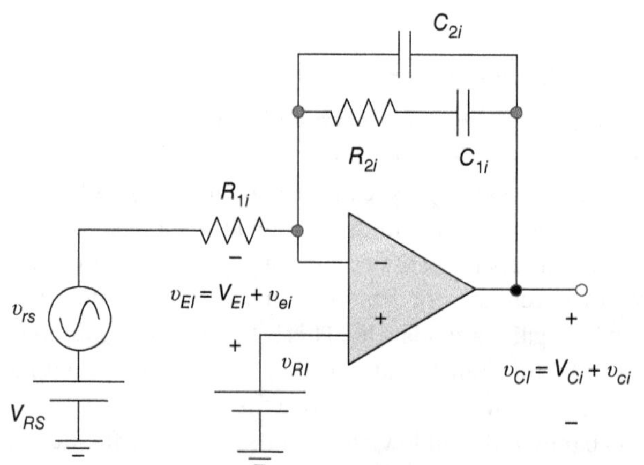

Figure 5.33 Circuit of the Type-II controller.

A detailed procedure to determine the transfer function of the controller to improve the DC gain (>20 dB) and achieve an inner-current loop phase margin $PM = 60°$ is explained in [45]. The remainder of this section details the controller transfer function design.

The magnitude of the uncompensated loop gain transfer function can be determined using

$$|T_{ki}(f)| = T_{kio}\frac{\sqrt{1+\left(\frac{f}{f_{zi}}\right)^2}}{\sqrt{1+\left(\frac{f}{f_{pf}}\right)^2}\sqrt{1-\left(\frac{f}{f_o}\right)^2+(4\xi^2\left(\frac{f}{f_o}\right)^2}}, \tag{5.178}$$

and the phase is

$$\phi_{Tk}(f) = -180° + \tan^{-1}\left(\frac{f}{f_{zi}}\right) - \tan^{-1}\left(\frac{f}{f_{pf}}\right) - \tan^{-1}\frac{2\xi\frac{f}{f_o}}{1-\left(\frac{f}{f_o}\right)^2}. \tag{5.179}$$

The desired crossover frequency for the loop gain is chosen as $f_c = 15$ kHz. At this frequency, the magnitude of the uncompensated loop gain transfer function is $|T_{ki}(f_c)| = 5.26$ V/V, and its phase is $\phi_{Tk}(f_c) = -101°$. The required phase margin is $PM = 60°$. Therefore, the phase boost required at f_c is

$$\phi_m = M - \phi_{Tk}(f_c) - 90° = 71°, \tag{5.180}$$

yielding the geometric mean between the controller corner frequencies as

$$K = \tan\left(\frac{\phi_m}{2} + 45°\right) = 6. \tag{5.181}$$

Assuming $R_{1i} = 1\ \text{k}\Omega/0.25\ W/2\%$, the value of C_{2i} is

$$C_{2i} = \frac{|T_k(f_c)|}{2\pi f_c K R_{1i}} = 9.34\ \text{nF}. \tag{5.182}$$

A 9.4 nF/25 V standard capacitor is selected. The expression for C_{1i} is

$$C_{1i} = C_{2i}(K^2 - 1) = 0.32\ \mu\text{F}. \tag{5.183}$$

A 0.33 μF/25 V standard capacitor is selected. The expression for R_{2i} is

$$R_{2i} = \frac{K}{2\pi f_c C_{1i}} = 196\ \Omega. \tag{5.184}$$

A 196 Ω/0.25 W/2% standard resistor is selected. The transfer function T_{ci} of the current-loop control circuit is

$$T_{ci}(s) = \frac{v_{ci}(s)}{v_{ei}(s)} = T_{cio}\frac{1+\frac{s}{\omega_{zci}}}{s\left(1+\frac{s}{\omega_{pci}}\right)}, \tag{5.185}$$

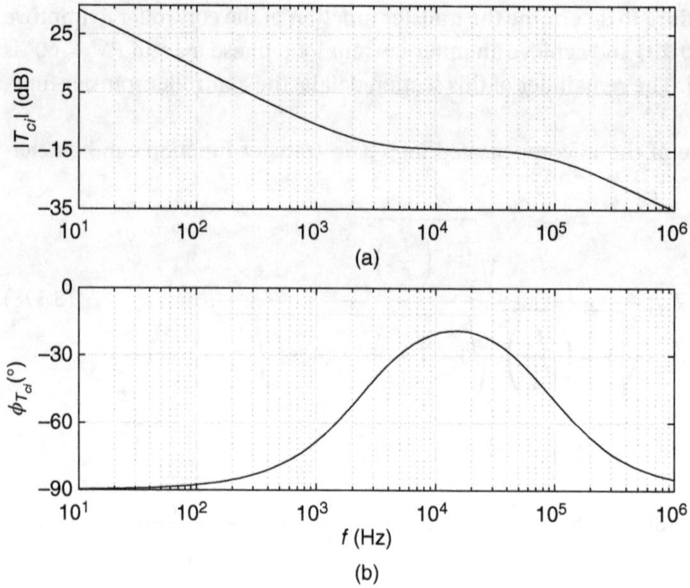

Figure 5.34 Theoretically obtained plots of controller transfer function T_{ci}. (a) Magnitude. (b) Phase.

where

$$T_{cio} = \frac{1}{R_{1i}(C_{1i} + C_{2i})} = 2.95 \text{ kV/V}, \tag{5.186}$$

$$\omega_{zci} = \frac{1}{R_{2i}C_{1i}} = 15.5 \text{ krad/s}, \tag{5.187}$$

and

$$\omega_{pci} = \frac{C_{1i} + C_{2i}}{R_{2i}C_{1i}C_{2i}} = 558 \text{ krad/s}. \tag{5.188}$$

Figure 5.34 shows the theoretically obtained magnitude and phase plots of the controller transfer function used in the inner-current loop. The DC gain is $T_{cio} =$ 69.4 dB, the frequency of the zero is $f_{zci} = \omega_{zci}/2\pi = 2.5$ kHz, and the frequency of the pole is $f_{pci} = \omega_{pci}/2\pi = 89$ kHz. The control transfer function is now incorporated into the critical path to determine the inner-loop gain and is discussed in Section 5.4.5.

5.4.5 Compensated Loop Gain T_i

Using the block diagram shown in Figure 5.35, the loop gain of the compensated inner-current loop can be determined as

$$T_i = \frac{v_{ei}}{v_{fi}} = T_{ki}T_{ci} = T_m T_{pi} R_s T_f T_{ci}, \tag{5.189}$$

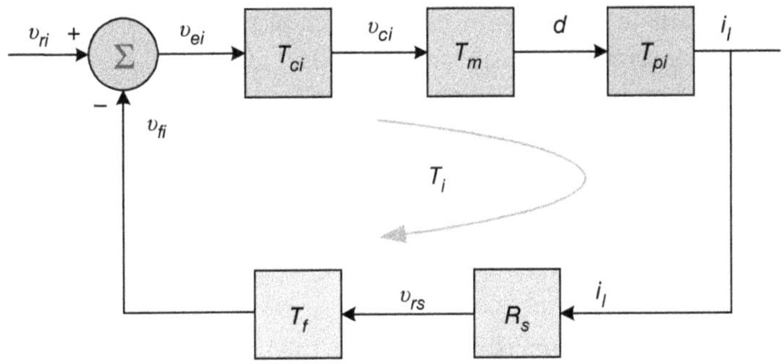

Figure 5.35 Block diagram of the inner-current loop.

yielding

$$T_i = T_{io} \frac{\left(1 + \frac{s}{\omega_{zi}}\right)\left(1 + \frac{s}{\omega_{zci}}\right)}{s\left(1 + \frac{s}{\omega_{pf}}\right)\left(1 + \frac{s}{\omega_{pci}}\right)\left(1 + \frac{2\xi s}{\omega_o} + \frac{s^2}{\omega_o^2}\right)}, \quad (5.190)$$

where T_{io} is the gain at $s = 0$ given as

$$T_{io} = T_{kio}T_{cio} = \frac{R_s}{V_{Tm}} \frac{nV_O}{D} \frac{1 + D}{n^2R_L(1 - D)^2 + r} \frac{1}{R_{1i}(C_{1i} + C_{2i})}. \quad (5.191)$$

Figure 5.36 shows the theoretically obtained magnitude and phase plots of compensated loop gain of the inner-current loop. The gain at DC is greater than 50 dB,

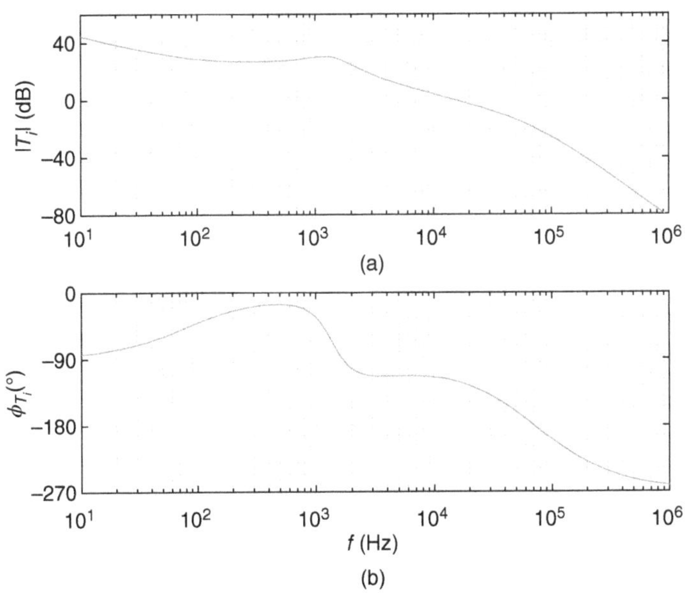

Figure 5.36 Theoretically obtained plots of compensated loop gain T_i of the inner current loop. (a) Magnitude. (b) Phase.

and the phase at DC starts at −90° due to the presence of the controller pole at the origin. The unity gain crossover frequency is 15 kHz, and the phase margin is $PM = 60°$. The inner-loop is stable. Section 5.5 describes the various important closed-loop transfer functions of the inner-current loop.

5.5 Closed-Loop Transfer Functions for Inner-Current Loop

The following closed-loop transfer functions relevant to the inner-loop are derived:

- Reference voltage to inductor current transfer function T_{icl} (control).
- Reference voltage to output voltage transfer function T_{picl} (control).
- Input voltage-to-inductor current transfer function M_{icl} (disturbance).
- Input voltage-to-output voltage transfer function M_{vicl} (disturbance).
- Input voltage-to-duty cycle transfer function M_{di} (disturbance).
- Input impedance Z_{iicl} (disturbance).
- Output impedance Z_{oicl} (disturbance).

The block diagram required to characterize the inner-current loop is shown in Figure 5.37. The small-signal current reference v_{ri} to the inner-loop is set by the outer loop. The focus of this section is to analyze the inner-current loop only. Therefore, the outer-voltage loop has not been shown in the block diagram. The error voltage v_{ei} is formed by the difference between the current reference v_{ri} and the feedback voltage v_{fi}. The feedback voltage v_{fi} is given by

$$v_{fi} = T_f R_s i_l, \tag{5.192}$$

where T_f is the transfer function of the low-pass filter and R_s is the sense resistor. The transfer function T_f is given by

$$T_f = \frac{v_{fi}}{v_{rs}} = \frac{1}{1 + \frac{s}{\omega_{pf}}}, \tag{5.193}$$

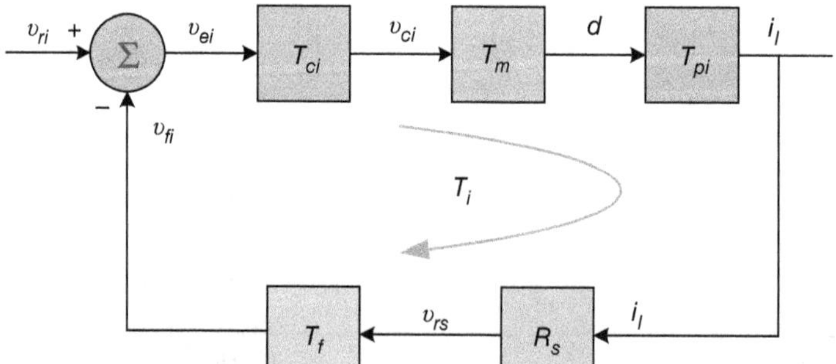

Figure 5.37 Block diagram used to derive the inner-loop control-to-inductor current transfer function T_{icl}.

where $\omega_{pf} = 2\pi f_{pf}$ is the filter cutoff frequency. The filter cutoff frequency is based on the allowable ripple in the sensed inductor current and is explained in Section 5.4.1. The gain of the low-pass filter is unity; therefore, it does not affect the DC gain of the loop gain. The transfer function T_{ci} is the voltage gain of the Type-II control circuit derived in Section 5.4.4. As the inductor current is controlled by the inner loop, the inductor current must be sensed, amplified by R_s, filtered by T_f, and compared with current reference v_{ri} to generate the small-signal error. This section performs a complete characterization of the inner-current loop.

5.5.1 Reference Voltage-to-Inductor Current Transfer Function T_{icl}

Using the block diagram shown in Figure 5.37, the closed-loop reference voltage-to-inductor current transfer function is

$$T_{icl}(s) = \left.\frac{i_l(s)}{v_{ri}(s)}\right|_{v_i=i_o=0}. \tag{5.194}$$

From Figure 5.37,

$$i_l = T_{ci} T_m T_{pi} v_{ei}. \tag{5.195}$$

Hence,

$$v_{ei} = \frac{i_l}{T_{ci} T_m T_{pi}} \tag{5.196}$$

and

$$v_{fi} = T_f R_s i_l. \tag{5.197}$$

Also, the error voltage is

$$\begin{aligned} v_{ei} &= v_{ri} - v_{fi}. \\ v_{ri} &= v_{ei} + v_{fi}. \end{aligned} \tag{5.198}$$

Substituting (5.196) and (5.197) into (5.198), we get

$$\begin{aligned} v_{ri} = v_{ei} + v_{fi} &= \frac{i_l}{T_{ci} T_m T_{pi}} + T_f R_s i_l = \left(\frac{1}{T_{ci} T_m T_{pi}} + T_f R_s\right) i_l \\ &= \left(\frac{1 + T_{ci} T_m T_{pi} T_f R_s}{T_{ci} T_m T_{pi}}\right) i_l = \left(\frac{1 + T_i}{T_{ci} T_m T_{pi}}\right) i_l. \end{aligned} \tag{5.199}$$

Hence,

$$T_{icl}(s) = \left.\frac{i_l(s)}{v_{ri}(s)}\right|_{v_i=i_o=0} = \frac{T_{ci} T_m T_{pi}}{1 + T_i}. \tag{5.200}$$

Figure 5.38 shows the theoretically obtained magnitude and phase plots of the inner-current loop reference voltage-to-inductor current transfer function. It can be observed that the bandwidth of the current loop seen by T_{icl} is high and equal

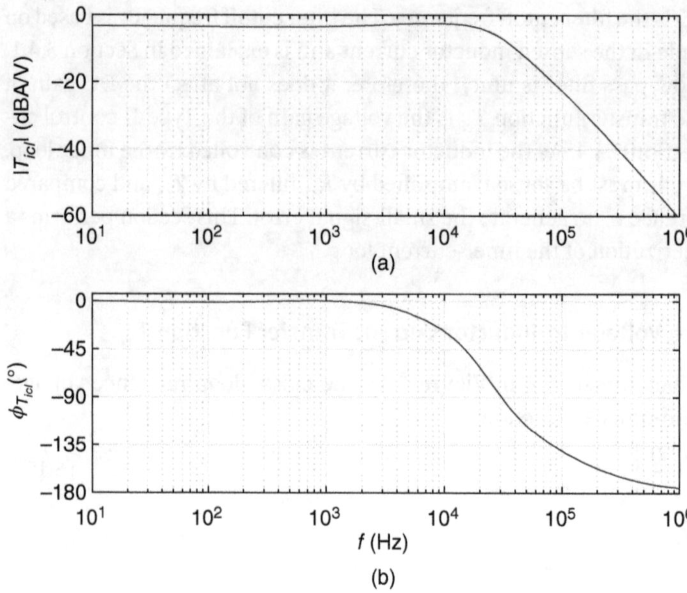

Figure 5.38 Theoretically obtained plots of inner-current loop reference voltage-to-inductor current transfer function T_{icl}. (a) Magnitude. (b) Phase.

to nearly 40 kHz. There is still a margin for bandwidth improvement but at the cost of reduced DC gain. The transfer function T_{icl} does not encompass the transfer function T_p (i.e. relevant to the outer-voltage loop). Therefore, maximum speed can be achieved by the inner-current loop. This means that for any input or load disturbances, the inner loop alone can sufficiently act against the disturbances and provide the necessary correction in the inductor current. Ideally, the inductor current is made to follow the fixed current reference v_{ri}; therefore, this form of the current-mode control can also be referred as *current-programmed control.*

The following approximation of (5.200) holds true as the loop gain $T_i \gg 1$:

$$T_{icl} = \frac{T_{ci}T_m T_{pi}}{1 + T_m T_{ci} T_{pi} T_f R_s} \approx \frac{T_{ci}T_m T_{pi}}{T_m T_{ci} T_{pi} T_f R_s} = \frac{1}{T_f R_s}. \tag{5.201}$$

5.5.2 Reference Voltage-to-Output Voltage Transfer Function T_{picl}

Using the block diagram shown in Figure 5.39, the closed-loop reference voltage-to-output voltage transfer function is

$$T_{picl}(s) = \left.\frac{v_o(s)}{v_{ri}(s)}\right|_{v_i=i_o=0} = \frac{T_{ci}T_m T_p}{1 + T_i}. \tag{5.202}$$

At low frequencies, $T_i \gg 1$. Therefore,

$$T_{picl} \approx \frac{T_{ci}T_m T_p}{T_i} = \frac{T_p}{T_f R_s}. \tag{5.203}$$

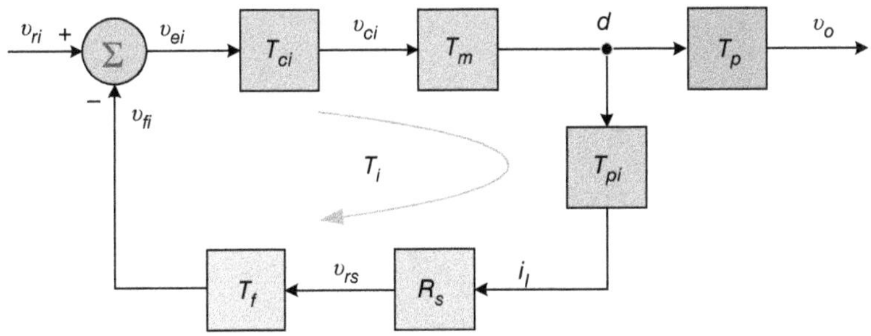

Figure 5.39 Block diagram required to derive the inner-loop control-to-output voltage transfer function T_{picl}.

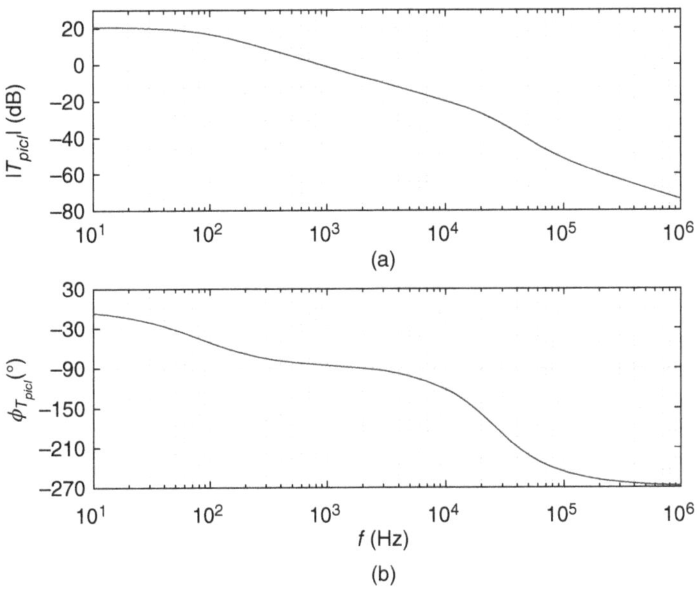

Figure 5.40 Theoretically obtained plots of reference voltage-to-output voltage transfer function T_{picl}. (a) Magnitude. (b) Phase.

Figure 5.40 shows the theoretically obtained magnitude and phase plots of the inner-current loop reference voltage-to-output voltage transfer function.

5.5.3 Input Voltage-to-Inductor Current Transfer Function M_{icl}

Using the block diagram shown in Figure 5.41, the closed-loop input voltage-to-inductor current transfer function is

$$M_{icl}(s) = \left.\frac{i_l(s)}{v_i(s)}\right|_{v_{ri}=i_o=0}. \tag{5.204}$$

From the block diagram,

$$i_l = i_l' + i_l'' = (-T_i)i_l + M_{vi}v_i \tag{5.205}$$

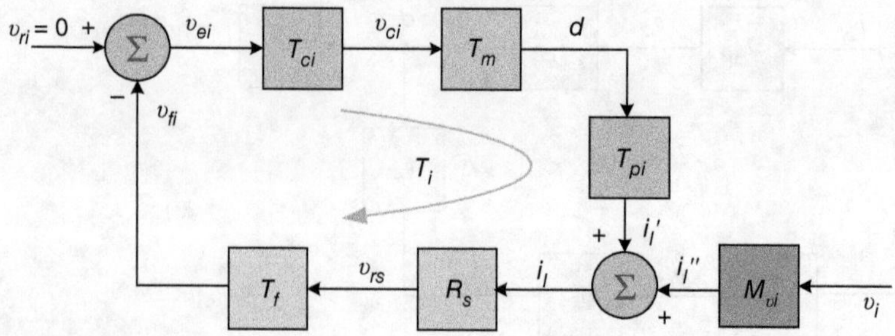

Figure 5.41 Block diagram required to derive the inner-loop input voltage-to-inductor current transfer function M_{icl}.

and

$$i_l(1 + T_i) = M_{vi}v_i. \tag{5.206}$$

Hence,

$$M_{icl}(s) = \frac{i_l(s)}{v_i(s)} = \frac{M_{vi}}{1 + T_i}. \tag{5.207}$$

Figure 5.42 shows the theoretically obtained magnitude and phase plots of the inner-current loop input voltage-to-inductor current transfer function. The gain

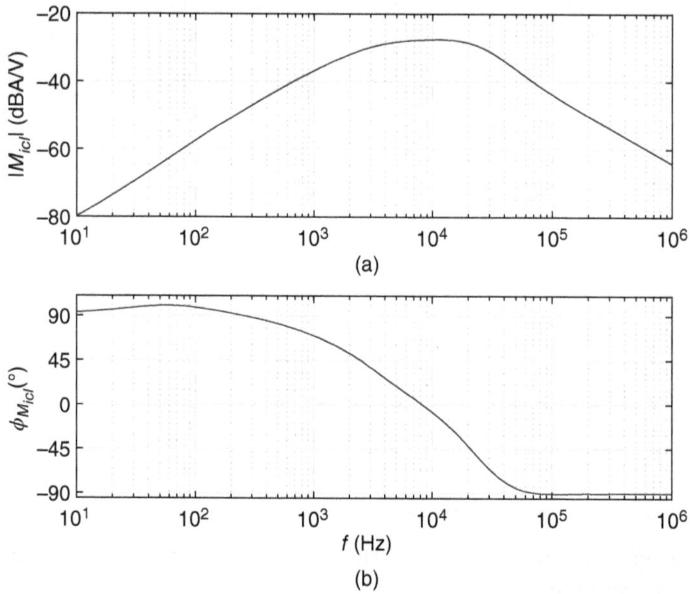

Figure 5.42 Theoretically obtained plots of input voltage-to-inductor current transfer function M_{icl}. (a) Magnitude. (b) Phase.

at DC and low frequencies is −50 dB or lower, indicating that the inner-loop can sufficiently attenuate the input voltage disturbance. The phase starts at 90° due to the presence of the pole at the origin in the loop gain transfer function T_i.

5.5.4 Input Voltage-to-Output Voltage Transfer Function M_{vicl}

Using the block diagram shown in Figure 5.43, the closed-loop input voltage-to-output voltage transfer function is

$$M_{vicl}(s) = \left.\frac{v_o(s)}{v_i(s)}\right|_{v_{ri}=i_o=0}. \tag{5.208}$$

From the block diagram,

$$i_l = i_l' + i_l'' = T_{pi}d + M_{vi}v_i \tag{5.209}$$

and

$$\frac{-d}{T_{ix}} = T_{pi}d + M_{vi}v_i, \tag{5.210}$$

where

$$T_{ix} = \frac{T_i}{T_{pi}} = T_{ci}T_mR_sT_f. \tag{5.211}$$

Rearranging (5.210) further, we get

$$-d\left(\frac{1}{T_{ix}} + T_{pi}\right) = M_{vi}v_i \tag{5.212}$$

yielding

$$d = -\frac{M_{vi}T_{ix}}{1+T_i}v_i. \tag{5.213}$$

From the block diagram in Figure 5.43,

$$v_o = v_o' + v_o'' = T_pd + M_vv_i. \tag{5.214}$$

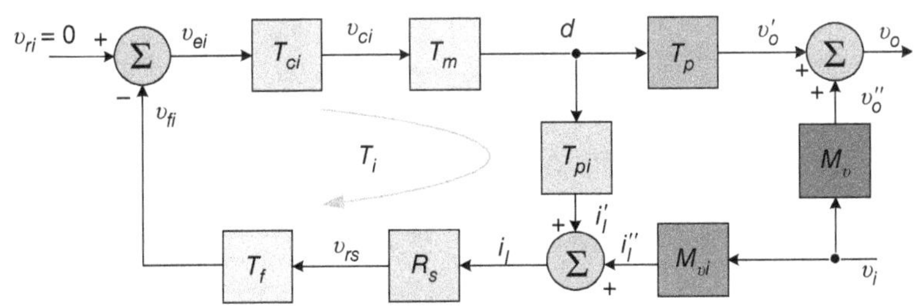

Figure 5.43 Block diagram required to derive the inner-loop input voltage-to-output voltage transfer function M_{vicl}.

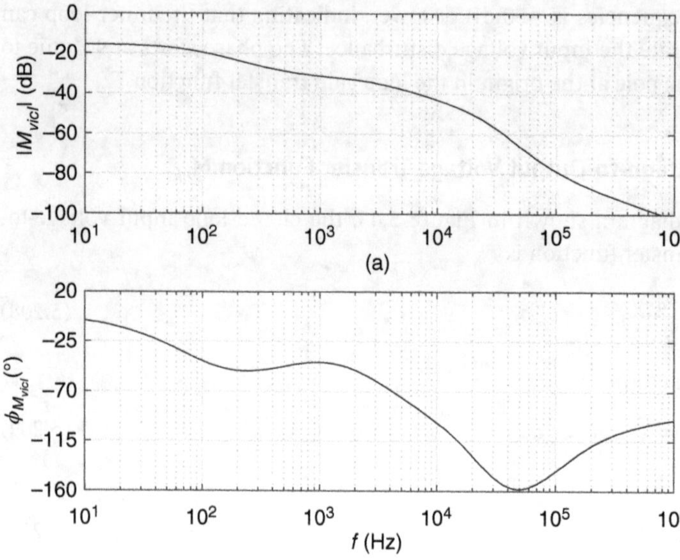

Figure 5.44 Theoretically obtained plots of input voltage-to-output voltage transfer function M_{vicl}. (a) Magnitude. (b) Phase.

Substituting d from (5.213) into (5.214), we get

$$v_o = T_p\left(-\frac{M_{vi}T_{ix}}{1+T_i}v_i\right) + M_v v_i = \left(M_v - \frac{T_p M_{vi} T_{ix}}{1+T_i}\right) v_i. \tag{5.215}$$

The closed-loop input-to-output transfer function is

$$M_{vicl}(s) = \frac{v_o(s)}{v_i(s)} = M_v - \frac{T_p M_{vi} T_{ix}}{1+T_i}. \tag{5.216}$$

Figure 5.44 shows the theoretically obtained magnitude and phase plots of the inner-current loop input voltage-to-output voltage transfer function. Unlike other disturbance transfer functions, the M_{vicl} transfer function obtained between the output and the input in the presence of the inner-current loop depends on T_i instead of $1/T_i$.

5.5.5 Input Voltage-to-Duty Cycle Transfer Function M_{di}

Using the block diagram shown in Figure 5.43, the closed-loop input voltage-to-duty cycle transfer function is

$$M_{di}(s) = \left.\frac{d(s)}{v_i(s)}\right|_{v_{ri}=i_o=0}. \tag{5.217}$$

From (5.213),

$$d = -\frac{M_{vi}T_{ix}}{1+T_i}v_i. \tag{5.218}$$

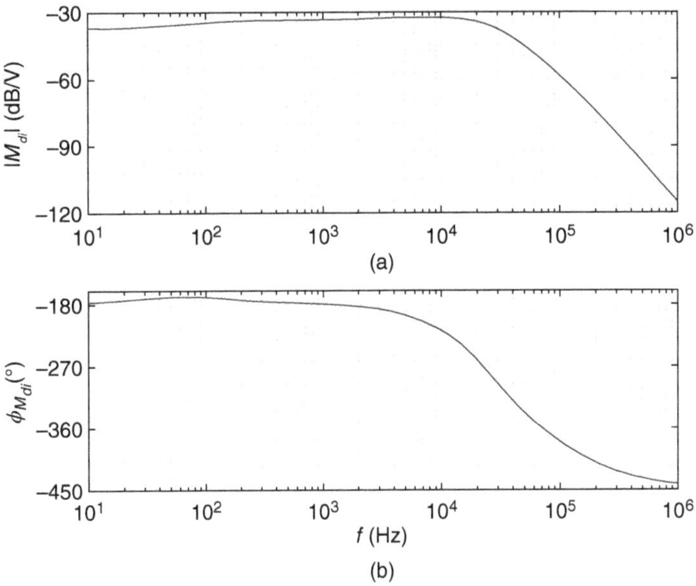

Figure 5.45 Theoretically obtained plots of input voltage-to-duty cycle transfer function M_{di}. (a) Magnitude. (b) Phase.

Hence, the closed-loop input voltage-to-duty cycle transfer function is

$$M_{di}(s) = \frac{d(s)}{v_i(s)} = -\frac{M_{vi}T_{ix}}{1 + T_i} = -\frac{M_{vi}}{T_{pi}}\frac{T_i}{1 + T_i}. \tag{5.219}$$

Figure 5.45 shows the theoretically obtained magnitude and phase plots of the inner-current loop input voltage-to-control transfer function. This transfer function determines the magnitude of change in duty cycle offered by the inner loop and outer loop for small-signal perturbations in the input voltage. From the Bode plot, the magnitude of the transfer function at DC is nearly $M_{dio} = 0.014$ 1/V = −37.1 dB/V, i.e. for a step change in input voltage by 1 V, the duty cycle changes by nearly 0.014 to provide the correction for disturbance. For an example, assuming that the steady-state input voltage of the designed flyback converter changes from 12 to 14 V, then the corresponding change in the duty cycle needed to maintain the output voltage V_O constant is $\Delta D = 0.04$. However, the current loop is able to provide only $\Delta D = M_{dio} \times \Delta V_I = 0.014 \times 2 = 0.028$. This means a deficit of $D = 0.012$ is needed to be provided by the outer-voltage loop.

5.5.6 Input Impedance Z_{iicl}

Using the block diagram shown in Figure 5.46, the closed-inner-loop input impedance is

$$Z_{iicl}(s) = \left.\frac{v_i(s)}{i_i(s)}\right|_{i_o=d=0}. \tag{5.220}$$

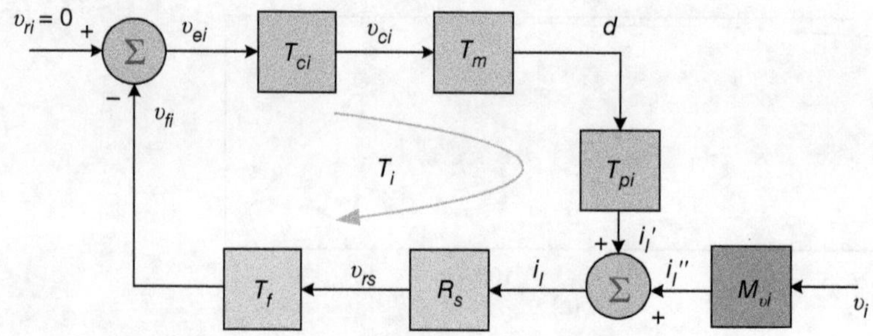

Figure 5.46 Block diagram used to derive the closed-inner-loop input impedance Z_{iicl}.

From (5.213),

$$d = -\frac{M_{vi}T_{ix}}{1+T_i}v_i. \tag{5.221}$$

Substituting (5.221) into (5.80) gives

$$i_i = I_L d + D i_l = -\frac{I_L M_{vi} T_{ix}}{1+T_i}v_i + D i_l. \tag{5.222}$$

Using (5.209) and (5.221),

$$\begin{aligned} i_l &= T_{pi} d + M_{vi} v_i = -\frac{M_{vi}T_{ix}T_{pi}}{1+T_i}v_i + M_{vi}v_i \\ &= \left[M_{vi} - \frac{M_{vi}T_i}{1+T_i}\right]v_i = \frac{M_{vi}}{1+T_i}v_i. \end{aligned} \tag{5.223}$$

Substituting (5.223) into (5.222),

$$\begin{aligned} i_i &= -\frac{I_L M_{vi} T_{ix}}{1+T_i}v_i + \frac{DM_{vi}}{1+T_i}v_i \\ &= \left[\frac{DM_{vi}}{1+T_i} - \frac{I_L M_{vi} T_{ix}}{1+T_i}\right]v_i. \end{aligned} \tag{5.224}$$

Hence, the closed-inner-loop input admittance is

$$Y_{iicl}(s) = \frac{i_i(s)}{v_i(s)} = \frac{DM_{vi}}{1+T_i} - \frac{I_L M_{vi} T_{ix}}{1+T_i}. \tag{5.225}$$

The closed-loop input impedance is

$$Z_{iicl}(s) = \frac{v_i(s)}{i_i(s)} = \frac{1}{Y_{iicl}(s)}. \tag{5.226}$$

Figure 5.47 shows the theoretically obtained magnitude and phase plots of the closed-inner-loop input impedance.

5.5.7 Output Impedance Z_{oicl}

Using the block diagram shown in Figure 5.48, the closed-loop output impedance is

$$Z_{oicl}(s) = \left.\frac{v_o(s)}{i_o(s)}\right|_{v_{ri}=v_i=0}. \tag{5.227}$$

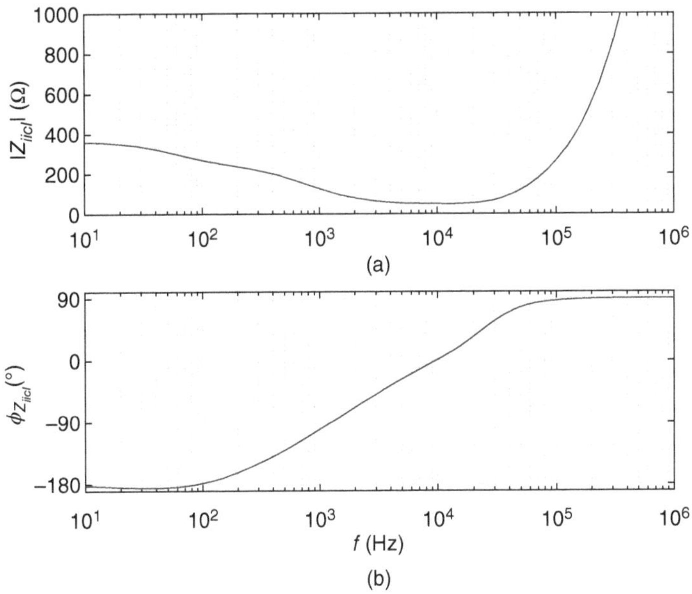

Figure 5.47 Theoretically obtained plots of closed-inner-loop input impedance Z_{iicl}. (a) Magnitude. (b) Phase.

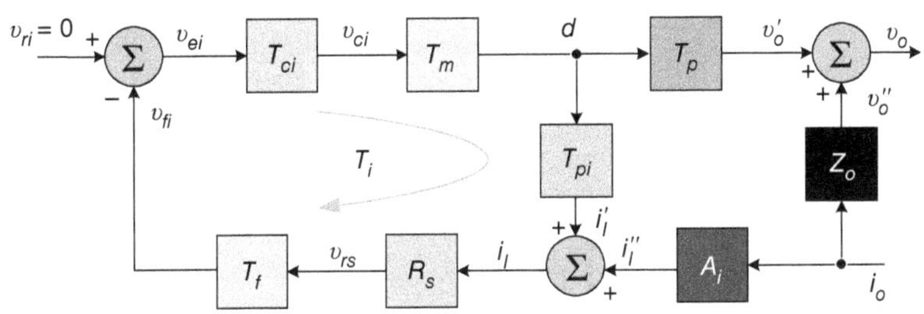

Figure 5.48 Block diagram required to derive the closed-inner-loop output impedance Z_{oicl}.

From the block diagram,

$$i_l = i_l' + i_l'' = T_{pi}d + A_i i_o \tag{5.228}$$

and

$$\frac{-d}{T_{ix}} = T_{pi}d + A_i i_o, \tag{5.229}$$

where

$$T_{ix} = \frac{T_i}{T_{pi}} = T_{ci}T_m R_s T_f. \tag{5.230}$$

Rearranging (5.229) further, we get

$$-d\left(\frac{1}{T_{ix}} + T_{pi}\right) = A_i i_o. \tag{5.231}$$

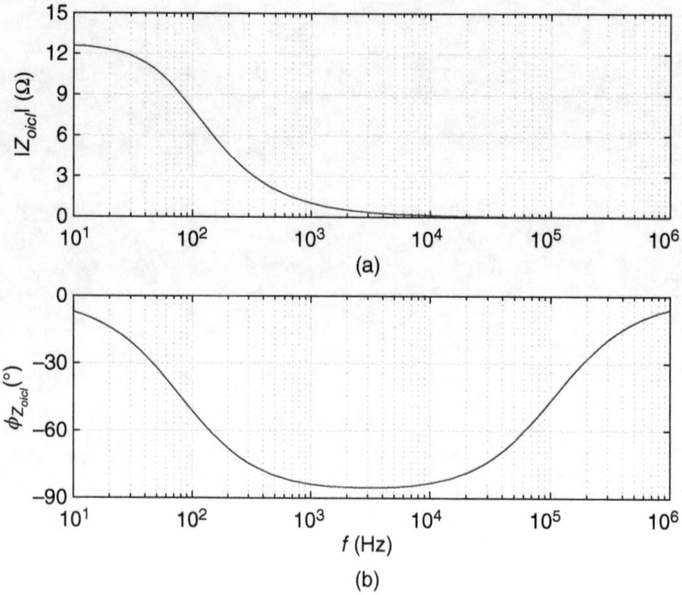

Figure 5.49 Theoretically obtained plots of closed-inner-loop output impedance Z_{oicl}. (a) Magnitude. (b) Phase.

Hence,

$$d = -\frac{A_i T_{ix}}{1 + T_i} i_o. \tag{5.232}$$

From the block diagram in Figure 5.48,

$$v_o = v_o' + v_o'' = T_p d + Z_o i_o. \tag{5.233}$$

Substituting d from (5.232) into (5.233), we get

$$v_o = T_p \left(-\frac{A_i T_{ix}}{1 + T_i} i_o \right) + Z_o i_o = \left(Z_o - \frac{A_i T_p T_{ix}}{1 + T_i} \right) i_o. \tag{5.234}$$

The closed-loop output impedance is

$$Z_{oicl}(s) = \frac{v_o(s)}{i_o(s)} = Z_o - \frac{A_i T_p T_{ix}}{1 + T_i}. \tag{5.235}$$

Figure 5.49 shows the theoretically obtained magnitude and phase plots of the closed-inner-loop output impedance.

5.6 Outer-Voltage Loop

The outer-voltage loop, i.e. the major loop, is essential in a converter to act against the nonlinear disturbances in the load. With change in the load resistance, the load current increases or decreases causing the inductor current to change. However, to maintain the output voltage constant, it is essential to introduce the outer-voltage loop with a set reference voltage V_{RV} that regulates the output voltage. In a typical two-loop system, the reference voltage to the outer loop is a fraction of the rated output voltage. The fraction is determined by the feedback network, whose voltage gain β is determined by a network of resistors. The reference voltage must be positive; however, the output voltage is negative. Therefore, the feedback network gain β must be negative to have a zero phase shift in the uncompensated loop gain at DC and low frequencies. The negative feedback factor can be achieved by various means, mostly depending upon the application and the type of control (analog or digital).

Figure 5.50 shows a complete block diagram of the average current-mode controlled flyback converter with two-loop control. Moreover, Figure 5.51 shows the circuit schematic of the flyback converter with average current-mode controlled inner-current loop and outer-voltage loop. It must be noted that the circuit schematic already includes the compensation scheme needed for the outer loop. Complete derivations of the transfer functions are provided in this section. The feedback gain is negative to account for the negative output voltage. The reference voltage V_{RI} for the current loop is set by the outer loop. Typically, the bandwidth of the outer loop is slow, and that of inner loop is fast.

The procedure to design the outer loop is identical to that involved in designing the inner-loop. The flyback power stage served as the plant for the inner-current loop. However, in the case of outer-loop design, the power stage incorporated within the inner loop serves as the plant to be controlled. Hence, the path between the current reference v_{ri} and the duty cycle d serves as the plant for the outer-voltage loop. The transfer function between v_{ri} and d is a sixth-order transfer function. Therefore, it is cumbersome to express the transfer functions in individual pole–zero form. The complex-conjugate poles in T_p, the RHP zero in T_p, the low-frequency LHP zero in T_{pi}, and the inner-current loop controller poles are the poles that may exist within the crossover frequency of the voltage loop gain. Therefore, attention must be paid in deciding the appropriate outer-loop gain crossover frequency.

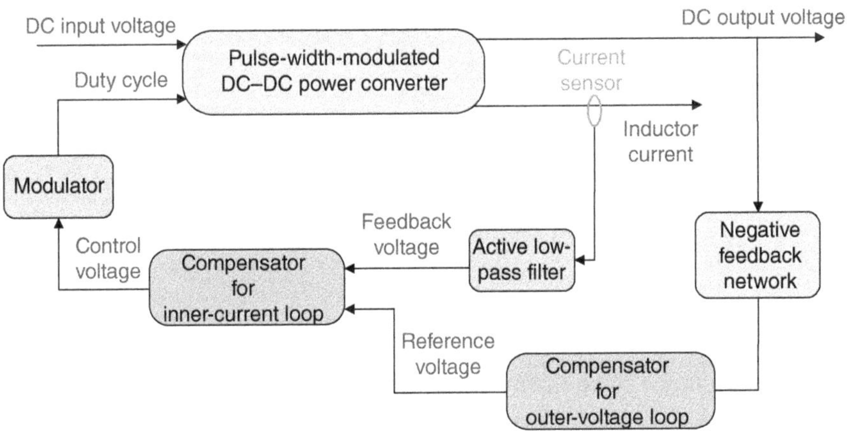

Figure 5.50 Architecture of the complete average current-mode controlled flyback converter with voltage-mode control.

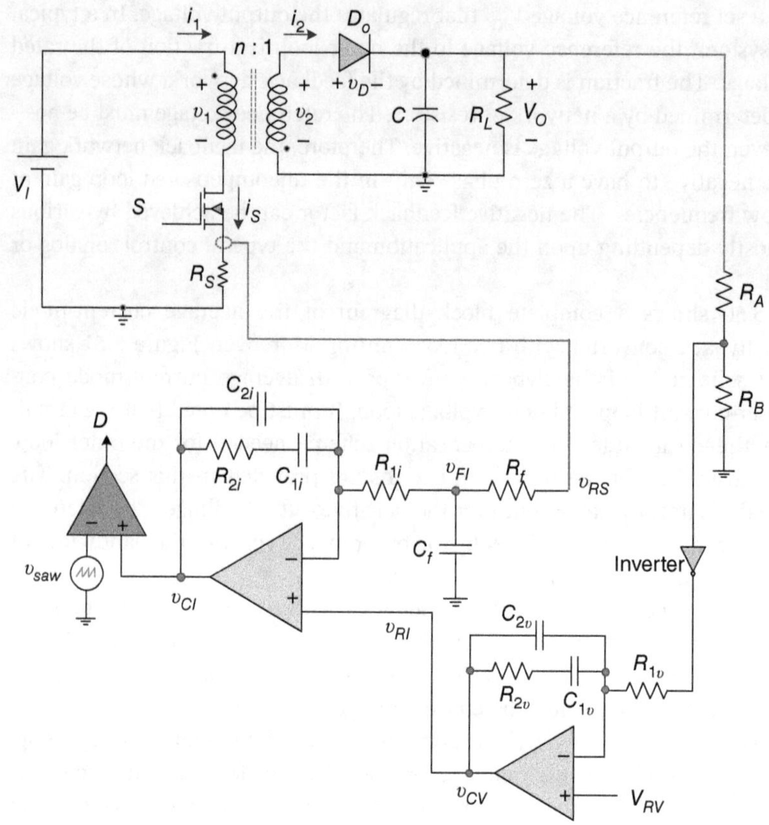

Figure 5.51 Complete circuit of the average current-mode controlled flyback converter with voltage-mode control.

5.6.1 Transfer Function of Feedback Network β

The transfer function of the feedback network is given as

$$\beta = \frac{V_{RV}}{V_O}, \tag{5.236}$$

where V_{RV} is the reference voltage to the outer-voltage loop. Assuming the reference voltage $V_{RV} = 2.5$ V, the transfer function of the feedback network is

$$\beta = \frac{V_{RV}}{V_O} = 2.5/5 = 0.5. \tag{5.237}$$

5.6.2 Uncompensated Loop Gain T_{kv}

The natural behavior of the outer-voltage loop can be determined using the uncompensated loop gain as

$$T_{kv} = \frac{v_{fv}}{v_{ev}} = \frac{\beta v_o}{v_{ri}} = \beta T_{picl} = \frac{\beta T_{ci} T_m T_p}{1 + Ti}. \tag{5.238}$$

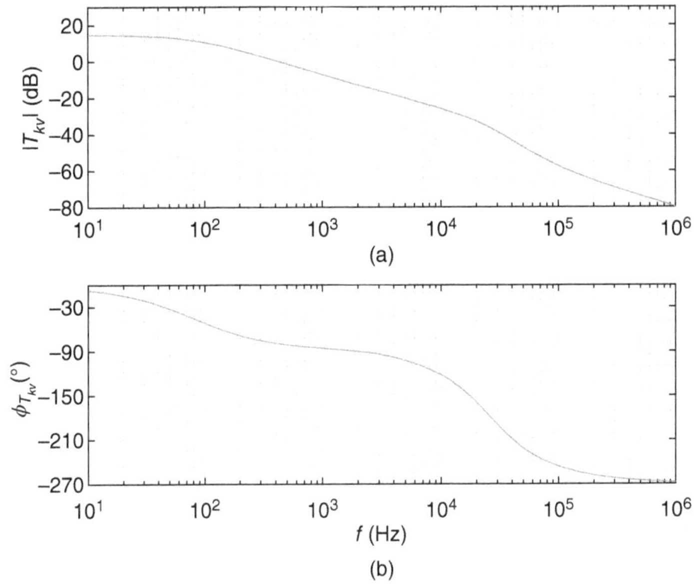

Figure 5.52 Theoretically obtained plots of uncompensated loop gain T_{kv}. (a) Magnitude. (b) Phase.

Figure 5.52 shows the theoretically obtained magnitude and phase plots of the outer-voltage uncompensated loop gain T_{kv}. The gain is low at DC and is nearly $T_{kvo} = 5.4$ V/V = 14.7 dB. The crossover frequency is nearly 402 Hz. The phase margin at the crossover frequency is nearly 100°. It is essential to provide a high gain at DC to reduce the steady-state error. The crossover frequency must be high for improved bandwidth; however, it cannot be higher than the frequency of the RHP zero present in the T_p transfer function. A significant phase boost must also be maintained at the crossover frequency to obtain a phase margin of 60°. Therefore, high-order control schemes such as Type II or Type III is preferred for this application. A Type-II (single-lead integral) control circuit is utilized in this analysis. The design was mentioned previously also in the design of the inner-current loop. Figure 5.53 shows the control circuit used in the outer-voltage loop.

5.6.3 Transfer Function of Compensation Circuit T_{cv}

Figure 5.53 shows the circuit of a single-lead integral (Type-II) controller. A single-lead integral control circuit introduces a pole at the origin, which is required to boost the DC gain and a pole–zero pair, which adjusts the crossover frequency f_c, thus maintaining the required PM. A detailed procedure to determine the transfer function of the controller to improve the DC gain (>20 dB) and achieve an inner-current-loop phase margin $PM = 60°$ is explained in [45].

From the uncompensated loop gain transfer function T_{kv}, the crossover frequency is $f_c = 402$ Hz. A crossover frequency of $f_c = 1$ kHz is assumed in the following discussion. At the new value of f_c, the magnitude of $T_{kv}(f_c) = 0.427 = -7.2$ dB, and the

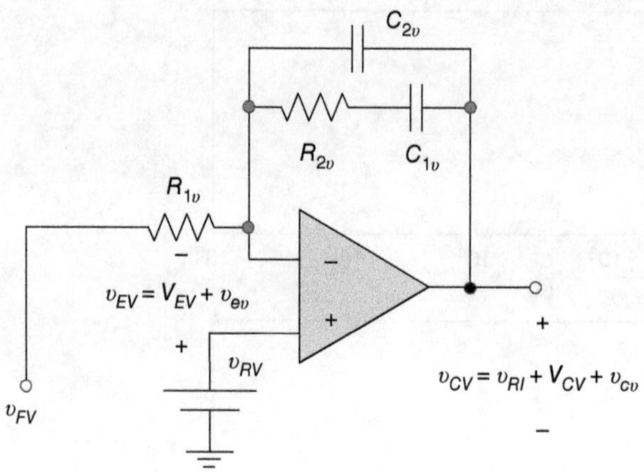

Figure 5.53 Circuit of Type-II controller used in outer-voltage loop.

phase is $\phi_{Tkv}(f_c) = -86°$. The required phase margin is $PM = 60°$. Therefore, the phase boost required at f_c is

$$\phi_m = PM - \phi_{Tk}(f_c) - 90° = 56°, \tag{5.239}$$

yielding the geometric mean between the controller corner frequencies as

$$K = \tan^{-1}\left(\frac{\phi_m}{2} + 45°\right) = 3.3. \tag{5.240}$$

Assuming $R_{1v} = 1\ \text{k}\Omega/0.25\ \text{W}/2\%$, the value of C_{2v} is

$$C_{2v} = \frac{|T_k(f_c)|}{2\pi f_c K R_{1v}} = 22\ \text{nF}. \tag{5.241}$$

A 22 nF/25 V standard capacitor is selected. The expression for C_{1v} is

$$C_{1v} = C_{2v}(K^2 - 1) = 0.2\ \mu\text{F}. \tag{5.242}$$

A 0.2 μF/20 V standard capacitor is selected. The expression for R_{2v} is

$$R_{2v} = \frac{K}{2\pi f_c C_{1v}} = 2.5\ \text{k}\Omega. \tag{5.243}$$

A 2.5 kΩ/0.25 W/5% standard resistor is selected. The transfer function T_{cv} of the voltage-loop control circuit is

$$T_{cv} = \frac{v_{cv}(s)}{v_{ev}(s)} = T_{cvo}\frac{1 + \dfrac{s}{\omega_{zcv}}}{s\left(1 + \dfrac{s}{\omega_{pcv}}\right)}, \tag{5.244}$$

where

$$T_{cvo} = \frac{1}{R_{1v}(C_{1v} + C_{2v})} = 4.5\ \text{kV/V}, \tag{5.245}$$

$$\omega_{zcv} = \frac{1}{R_{2v}C_{1v}} = 2\ \text{krad/s}, \tag{5.246}$$

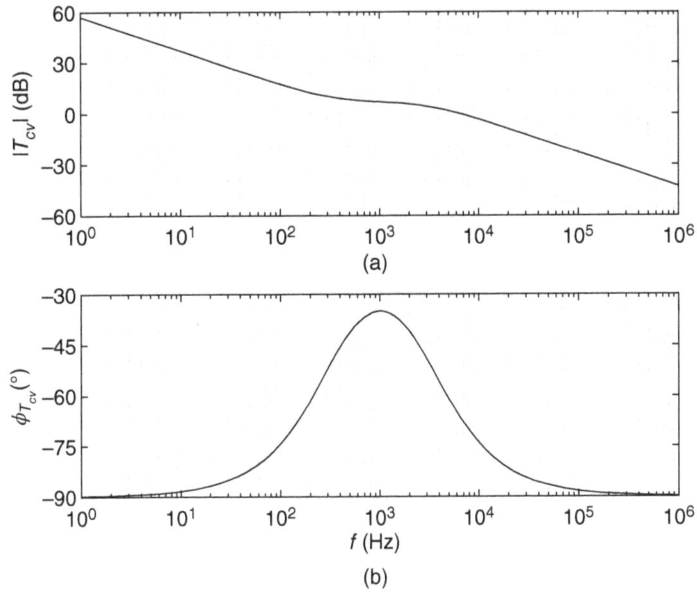

Figure 5.54 Theoretically obtained plots of controller transfer function T_{cv} of the outer-voltage loop. (a) Magnitude. (b) Phase.

and

$$\omega_{pcv} = \frac{C_{1v} + C_{2v}}{R_{2v}C_{1v}C_{2v}} = 20 \text{ krad/s}. \tag{5.247}$$

Figure 5.54 shows the theoretically obtained magnitude and phase plots of the controller transfer function used in the outer-voltage loop. The poles are located at 0 Hz and 3.2 kHz, and the zero is located at 318 Hz. A phase boost of nearly 40° is provided at a crossover frequency of 1 kHz.

5.6.4 Compensated Loop Gain T_v

The loop gain of the compensated outer-voltage loop is

$$T_v = \frac{v_{ev}}{v_{fv}} = T_{kv}T_{cv} = \beta T_{picl}T_{cv} = \frac{\beta T_{ci}T_m T_p T_{cv}}{1 + T_i}. \tag{5.248}$$

Figure 5.55 shows the theoretically obtained magnitude and phase plots of the loop gain of the compensated outer-voltage loop. The outer-voltage loop gain has a phase margin of 60°. The gain at DC is higher than 60 dB.

5.7 Closed-Loop Transfer Functions for Outer-Voltage Loop

The following closed-loop transfer functions are derived:

- Reference voltage-to-output voltage transfer function T_{pcl} (control).

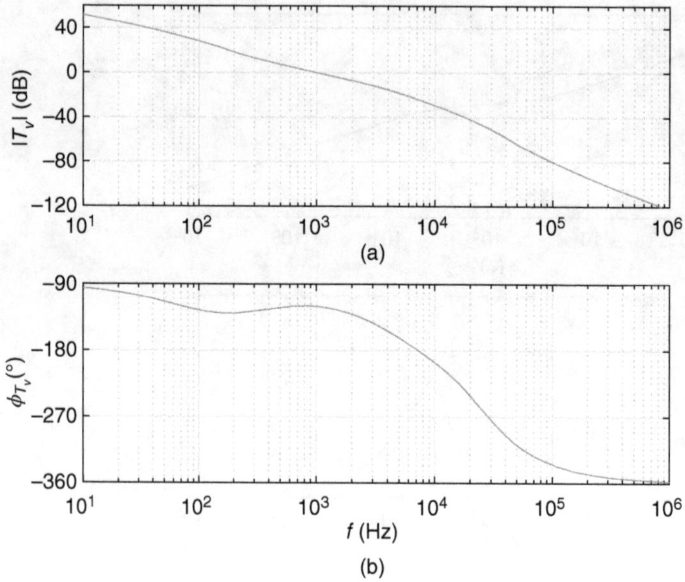

Figure 5.55 Theoretically obtained plots of loop gain T_v of the compensated outer-voltage loop. (a) Magnitude. (b) Phase.

- Input voltage-to-output voltage transfer function M_{vcl} (disturbance).
- Input voltage-to-duty cycle transfer function M_{dv} (disturbance).
- Input impedance Z_{ivcl} (disturbance).
- Output impedance Z_{ovcl} (disturbance).

5.7.1 Reference Voltage-to-Output Voltage Transfer Function T_{pcl}

Using the block diagram shown in Figure 5.56, the closed-loop reference voltage-to-output voltage transfer function is

$$T_{pcl}(s) = \left.\frac{v_o(s)}{v_{rv}(s)}\right|_{v_i=i_o=0} = \frac{T_{cv}T_{picl}}{1+T_v}. \tag{5.249}$$

Figure 5.57 shows the theoretically obtained magnitude and phase plots of the reference voltage-to-output voltage transfer function.

5.7.2 Input Voltage-to-Duty Cycle Transfer Function M_{dv}

Using the block diagram shown in Figure 5.58, the closed-loop input voltage-to-duty cycle transfer function is

$$M_{dv}(s) = \left.\frac{d(s)}{v_i(s)}\right|_{v_{rv}=i_o=0}. \tag{5.250}$$

From the block diagram,

$$v_o = v_o' + v_o'' = T_p d + M_v v_i. \tag{5.251}$$

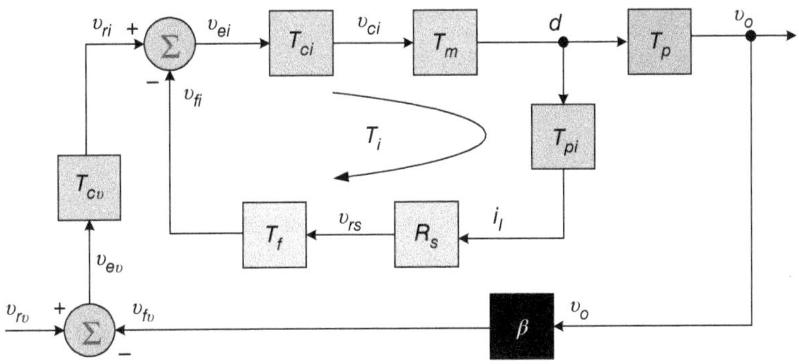

Figure 5.56 Block diagram used to derive the control-to-output voltage transfer function T_{pcl}.

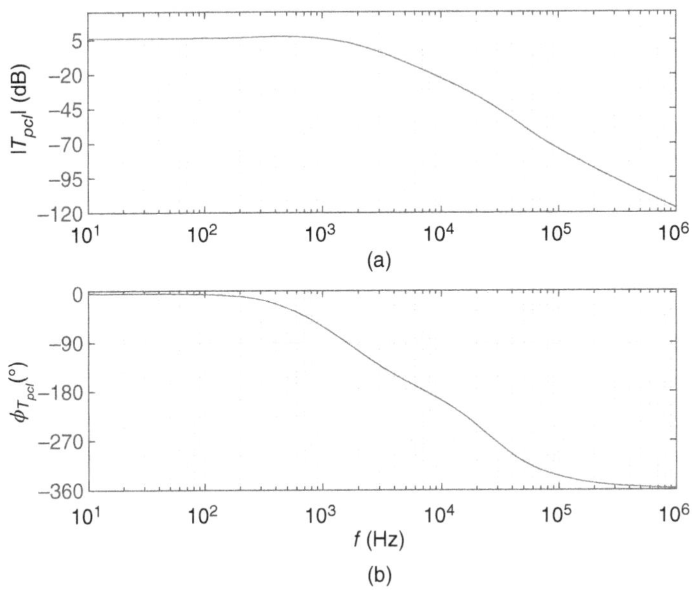

Figure 5.57 Theoretically obtained plots of the reference voltage-to-output voltage T_{pcl}. (a) Magnitude. (b) Phase.

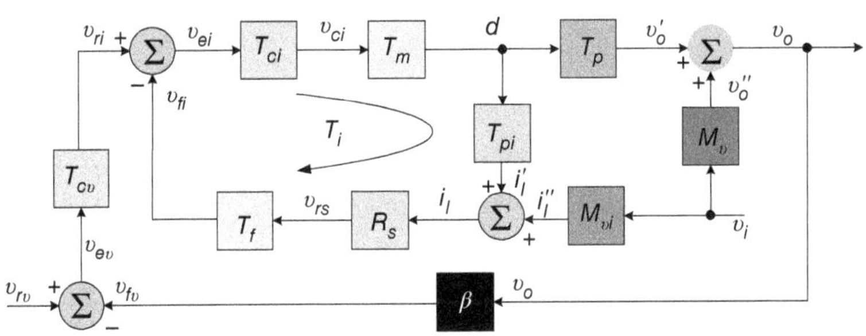

Figure 5.58 Block diagram used to derive the input voltage-to-duty cycle transfer function M_{dv}.

Also,

$$v_{fv} = -\beta v_o \tag{5.252}$$

and

$$v_{cv} = v_{ri} = -\beta T_{cv} v_o. \tag{5.253}$$

The error voltage v_{ei} is

$$v_{ei} = v_{ri} - v_{fi} = -T_{cv}\beta v_o - R_s T_f i_l \tag{5.254}$$

and

$$v_{ei} = \frac{d}{T_m T_{ci}}. \tag{5.255}$$

The inductor current is

$$i_l = T_{pi} d + M_{vi} v_i. \tag{5.256}$$

Substituting (5.256) and (5.255) in (5.254) yields

$$\frac{d}{T_m T_{ci}} = -T_{cv}\beta v_o - R_s T_f (T_{pi} d + M_{vi} v_i). \tag{5.257}$$

Rearranging gives

$$v_o = -\frac{d}{T_{cv}\beta}\left(\frac{1}{T_m T_{ci}} + R_s T_f T_{pi}\right) - \frac{v_i}{T_{cv}\beta}\left(R_s T_f M_{vi}\right). \tag{5.258}$$

Equating (5.251) and (5.258) and rearranging, we get

$$v_i\left(-\frac{M_{vi} R_s T_f}{T_{cv}\beta} - M_v\right) = d\left(T_p + \frac{1}{T_m T_{ci} T_{cv}\beta} + \frac{R_s T_f T_{pi}}{T_{cv}\beta}\right). \tag{5.259}$$

Simplifying further gives

$$-\frac{v_i}{T_{cv}\beta}\left(M_{vi} R_s T_f + M_v T_{cv}\beta\right) = \frac{d}{T_{cv}\beta}\left(T_p T_{cv}\beta + \frac{1 + R_s T_f T_{pi} T_m T_{ci}}{T_m T_{ci}}\right) \tag{5.260}$$

and

$$-v_i\left(M_{vi} R_s T_f + M_v T_{cv}\beta\right) = d\left(T_p T_{cv}\beta + \frac{1 + R_s T_f T_{pi} T_m T_{ci}}{T_m T_{ci}}\right). \tag{5.261}$$

Hence,

$$M_{dv}(s) = \frac{d(s)}{v_i(s)} = -\frac{M_{vi} R_s T_f + \beta M_v T_{cv}}{\beta T_p T_{cv} + \dfrac{1 + T_i}{T_m T_{ci}}} = -\frac{T_{ci} T_m (M_{vi} R_s T_f + \beta M_v T_{cv})}{\beta T_p T_{cv} T_{ci} T_m + 1 + T_i}. \tag{5.262}$$

Figure 5.59 shows the theoretically obtained magnitude and phase plots of the input voltage-to-duty cycle transfer function. From the Bode plot, the magnitude of the transfer function at DC is nearly $M_{dvo} = 0.022$ 1/V = −33.15 dB/V, i.e. for a step change in input voltage by 1 V, the duty cycle changes by nearly 0.022 to correct the disturbance. With only the inner-loop control, the change in duty cycle is merely $M_{dio} = 0.014$. Let us consider a case where the steady-state input

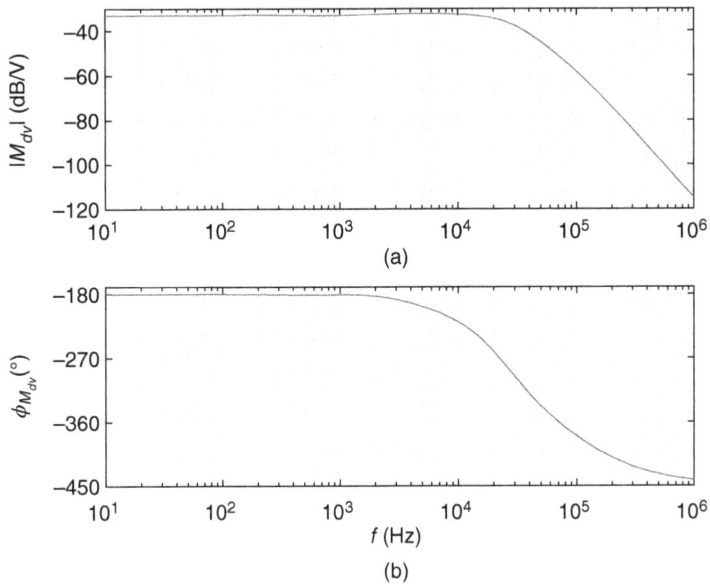

Figure 5.59 Theoretically obtained plots of the input voltage-to-duty cycle transfer function M_{dv}. (a) Magnitude. (b) Phase.

voltage of the designed flyback converter changes from 12 to 14 V. Then the corresponding change in the duty cycle needed to maintain the output voltage V_O constant is $\Delta D = 0.038$. However, the current loop is able to provide only $\Delta D = M_{dio} \times \Delta V_I = 0.014 \times 2 = 0.028$. The outer loop is able to achieve the remaining required compensation for any change in the input voltage and also achieve tightly controlled output voltage maintained at $V_O = 5$ V. In addition, for a change in the supply voltage in the positive direction, the duty cycle has to decrease to maintain the output voltage constant. Therefore, the phase difference between input voltage and duty cycle is −180°.

5.7.3 Input Voltage-to-Output Voltage Transfer Function M_{vcl}

Using the block diagram shown in Figure 5.60, the closed-loop input voltage-to-output voltage transfer function with outer-voltage loop is

$$M_{vcl}(s) = \left. \frac{v_o(s)}{v_i(s)} \right|_{v_{rv}=i_o=0}. \tag{5.263}$$

From (5.250),

$$d = M_{dv} v_i. \tag{5.264}$$

Substituting (5.264) into (5.251),

$$v_o = v_o' + v_o'' = T_p d + M_v v_i = T_p M_{dv} v_i + M_v v_i. \tag{5.265}$$

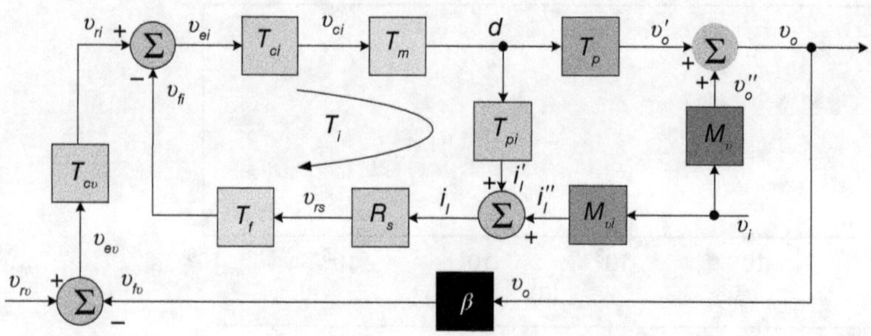

Figure 5.60 Block diagram used to derive the input voltage-to-output voltage transfer function M_{vcl}.

Rearrangement of (5.265) yields

$$M_{vcl}(s) = \frac{v_o(s)}{v_i(s)} = T_p M_{dv} + M_v. \tag{5.266}$$

Figure 5.61 shows the theoretically obtained magnitude and phase plots of the input voltage-to-output voltage transfer function M_{vcl}. By comparing the results of M_{vicl} and M_{vcl}, it can be observed that the outer-voltage loop provides a stronger attenuation to disturbances in the input voltage than that offered by the inner-current loop alone. The attenuation is stronger at low DC and low frequencies. However, at high frequencies, approximately beyond f_c, the transfer function takes the form of M_{vicl}, where the negative feedback is almost ineffective.

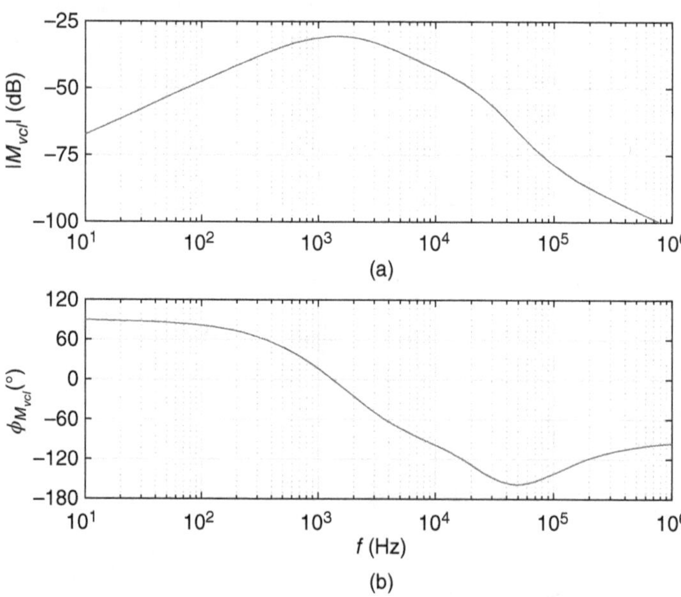

Figure 5.61 Theoretically obtained plots of the input voltage-to-output voltage transfer function M_{vcl}. (a) Magnitude. (b) Phase.

5.7.4 Input Impedance Z_{ivcl}

Using the block diagram shown in Figure 5.62, the closed-loop input impedance with outer-voltage loop is

$$Z_{ivcl}(s) = \left. \frac{v_i(s)}{i_l(s)} \right|_{v_{rv}=i_o=0}. \tag{5.267}$$

From Figure 5.62,

$$v_o = v_o' + v_o'' = T_p d + M_v v_i \tag{5.268}$$

and

$$i_l = i_l' + i_l'' = T_{pi} d + M_{vi} v_i. \tag{5.269}$$

Also,

$$v_{fv} = \beta v_o, \tag{5.270}$$

$$v_{ev} = v_{rv} - v_{fv} = 0 - \beta v_o = -\beta v_o, \tag{5.271}$$

and

$$v_{cv} = v_{ri} = T_{cv} v_{ev} = -\beta T_{cv} v_o. \tag{5.272}$$

The error voltage v_{ei} is

$$v_{ei} = \frac{d}{T_{ci} T_m}. \tag{5.273}$$

The feedback voltage v_{fi} is

$$v_{fi} = T_f R_s i_l. \tag{5.274}$$

The reference voltage to inner-current loop is

$$v_{ri} = v_{ei} + v_{fi}. \tag{5.275}$$

Substituting (5.272), (5.273), and (5.274) into (5.275),

$$-\beta T_{cv} v_o = \frac{d}{T_{ci} T_m} + T_f R_s i_l. \tag{5.276}$$

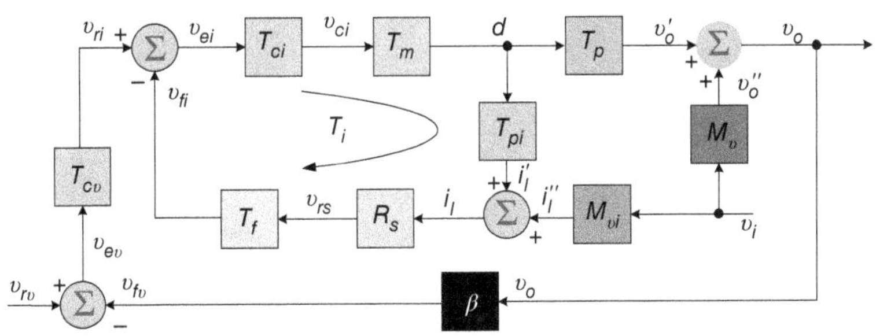

Figure 5.62 Block diagram used to derive the closed-outer-voltage-loop input impedance Z_{ivcl}.

Rearranging gives

$$v_o = -\frac{d}{\beta T_{cv} T_{ci} T_m} - \frac{T_f R_s i_l}{\beta T_{cv}}. \tag{5.277}$$

Substituting (5.268) into (5.277),

$$T_p d + M_v v_i = -\frac{d}{\beta T_{cv} T_{ci} T_m} - \frac{T_f R_s i_l}{\beta T_{cv}}. \tag{5.278}$$

Rearranging yields

$$d\left(T_p + \frac{1}{\beta T_{cv} T_{ci} T_m}\right) = -\frac{T_f R_s i_l}{\beta T_{cv}} - M_v v_i. \tag{5.279}$$

Substituting (5.269) into (5.279),

$$\begin{aligned} d\left(T_p + \frac{1}{\beta T_{cv} T_{ci} T_m}\right) &= -\frac{T_f R_s (T_{pi} d + M_{vi} v_i)}{\beta T_{cv}} - M_v v_i \\ d\left(T_p + \frac{1}{\beta T_{cv} T_{ci} T_m} + \frac{T_f R_s T_{pi}}{\beta T_{cv}}\right) &= \left(-\frac{T_f R_s M_{vi}}{\beta T_{cv}} - M_v\right) v_i \\ d\left(\frac{\beta T_p T_{ci} T_m T_{cv} + 1 + T_f R_s T_{pi} T_{ci} T_m}{\beta T_{ci} T_m T_{cv}}\right) &= \left(\frac{-T_f R_s M_{vi} - \beta T_{cv} M_v}{\beta T_{cv}}\right) v_i. \end{aligned} \tag{5.280}$$

Further solving for d, we get

$$d = -\frac{T_{ci} T_m (T_f R_s M_{vi} + \beta M_v T_{cv})}{\beta T_p T_{ci} T_m T_{cv} + 1 + T_f R_s T_{pi} T_{ci} T_m} v_i. \tag{5.281}$$

Substituting (5.269) into (5.80),

$$i_i = I_L d + D(T_{pi} d + M_{vi} v_i) = (I_L + D T_{pi}) d + D M_{vi} v_i. \tag{5.282}$$

Substituting (5.281) into (5.282),

$$\begin{aligned} i_i &= (I_L + D T_{pi}) d + D M_{vi} v_i \\ &= (I_L + D T_{pi}) \left[-\frac{T_{ci} T_m (T_f R_s M_{vi} + \beta M_v T_{cv})}{\beta T_p T_{ci} T_m T_{cv} + 1 + T_f R_s T_{pi} T_{ci} T_m} v_i\right] + D M_{vi} v_i \\ &= \left\{(I_L + D T_{pi}) \left[-\frac{T_{ci} T_m (T_f R_s M_{vi} + \beta M_v T_{cv})}{\beta T_p T_{ci} T_m T_{cv} + 1 + T_f R_s T_{pi} T_{ci} T_m}\right] + D M_{vi}\right\} v_i. \end{aligned} \tag{5.283}$$

Hence, the closed-loop input admittance for the outer-voltage loop is

$$\begin{aligned} Y_{ivcl}(s) &= \frac{i_i(s)}{v_i(s)} \\ &= \left\{(I_L + D T_{pi}) \left[-\frac{T_{ci} T_m (T_f R_s M_{vi} + \beta M_v T_{cv})}{\beta T_p T_{ci} T_m T_{cv} + 1 + T_f R_s T_{pi} T_{ci} T_m}\right] + D M_{vi}\right\}. \end{aligned} \tag{5.284}$$

The closed-loop input impedance for the outer-voltage loop is

$$Z_{ivcl}(s) = \frac{v_i(s)}{i_i(s)} = \frac{1}{Y_{ivcl}(s)}. \tag{5.285}$$

Figure 5.63 shows the theoretically obtained magnitude and phase plots of the closed-loop input impedance Z_{ivcl}.

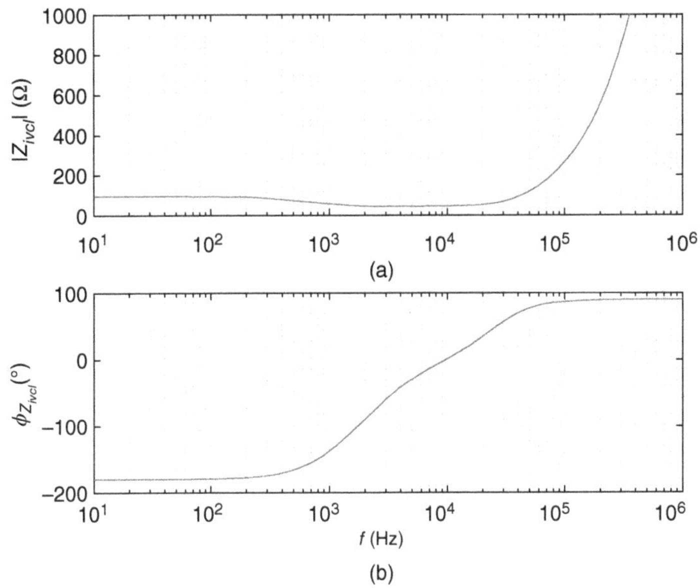

Figure 5.63 Theoretically obtained plots of the closed-outer-voltage loop input impedance Z_{ivcl}. (a) Magnitude. (b) Phase.

5.7.5 Output Impedance Z_{ovcl}

Using the block diagram shown in Figure 5.64, the closed-loop output impedance with outer-voltage loop is

$$Z_{ovcl}(s) = \left.\frac{v_o(s)}{i_o(s)}\right|_{v_{rv}=v_i=0}. \tag{5.286}$$

From Figure 5.64,

$$v_o = v_o' + v_o'' = T_p d + Z_o i_o \tag{5.287}$$

and

$$i_l = i_l' + i_l'' = A_i i_o + T_{pi} d. \tag{5.288}$$

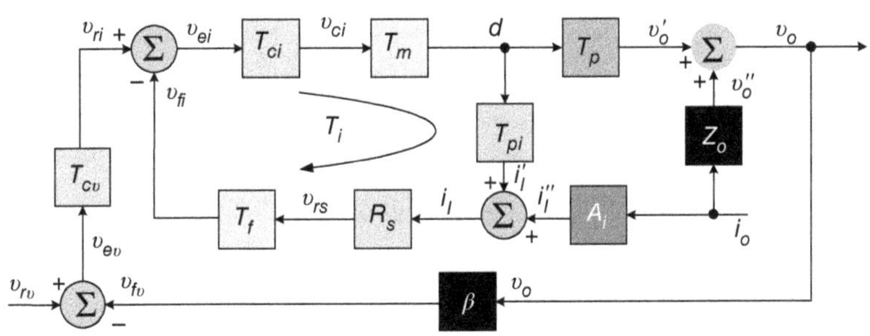

Figure 5.64 Block diagram used to derive the closed-outer-voltage-loop output impedance Z_{ovcl}.

Also,

$$v_{fv} = \beta v_o \tag{5.289}$$

and

$$v_{cv} = v_{ri} = -\beta T_{cv} v_o. \tag{5.290}$$

The error voltage v_{ei} is

$$v_{ei} = v_{ri} - v_{fi} = -T_{cv}\beta v_o - R_s T_f i_l \tag{5.291}$$

and

$$v_{ei} = \frac{d}{T_m T_{ci}}. \tag{5.292}$$

Substituting (5.288) and (5.292) into (5.291) yields

$$\frac{d}{T_m T_{ci}} = -T_{cv}\beta v_o - R_s T_f (A_i i_o + T_{pi} d). \tag{5.293}$$

Rearranging (5.293),

$$d = -\frac{(T_{cv}\beta v_o + R_s T_f A_i i_o) T_{ci} T_m}{1 + T_i}. \tag{5.294}$$

Substituting (5.294) into (5.287) yields

$$v_o = T_p \left[-\frac{(T_{cv}\beta v_o + R_s T_f A_i i_o) T_{ci} T_m}{1 + T_i} \right] + Z_o i_o. \tag{5.295}$$

Rearranging gives

$$v_o \left(1 + \frac{\beta T_p T_{cv} T_{ci} T_m}{1 + T_i} \right) = i_o \left(Z_o - \frac{T_p R_s T_f A_i T_{ci} T_m}{1 + T_i} \right). \tag{5.296}$$

Hence, the closed-loop output impedance for the outer-voltage loop is

$$Z_{ovcl}(s) = \frac{v_o(s)}{i_o(s)} = \frac{Z_o(1 + T_i) - A_i T_p T_{ix}}{1 + T_i + T_p T_{cv} \beta T_{ci} T_m}. \tag{5.297}$$

Figure 5.65 shows the theoretically obtained magnitude and phase plots of the closed-outer-voltage-loop output impedance Z_{ovcl}.

5.8 Comparison of Closed-Loop and Open-Loop Step Responses

5.8.1 Response of Output Voltage to Step Change in Input Voltage

Figure 5.66 compares the responses of the output voltage for unit step changes in input voltage by 1 V for (i) open loop (M_v), (ii) closed-inner-current loop (M_{vicl}), and (iii) closed-outer-voltage loop in the presence of the inner-current loop (M_{vcl}). In the open-loop configuration, a step change $\Delta V_I = 1$ V in the input voltage causes the output voltage to achieve a new steady-state value $V_O(\infty) = 5.44$ V. Therefore, the step

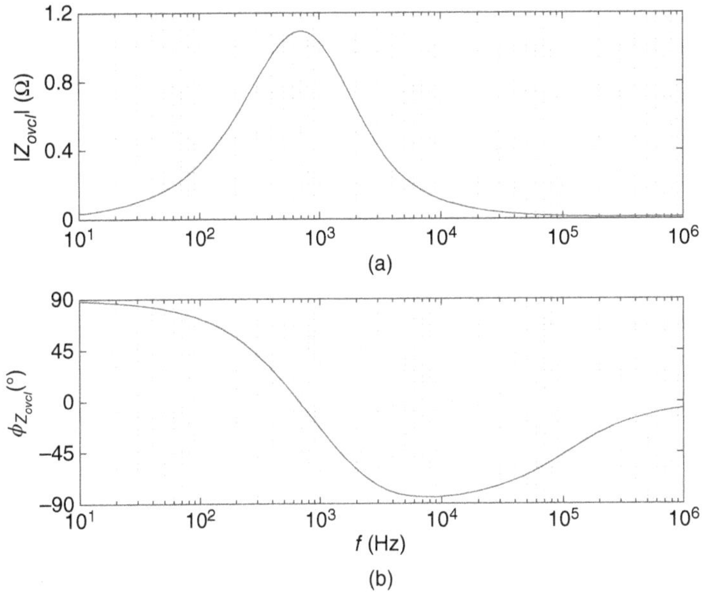

Figure 5.65 Theoretically obtained plots of the closed-outer-voltage loop output impedance Z_{ovcl}. (a) Magnitude. (b) Phase.

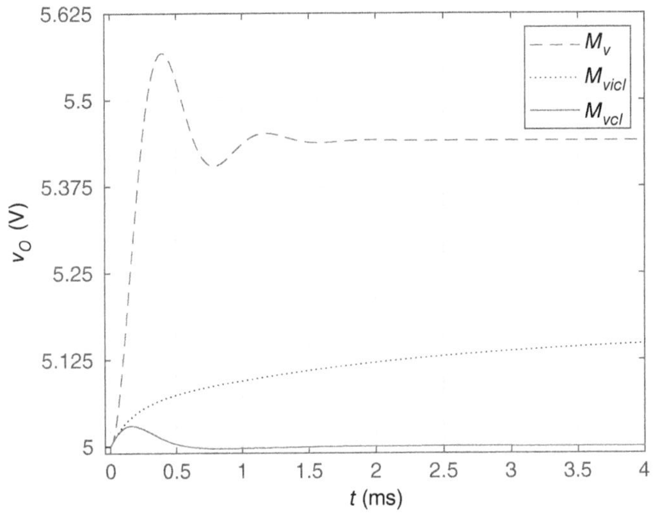

Figure 5.66 Comparison of responses in output voltage for step changes in input voltage by 1 V obtained using M_v, M_{vicl}, and M_{vcl} transfer functions, respectively.

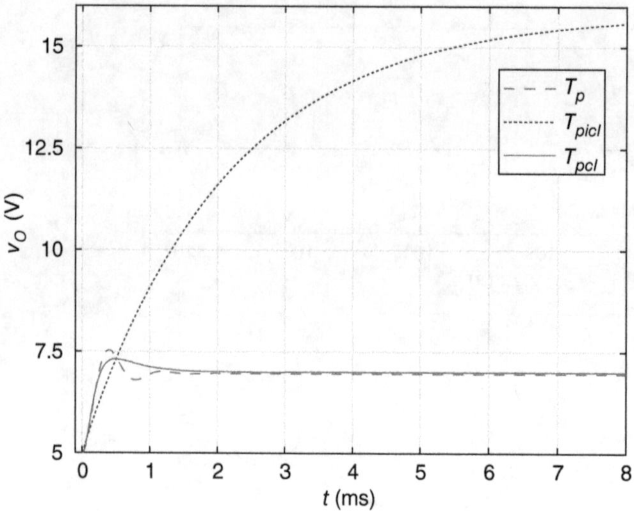

Figure 5.67 Comparison of responses in output voltage for step changes in duty cycle by 0.1, current-loop reference voltage by 1 V, and voltage-loop reference voltage by 1 V, obtained using T_p, T_{picl}, and T_{pcl} transfer functions, respectively.

change in the output voltage is $\Delta V_O = V_O(\infty) - V_O(0) = 5.44 - 5 = 0.44$ V, yielding $M_{vo} = (\Delta V_O)/(\Delta V_I) = 0.44/1 = 0.44$ V/V, which was also observed in the magnitude response of M_v. However, for closed-loop responses determined by current-loop transfer function M_{vicl}, a step changes in input voltage causes the output voltage to rise to a new steady-state value $V_O(\infty) \approx 5.16$ V within 4 ms. Therefore, the step change in the output voltage is $\Delta V_O = V_O(\infty) - V_O(0) = 5.16 - 5 = 0.16$ V, yielding $M_{viclo} = (\Delta V_O)/(\Delta V_I) = 0.16/1 = 0.16$ V/V, which was also observed in the magnitude response of M_{vicl}. For the voltage-loop transfer function M_{vcl}, step changes in input voltage cause the output voltage to return to its steady-state value $V_O(\infty) = 5$ V within 1 ms.

5.8.2 Response of Output Voltage to Step Change in Duty Cycle, Current-Loop Reference Voltage, and Voltage-Loop Reference Voltage

Figure 5.67 compares the responses of the output voltage for step changes in (i) duty cycle for open loop by 0.1 (T_p), (ii) current-loop reference voltage for the closed-inner-current loop by 1 V (T_{picl}), and (iii) voltage-loop reference voltage for the closed-outer-voltage loop by 1 V (T_{pcl}). In the open-loop configuration, a step change in duty cycle $\Delta D = 0.1$ causes the output voltage to achieve a new steady-state value $V_O(\infty) = 6.96$ V. Therefore, the step change in the output voltage is $\Delta V_O = V_O(\infty) - V_O(0) = 6.96 - 5 = 1.96$ V, yielding $T_{po} = (\Delta V_O)/\Delta D = 1.96/0.1 = 19.6$ V, which was also observed in the magnitude plot of T_p. The step response shows a settling time of $\approx$2 ms. The output voltage response for the closed-inner current loop achieves a steady-state value

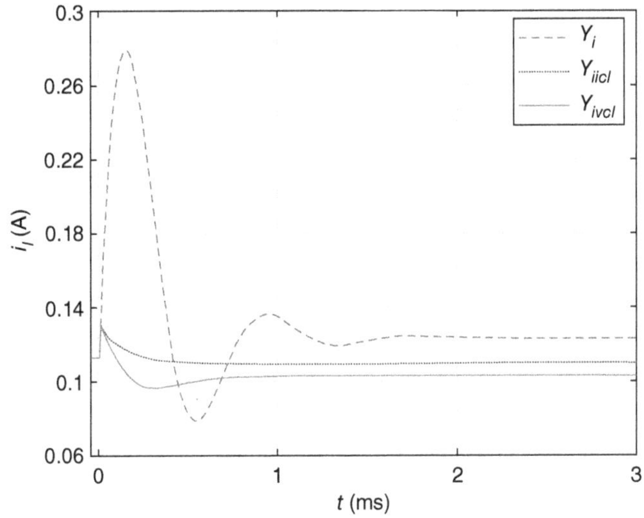

Figure 5.68 Comparison of responses in input current for step changes in the input voltage by 1 V obtained using Y_i, Y_{iicl}, and Y_{ivcl} transfer functions.

$V_O(\infty) = 15.7$ V in a settling time of more than 8 ms. Therefore, the step change in the output voltage is $\Delta V_O = V_O(\infty) - V_O(0) = 15.7 - 5 = 10.7$ V, yielding $T_{piclo} = (\Delta V_O)/\Delta D = 10.7/1 = 10.7$ V, which was also observed in the magnitude plot of T_{picl}. The output voltage response for the closed-outer voltage loop achieves a steady-state value $V_O(\infty) = 7$ V in a rapid settling time of nearly 2 ms. Therefore, the step change in the output voltage is $\Delta V_O = V_O(\infty) - V_O(0) = 7 - 5 = 2$ V, yielding $T_{pclo} = (\Delta V_O)/\Delta D = 2/1 = 2$ V, which was also observed in the magnitude plot of T_{pcl}.

5.8.3 Response of Input Current to Step Change in Input Voltage

Figure 5.68 compares the responses of the input current for unit step change in input voltage by 1 V for (i) open loop (Y_i), (ii) closed-inner-current loop (Y_{iicl}), and (iii) closed-outer-voltage loop (Y_{ivcl}). In the open-loop configuration, a step change in input voltage $\Delta V_I = 1$ V causes the input current to achieve a new steady-state value $I_I(\infty) = 0.123$ A. Therefore, the step change in the input current is $\Delta I_I = I_I(\infty) - I_I(0) = 0.123 - 0.1132 = 0.01$ A, yielding $Z_{io} = R_i = (\Delta V_I)/(\Delta I_I) = 1/0.01 = 100\ \Omega$, which was also observed in the magnitude plot of Z_i. The step response shows a settling time of 2 ms. The input current response for the closed-inner current loop achieves a steady-state value $I_I(\infty) = 0.11$ A in a settling time of 0.5 ms. The input current response for the closed-outer voltage loop achieves a steady-state value $I_I(\infty) = 0.10$ A in a rapid settling time of nearly 0.5 ms. The reduction in the input current in negative direction for positive step change in output voltage is due to the negative input resistance caused at DC by the inner-current and outer-voltage loops.

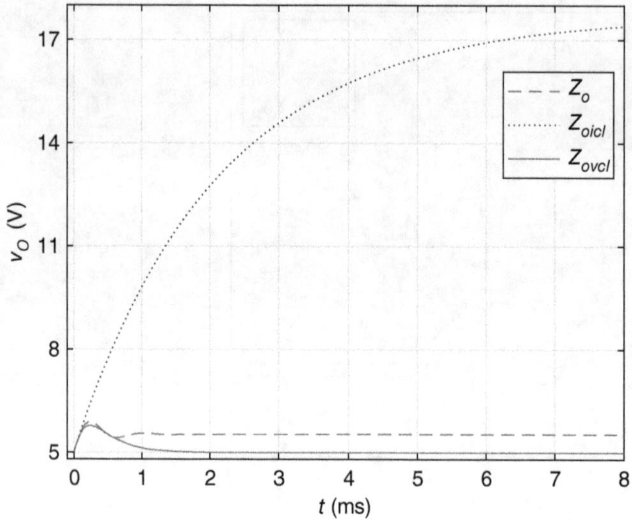

Figure 5.69 Comparison of responses in the output voltage for step changes in the output current by 1 A obtained using Z_o, Z_{oicl}, and Z_{ovcl} transfer functions.

5.8.4 Response of Output Voltage to Step Change in Load Current

Figure 5.69 compares the responses of the output voltage for unit step changes in output current by 1 A for (i) open loop (Z_o), (ii) closed-inner-current loop (Z_{oicl}), and (iii) closed-outer-voltage loop (Z_{ovcl}). In the open-loop configuration, a step change in output current $\Delta I_O = 1$ A causes the output voltage to achieve a new steady-state value $V_O(\infty) = 5.5$ V. Therefore, the step change in the output voltage is $\Delta V_O = V_O(\infty) - V_O(0) = 5.5 - 5 = 0.5$ V, yielding $Z_{oo} = R_o = (\Delta V_O)/(\Delta I_O) = 0.5/1 = 0.5$ V/A, which was also observed in the magnitude plot of Z_o. The step response shows a settling time of 1 ms. The output voltage response for the closed-inner current loop achieves a steady-state value $V_O(\infty) = 17.4$ V in a settling time of more than 8 ms. The output voltage for the closed-outer voltage loop returns to the steady-state value $V_O(\infty) = V_O(0) = 5$ V in a settling time of 1 ms.

5.9 Summary

1. The small-signal, linear model of the flyback converter in CCM consists of a current-controlled current source representing the transistor and a voltage-controlled voltage source representing the diode, in addition to the input AC voltage source, inductor, and filter capacitor.
2. The small-signal duty cycle-to-output voltage transfer function of a lossy flyback converter is a second-order transfer function constituting an LHP real zero, an RHP real zero and a pair of complex-conjugate poles.

3. The small-signal duty cycle-to-inductor transfer function of a lossy flyback converter is a second-order transfer with a real LHP zero and a pair of complex-conjugate LHP poles.
4. The small-signal input-to-output voltage transfer function of a lossy flyback converter is a second-order transfer exhibiting an LHP zero and a pair of complex-conjugate LHP poles. The location of the LHP zero and complex poles is identical to that of the duty cycle-to-output voltage transfer function.
5. The small-signal input voltage-to-inductor transfer function of a lossy flyback converter is a second-order transfer exhibiting an LHP zero and a pair of complex-conjugate LHP poles.
6. The input impedance of a lossy flyback converter exhibits a pair of LHP complex-conjugate zeros and an LHP real zero. The input resistance is a function of the load resistance and the duty cycle. The location of the complex zero is identical to the complex pole produced by the duty cycle-to-output voltage transfer function.
7. The output impedance of a lossy flyback converter has two LHP real zeros and an LHP complex-conjugate poles. The output resistance is a function of the load resistance. The location of the complex pole is identical to that produced by the duty cycle-to-output voltage transfer function.
8. The current-mode control of a flyback converter can be achieved using an integral single-lead controller and can effectively reduce the DC steady-state error and improve the closed-loop bandwidth. This is a major advantage of the voltage-mode control, where the presence of RHP zero in the duty cycle-to-output voltage transfer function demands double-lead or higher forms of control scheme for effective compensation.
9. The average current-mode control methodology can be extended for the inverting flyback converter.
10. The effects of the transformer leakage inductances on the flyback dynamic performance should be considered.

References

1 T. A. Froeschle, "Two-state modulation techniques for power systems," Semi-annual reports prepared by Bose Corporation for U.S Army Electronics Command, Natick, MA, Contract DA28-043-AMC-022812(E), Access Number AD 815603, June 1967, Access Number AD 823239, November 1967, Access Number AD 831768, April 1968, and Access Number AD 844635, November 1968.

2 C. W. Deisch, "Simple switching control method changes power converter into a current source," *IEEE Power Electronics Specialists Conference*, Syracuse, NY, USA, 1978, pp. 300–306.

3 R. D. Middlebrook, "Topics in multiple-loop regulators and current-mode programming," *IEEE Transactions on Power Electronics*, vol. PE-2, no. 2, pp. 109–125, April 1987.

4 R. D. Middlebrook, "Modeling current-programmed buck and boost regulators," *IEEE Transactions on Power Electronics*, vol. 4, no. 1, pp. 36–52, January 1989.

5 R. D. Middlebrook, "The general feedback theorem: a final solution for feedback systems," *IEEE Microwave Magazine*, vol. 7, no. 2, pp. 50–63, April 2006.

6 S. P. Hsu, A. Brown, L. Rensink, and R. D. Middlebrook, "Modelling and analysis of switching DC-to-DC converters in constant-frequency current-programmed mode," *IEEE Power Electronics Specialists Conference*, San Diego, CA, USA, 1979, pp. 284–301.

7 L. H. Dixon, "Average current-mode control of switching power supplies," *Unitrode Power Supply Design Seminar Manual*, 1990.

8 L. H. Dixon, "Current-mode control of switching power power supplies," Unitrode Power Supply Design Notebook, SEM-400, 1985.

9 R. B. Ridley, B. H. Cho, and F. C. Y. Lee, "Analysis and interpretation of loop gains of multi-loop-controlled switching regulators," *IEEE Transactions on Power Electronics*, vol. 3, no. 4, pp. 489–498, October 1988.

10 R. B. Ridley, "A new, continuous-time model for current-mode control," *Power Conversion and Intelligent Motion Conference*, pp. 16–19, October 1989.

11 W. Tang, F. C. Lee, and R. B. Ridley, "Small-signal modeling of average current-mode control," *IEEE Transactions on Power Electronics*, vol. 8, no. 2, pp. 112–119, April 1993.

Average Current-Mode Control of DC-DC Power Converters, First Edition.
Marian K. Kazimierczuk, Dalvir K. Saini, and Agasthya Ayachit.

12 J. Sun and R. M. Bass, "Modeling and practical design issues for average current control," *Proceedings of Applied Power Electronics Conference and Exposition*, Dallas, TX, March 1999, vol. 2, pp. 980–986.

13 J. Sun and R. M. Bass, "Modeling and practical design issues for average current control," *Proceedings of Applied Power Electronics Conference and Exposition*, Dallas, TX, March 14–18, 1999, vol. 2, pp. 980–986.

14 R. Redl and I. Novak, "Instabilities in current-mode controlled switching voltage regulators," *Proceedings of Power Electronics Specialists Conference*, pp. 17–28, 1981.

15 Y. W. Lu, G. Feng, and Y.-F. Liu, "A large signal dynamic model for DC-to-DC converters with average current control," *Applied Power Electronics Conference and Exposition*, 2004, vol. 2, pp. 797–803.

16 J. Leyva-Ramos, M. G. Ortiz-Lopez, L. H. Diaz-Saldierna, and M. Martinez-Cruz, "Average current controlled switching regulators with cascade boost converters," *IET Power Electronics*, vol. 4, no. 1, pp. 1–10, January 2011.

17 J. A. Morales-Saldana, J. Leyva-Ramos, E. E. Carbajal-Gutierrez and M. G. Ortiz-Lopez, "Average current-mode control scheme for a quadratic buck converter with a single switch," *IEEE Transactions on Power Electronics*, vol. 23, no. 1, pp. 485–490, January 2008.

18 C. Sun and B. Lehman, "Discussions on control loop design in average current mode control [PWM DC/DC power convertors]," *Proceedings of IEEE Industry Applications Conference*, Rome, Italy, 2000, vol. 4, pp. 2411–2417.

19 B. Bryant and M. K. Kazimierczuk, "Open-loop power-stage transfer functions relevant to current-mode control of boost PWM converter operating in CCM," *IEEE Transactions on Circuits and Systems*, vol. 52, no. 10, pp. 2158–2164, October 2005.

20 B. Bryant and M. K. Kazimierczuk, "Modeling the closed-current loop of PWM boost DC–DC converters operating in CCM with peak current-mode control," *IEEE Transactions on Circuits and Systems*, vol. 52, no. 11, pp. 2404–2412, November 2005.

21 B. Bryant and M. K. Kazimierczuk, "Voltage-loop of boost PWM DC–DC converters with peak current-mode control," *IEEE Transactions on Circuits and Systems*, vol. 53, no. 1, pp. 99–105, January 2006.

22 B. Bryant and M. K. Kazimierczuk, "Voltage-loop power-stage transfer functions with MOSFET delay for boost PWM converter operating in CCM," *IEEE Transactions on Industrial Electronics*, vol.54, no. 1, pp. 347–353, February 2007.

23 B. Bryant and M. K. Kazimierczuk, "Sample and hold effect in PWM DC–DC converters with peak current-mode control," *IEEE International Symposium on Circuits and Systems*, Vancouver, BC, Canada, May 23–26, 2004, Vol. V, pp. 860–863.

24 B. Bryant and M. K. Kazimierczuk, "Small-signal duty cycle to inductor current transfer function for boost PWM DC–DC converter in continuous conduction mode," *IEEE International Symposium on Circuits and Systems*, Vancouver, BC, Canada, May 23–26, 2004, Vol. V, pp. 856–859.

25 N. Kondrath and M. K. Kazimierczuk, "Audio-susceptibility of the inner-loop of peak current-mode controlled PWM DC–DC buck converter in CCM," *Proceedings of IEEE Industrial Electronics Conference*, Montreal, QC, 2012, pp. 250–255.

26 N. Kondrath and M. K. Kazimierczuk, "Comparison of wide- and high-frequency duty-ratio-to-inductor-current transfer functions of DC–DC PWM buck converter in CCM," *IEEE Transactions on Industrial Electronics*, vol. 59, no. 1, pp. 641–643, January 2012.

27 N. Kondrath and M. K. Kazimierczuk, "Control-to-output and duty ratio-to-inductor current transfer functions of peak current-mode controlled DC–DC PWM buck converter in CCM," *IEEE International Symposium on Circuits and Systems*, 2010, Paris, France, pp. 2734–2737.

28 N. Kondrath and M. K. Kazimierczuk, "Control current and relative stability of peak current-mode controlled pulse-width modulated DC–DC converters without slope compensation," *IET Power Electronics*, vol. 3, no. 6, pp. 936–946, November 2010.

29 N. Kondrath and M. K. Kazimierczuk, "Loop gain and margins of stability of inner-current loop of peak current-mode controlled PWM DC–DC converters in CCM," *IET Proceedings Power Electronics*, vol. 4, no. 6, pp. 701–707, 2011.

30 N. Kondrath and M. K. Kazimierczuk, "Control-to-output transfer function of peak current-mode controlled PWM DC–DC boost converter in CCM," *IET Electronics Letters*, vol. 47, no. 17, pp. 991–993, August 2011.

31 N. Kondrath and M. K. Kazimierczuk, "Control-to-output transfer function of a DC–DC pulse-width-modulated buck converter in continuous conduction mode," *IET Power Electronics*, vol. 5, no. 5, pp. 582–590, May 2012.

32 N. Kondrath and M. K. Kazimierczuk, "Unified model to derive control-to-output transfer function of peak current-mode controlled PWM DC–DC converters in CCM," *IET Power Electronics*, vol. 5, no. 9, pp. 1706–1713, November 2012.

33 N. Kondrath and M. K. Kazimierczuk, "Margins of stability of inner-current loop of current-mode controlled PWM DC–DC converters," *IEEE International Symposium on Circuits and Systems*, Taipei, Taiwan, May 24–27, 2009, pp. 1985–1988.

34 D. K. Saini and M. K. Kazimierczuk, "Average current-mode control of buck DC–DC converter with reduced control voltage ripple," *Proceedings of IEEE Industrial Electronics Conference*, Florence, Italy, October 2016.

35 D. K. Saini and M. K. Kazimierczuk, "Audio-susceptibility of inner-loop of true-average current-mode controlled buck DC–DC converter," *IEEE Midwest Symposium on Circuits and Systems*, Boston, Massachusetts, August 6-9, 2017, pp. 460–463.

36 D. Czarkowski and M. K. Kazimierczuk, "Energy-conservation approach to modeling PWM DC–DC converters," *IEEE Transactions on Aerospace and Electronic Systems*, vol. 29, no. 3, pp. 1059–1063, July 1993.

37 M. K. Kazimierczuk and D. Czarkowski, "Application of the principle of energy conservation to modeling the PWM converters," *Proceedings of the 2nd IEEE Conference on Control Applications*, Vancouver, Canada, September 13–16, 1993, pp. 291–296.

38 D. Czarkowski and M. K. Kazimierczuk, "Static- and dynamic-circuit models of PWM buck-derived DC–DC converters," *IEEE Proceedings Part G, Circuits, Devices and Systems*, vol. 139, pp. 669–679, December 1992.

39 A. Reatti and M. K. Kazimierczuk, "Small-signal model of PWM converters for discontinuous conduction mode and its application for boost converter," *IEEE Transactions on Circuits and Systems I: Fundamental Theory and Applications*, vol. 50, no. 1, pp. 65–73, January 2003.

40 V. P. Galigekere and M. K. Kazimierczuk, "Small-signal modeling of Z-source DC-to-DC converter using circuit averaging technique," *IEEE Transactions on Power Electronics*, vol. 28, no. 3, pp. 1286–1296, March 2013.

41 M. K. Kazimierczuk and A. Ayachit, *Laboratory Manual for Pulse-Width Modulated DC–DC Power Converters*, John Wiley & Sons, Chichester, UK, 2016.

42 Y. P. Siwakoti, F. Blaabjerg, V. P. Galigekere, A. Ayachit, and M. K. Kazimierczuk, "A-source impedance network," *IEEE Transactions on Power Electronics*, vol. 31, no. 12, pp. 8081–8087, December 2016.

43 A. Ayachit, Y. P. Siwakoti, V. P. N. Galigekere, M. K. Kazimierczuk, and F. Blaabjerg, "Steady-state and small-signal analysis of A-source converter," *IEEE Transactions on Power Electronics*, vol. 33, no. 8, pp. 7118–7131, August 2018.

44 A. Chadha and M. K. Kazimierczuk, "Small-signal modeling of open-loop PWM tapped-inductor buck DC–DC converter in CCM," *IEEE Transactions on Industrial Electronics*, vol. 68, no. 7, pp. 5765–5775, July 2021.

45 M. K. Kazimierczuk, *Pulse-width Modulated DC–DC Power Converters*, 2nd Ed., Wiley, Chichester, UK, 2015.

46 A. Reatti, F. Corti, A. Tesi, A. Torali, and M. K. Kazimierczuk, "Effects of parasitic components on dynamic performance of power stages of DC–DC PWM buck and boost converters in CCM," *IEEE International Symposium on Circuits and Systems*, Sapporo, Japan, May 26–29, 2019.

47 H. D. Venable, "The K Factor: a new mathematical tool for stability analysis and synthesis," *Proceedings of POWERCON 10*, San Diego, California, March 1983.

48 A. Reatti, F. Corti, A. Tesi, A. Torali, and M. K. Kazimierczuk, "Nonlinear exact analysis and solution of power stage of DC–DC PWM boost converter," *IEEE International Symposium on Circuits and Systems*, Sapporo, Japan, May 26–29, 2019.

49 J. J. Lee and M. K. Kazimierczuk, "Effects of load changes on the control-to-output transfer function of a buck-boost converter in CCM," *How2Power*, May 2010.

50 B. Bryant and M. K. Kazimierczuk, "Derivation of PWM DC–DC buck-boost converter topology," *IEEE International Symposium on Circuits and Systems*, May 2002, Scottsdale, AZ, pp. 841–844.

51 W. K. Ho, O. P. Gan, E. B. Tay, and E. L. Ang, "Performance and gain and phase margins of well-known PID tuning formulas," *IEEE Transactions on Control Systems Technology*, vol. 4, no. 4, pp. 473–477, July 1996.

52 D. K. Saini, A. Reatti, and M. K. Kazimierczuk, "Average current-mode control of buck DC–DC converter with reduced control voltage ripple," *Proc. IEEE Industrial Electronics Conference*, Florence, Italy, October 2016.

53 M. K. Kazimierczuk, "Transfer function of current modulator in PWM converters with current-mode control," *IEEE Transactions on Circuits and Systems I: Fundamental Theory and Applications*, vol. 47, no. 9, pp. 1407–1412, September 2000.

Appendix A

Design Equations for Continuous-Conduction Mode

A.1 Common Equations Needed for the Design of Converters

A.1.1 DC Output Power

$$P_{Omin} = V_O I_{Omin} = \frac{V_O^2}{R_{Lmax}}. \tag{A.1}$$

$$P_{Omax} = V_O I_{Omax} = \frac{V_O^2}{R_{Lmin}}. \tag{A.2}$$

A.1.2 DC Voltage Transfer Function

$$M_{VDCmin} = \frac{V_O}{V_{Imax}}. \tag{A.3}$$

$$M_{VDCnom} = \frac{V_O}{V_{Inom}}. \tag{A.4}$$

$$M_{VDCmax} = \frac{V_O}{V_{Imin}}. \tag{A.5}$$

A.2 Specific Expressions for the Design of Converters in CCM

The equations for estimating the values of the different parameters and components of the buck, boost, and buck–boost DC–DC converters in continuous-conduction mode (CCM) are provided in Table A.1.

Average Current-Mode Control of DC-DC Power Converters, First Edition.
Marian K. Kazimierczuk, Dalvir K. Saini, and Agasthya Ayachit.

Table A.1 Steady-state design equations for buck, boost, and buck–boost DC–DC converters in CCM.

Parameter/topology	Buck	Boost	Buck-boost
Min. duty cycle D_{min}	$\frac{M_{VDCmin}}{\eta}$	$1 - \frac{\eta}{M_{VDCmin}}$	$\frac{M_{VDCmin}}{M_{VDCmin} + \eta}$
Nom. duty cycle D_{nom}	$\frac{M_{VDCnom}}{\eta}$	$1 - \frac{\eta}{M_{VDCnom}}$	$\frac{M_{VDCnom}}{M_{VDCnom} + \eta}$
Max. duty cycle D_{max}	$\frac{M_{VDCmax}}{\eta}$	$1 - \frac{\eta}{M_{VDCmax}}$	$\frac{M_{VDCmax}}{M_{VDCmax} + \eta}$
Min. inductance L_{min}	$\frac{R_{Lmax}(1 - D_{min})}{2f_s}$	$\frac{2}{27}\frac{R_{Lmax}}{f_s}$ for $D_{min} < 1/3$ $\frac{R_{Lmax}D_{min}(1 - D_{min})^2}{2f_s}$ for $D_{min} > 1/3$	$\frac{R_{Lmax}(1 - D_{min})^2}{2f_s}$
Max. inductor current ripple Δi_{Lmax} (where $L > L_{min}$)	$\frac{V_O(1 - D_{min})}{f_s L}$	$\frac{V_O D_{min}(1 - D_{min})}{f_s L}$	$\frac{V_O(1 - D_{max})}{2f_s}$
Max. ESR of filter capacitor r_{Cmax}	$\frac{V_r}{\Delta i_{Lmax}}$	$\frac{V_r}{2I_{DMmax}}$	$\frac{V_r}{2I_{DMmax}}$
Min. filter capacitance C_{min} (where $r_C < r_{Cmax}$)	$\max\left[\frac{D_{max}}{2f_s r_C}, \frac{1 - D_{min}}{2f_s r_C}\right]$	$\frac{2D_{max}V_O}{f_s R_{Lmin} V_r}$	$\frac{2D_{max}V_O}{f_s R_{Lmin} V_r}$
Max. MOSFET voltage stress V_{SM}	V_{Imax}	V_O	$V_O + V_{Imax}$
Max. diode voltage stress V_{DM}	V_{Imax}	V_O	$V_O + V_{Imax}$
Max. MOSFET current stress I_{SM}	$I_{Omax} + \frac{\Delta i_{Lmax}}{2}$	$\frac{I_{Omax}}{1 - D_{max}} + \frac{V_O D_{max}(1 - D_{max})}{2f_s L}$	$I_{Imax} + I_{Omax} + \frac{\Delta i_{Lmax}}{2}$
Max. diode current stress I_{DM}	$I_{Omax} + \frac{\Delta i_{Lmax}}{2}$	$\frac{I_{Omax}}{1 - D_{max}} + \frac{V_O D_{max}(1 - D_{max})}{2f_s L}$	$I_{Imax} + I_{Omax} + \frac{\Delta i_{Lmax}}{2}$

Appendix B

MOSFET Parameters

Part number	V_{DSS} (V)	I_{SM} (A)	r_{DS} (mΩ)	C_o (pF)	V_{th} (V)
IRF150	100	40	0.055	100	4
IRF430	500	4.5	1.8	135	4
NTD3055	60	12	94	107	4
IRF142	100	24	0.11	100	4
IRF840	500	8	0.85	100	4
IRF740	400	10	0.55	100	4

Average Current-Mode Control of DC-DC Power Converters, First Edition.
Marian K. Kazimierczuk, Dalvir K. Saini, and Agasthya Ayachit.

Appendix C

Diode Parameters

Part number	V_{RRM} (V)	I_{Fave} (A)	R_F (mΩ)	V_F (V)
MBR1060	60	20	25	0.4
MUR1560	600	15	17.1	1.5
MBR10100	100	10	15	0.8
MBR10100	100	10	15	0.8
MUR2510	100	25	20	0.7
MUR2540	40	25	10	0.3
MR826	600	35	—	—
MBR4040	40	40	25	0.4

Average Current-Mode Control of DC-DC Power Converters, First Edition.
Marian K. Kazimierczuk, Dalvir K. Saini, and Agasthya Ayachit.

Appendix D

Selected MOSFETs' Spice Models

D.1 IRF430

```
*******************************************************************************
.model IRF430  NMOS(Level=3 Gamma=0 Delta=0 Eta=0 Theta=0 Kappa=0.2 Vmax=0 Xj=0
+ Tox=100n Uo=600 Phi=.6 Rs=2.514m Kp=20.56u W=.33 L=2u Vto=3.783
+ Rd=.944 Rds=2.222MEG Cbd=677.2p Pb=.8 Mj=.5 Fc=.5 Cgso=1.725n
+ Cgdo=136.7p Rg=1.159 Is=46.85p N=1 Tt=585n)
*******************************************************************************
```

D.2 IRF520

```
*******************************************************************************
.model IRF520  NMOS(Level=3 Gamma=0 Delta=0 Eta=0 Theta=0 Kappa=0.2 Vmax=0 Xj=0
+ Tox=100n Uo=600 Phi=.6 Rs=.1459 Kp=20.79u W=.73 L=2u Vto=3.59
+ Rd=80.23m Rds=444.4K Cbd=622.1p Pb=.8 Mj=.5 Fc=.5 Cgso=517.9p
+ Cgdo=137.3p Rg=6.675 Is=2.438p N=1 Tt=137n)
*******************************************************************************
```

D.3 IRF150

```
*******************************************************************************
.model IRF150 NMOS(Level=3 Gamma=0 Delta=0 Eta=0 Theta=0 Kappa=0 Vmax=0 Xj=0
+ Tox=100n Uo=600 Phi=.6 Rs=1.624m Kp=50u W=160u L=2u Vto=2
+ Rd=1.031m Rds=444.4K Cbd=3.229n Pb=.8 Mj=.5 Fc=.5 Cgso=9.027n
+ Cgdo=1.679n Rg=13.89 Is=194E-18 N=1 Tt=288n lambda=0.05)
*******************************************************************************
```

D.4 IRF142

```
*******************************************************************************
.model IRF142  NMOS(Level=3 Gamma=0 Delta=0 Eta=0 Theta=0 Kappa=0.2 Vmax=0 Xj=0
+ Tox=100n Uo=600 Phi=.6 Rs=21.94m Kp=20.61u W=.97 L=2u Vto=3.189
+ Rd=42.19m Rds=444.4K Cbd=2.398n Pb=.8 Mj=.5 Fc=.5 Cgso=1.113n
+ Cgdo=432.1p Rg=5.659 Is=2.823p N=1 Tt=142n)
*******************************************************************************
```

Average Current-Mode Control of DC-DC Power Converters, First Edition.
Marian K. Kazimierczuk, Dalvir K. Saini, and Agasthya Ayachit.

D.5 IRF840

```
******************************************************************************
.model IRF840  NMOS(Level=3 Gamma=0 Delta=0 Eta=0 Theta=0 Kappa=0.2 Vmax=0 Xj=0
+ Tox=100n Uo=600 Phi=.6 Rs=6.382m Kp=20.85u W=.68 L=2u Vto=3.879
+ Rd=.6703 Rds=2.222MEG Cbd=1.415n Pb=.8 Mj=.5 Fc=.5 Cgso=1.625n
+ Cgdo=133.4p Rg=.6038 Is=56.03p N=1 Tt=710n)
******************************************************************************
```

D.6 IRF740

```
******************************************************************************
.model IRF740  NMOS(Level=3 Gamma=0 Delta=0 Eta=0 Theta=0 Kappa=0.2 Vmax=0 Xj=0
+ Tox=100n Uo=600 Phi=.6 Rs=8.563m Kp=20.59u W=.78 L=2u Vto=3.657
+ Rd=.3915 Rds=1.778MEG Cbd=1.419n Pb=.8 Mj=.5 Fc=.5 Cgso=1.392n
+ Cgdo=146.6p Rg=.9088 Is=17.65p N=1 Tt=570n)
******************************************************************************
```

Appendix E

Selected Diodes' Spice Models

E.1 MUR1560

```
*****************************************************************************
.MODEL Dmur1560 d
+IS=1.22946e-07 RS=0.0276435 N=2 EG=1.3 XTI=4 BV=600 IBV=0.00001
+CJO=5.43205e-10 VJ=0.4 M=0.373483 FC=0.5 TT=7.14134e-08 KF=0 AF=1
*****************************************************************************
```

E.2 MBR10100

```
*****************************************************************************
.MODEL Dmbr10100 d
+IS=0.000132385 RS=0.0122186 N=2 EG=0.828762 XTI=3.80757 BV=100 IBV=0.0001
+CJO=1e-11 VJ=0.7 M=0.5 FC=0.5 TT=0 KF=0 AF=1
*****************************************************************************
```

E.3 MBR1060

```
*****************************************************************************
.MODEL Dmbr1060 d
+IS=0.000132385 RS=0.0122186 N=2 EG=0.828762 XTI=3.80757 BV=60 IBV=0.0001
+CJO=1e-11 VJ=0.7 M=0.5 FC=0.5 TT=0 KF=0 AF=1
*****************************************************************************
```

E.4 MUR2510

```
*****************************************************************************
.MODEL MUR2510  D(Is=67.39p Rs=4.351m Ikf=0 N=1 Xti=3 Eg=1.11 Cjo=243.6p
+ M=.1593 Vj=.75 Fc=.5 Isr=100.8n Nr=2 Tt=123.3n)
*****************************************************************************
```

E.5 MBR2540

```
*****************************************************************************
.model MBR2540  D(Is=21u Rs=5.011m Ikf=69.6 N=1 Xti=0 Eg=1.11 Cjo=3.507n
+ M=.5031 Vj=.75 Fc=.5 Isr=1.176m Nr=2)
*****************************************************************************
```

Average Current-Mode Control of DC-DC Power Converters, First Edition.
Marian K. Kazimierczuk, Dalvir K. Saini, and Agasthya Ayachit.

E.6 MBR4040

```
******************************************************************************
.model MBR4040  D(Is=27.51u Rs=5.244m Ikf=124.2 N=1 Xti=0 Eg=1.11 Cjo=3.468n
+ M=.5003 Vj=.75 Fc=.5 Isr=1.175m Nr=2)
******************************************************************************
```

Appendix F

Simulation Tools

F.1 SPICE Model of Power MOSFETs

Figure F.1 shows an SPICE large-signal model for n-channel enhancement MOSFETs. It is a model of integrated MOSFETs, which can be adopted to power MOSFETs. SPICE parameters of the large-signal model of enhancement-type n-channel MOSFETs are given in Table F.1. The diode currents are

$$i_{BD} = IS\left(e^{\frac{v_{BD}}{V_T}} - 1\right) \tag{F.1}$$

and

$$i_{BS} = IS\left(e^{\frac{v_{BS}}{V_T}} - 1\right). \tag{F.2}$$

The junction capacitances in the voltage range close to zero are

$$C_{BD} = \frac{(CJ)(AD)}{\left(1 - \frac{v_{BD}}{PB}\right)^{MJ}} \quad \text{for} \quad v_{BD} \le (FC)(PB) \tag{F.3}$$

and

$$C_{BS} = \frac{(CJ)(AS)}{\left(1 - \frac{v_{BS}}{PB}\right)^{MJ}} \quad \text{for} \quad v_{BS} \le (FC)(PB), \tag{F.4}$$

where CJ is the zero-bias junction capacitance per unit area, AD is the drain area, AS is the source area, PB is the built-in potential, and MJ is the grading coefficient.

The junction capacitances in the voltage range far from zero are

$$C_{BD} = \frac{(CJ)(AD)}{(1 - FC)^{1+MJ}}\left[1 - (1 + MJ)FC + MJ\frac{v_{BD}}{PB}\right] \quad \text{for} \quad v_{BD} \ge (FC)(PB) \tag{F.5}$$

and

$$C_{BS} = \frac{(CJ)(AS)}{(1 - FC)^{1+MJ}}\left[1 - (1 + MJ)FC + MJ\frac{v_{BD}}{PB}\right] \quad \text{for} \quad v_{BS} \ge (FC)(PB). \tag{F.6}$$

Average Current-Mode Control of DC-DC Power Converters, First Edition.
Marian K. Kazimierczuk, Dalvir K. Saini, and Agasthya Ayachit.

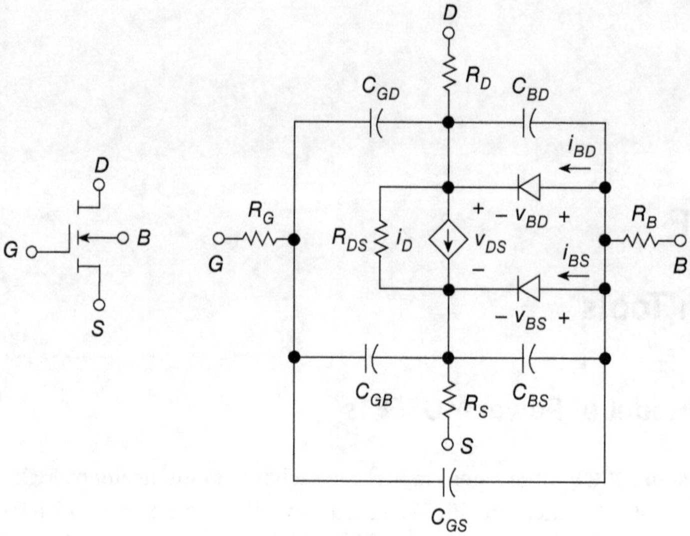

Figure F.1 SPICE large-signal model for *n*-channel MOSFET.

The typical values are as follows:

$C_{ox} = 3.45 \times 10^{-5}$ pF/μm
$t_{ox} = 4.1 \times 10^{-3}$ μm
$\epsilon_{ox(SiO_2)} = 3.9\epsilon_0$
$C_{j0} = 2 \times 10^{-4}$ F/m^2
$C_{jsw} = 10^{-9}$ F/m
$C_{GBO} = 2 \times 10^{-10}$ F/m
$C_{GDO} = C_{GSO} = 4 \times 10^{-11}$ F/m

F.1.1 SPICE NMOS Syntax

```
Mxxxx D G S B MOS-model-name L = xxx  W = yyy
```

Example:

```
M1 2 1 0 0 M1-FET L = 0.18um W = 1800um
```

F.1.2 SPICE NMOS Model Syntax

```
.model model-name NMOS (parameter=value ...)
```

Example:

```
.model M1-FET NMOS (Vto = 1V Kp = E-4)
```

Table F.1 Selected SPICE level-1 NMOS large-signal model parameters.

Sym.	SPICE S.	Model parameter	Default value	Typical value
V_{to}	VTO	Zero-bias threshold voltage	0 V	0.3–3 V
μC_{ox}	KP	Process constant	2×10^{-5} A/V^2	20–346 μA/V^2
λ	Lambda	Channel-length modulation	0 V^{-1}	0.5–10^{-5} V^{-1}
γ	Gamma	Body-effect V_t parameter	0 V$^{\frac{1}{2}}$	0.35 $V^{\frac{1}{2}}$
$2\phi_F$	PHI	Surface potential	0.6 V	0.7 V
R_D	RD	Drain series resistance	0 Ω	0.2 Ω
R_S	RS	Source series resistance	0 Ω	0.1 Ω
R_G	RG	Gate series resistance	0 Ω	1 Ω
R_B	RB	Body series resistance	0 Ω	1 Ω
R_{DS}	RDS	Drain–source resistance R	∞	1 MΩ
R_{SH}	RSH	Drain–source diffusion sheet resistance R	0	20 Ω/Sq.
I_S	IS	Saturation current	10^{-14} A	10^{-9} A
M_j	MJ	Grading coefficient	0.5	0.36
C_{j0}	CJ	Zero-bias bulk junction C/m^2	0 F/m^2	1 nF/m^2
V_{bi}	PB	Junction potential	1 V	0.72 V
M_{jsw}	MJSW	Grading coefficient	0.333	0.12
C_{j0sw}	CJSW	Zero-bias junc. perimeter C/m	0 F/m	380 pF/m
V_{BSW}	PBSW	Junction sidewell potential	1 V	0.42 V
C_{GDO}	CGDO	Gate–drain overlap C/m	0 F/m	220 pF/m
C_{GSO}	CGSO	Gate–source overlap C/m	0 F/m	220 pF/m
C_{GBO}	CGBO	Gate–bulk overlap C/m	0 F/m	700 pF/m
F_C	FC	Forward-biased C_J coefficient	0.5	0.5
t_{ox}	TOX	Oxide thickness	∞	4.1–100 nm
μ_{ns}	UO	Surface mobility	600 cm^2/V s	600 cm^2/V s
n_{sub}	NSUB	Substrate doping	0 cm^{-3}/V s	0 cm^{-3}/V s

F.1.3 SPICE PMOS Model Syntax

```
.model model-name PMOS (parameter=value ...)
```

F.1.4 SPICE Subcircuit Model Syntax

```
xname N1 N2 N3 model-name
```

Example:

```
x1 2 1 0 IRF840
```

Copy and paste the obtained device model:

```
.SUBCKT  IRF840 1 2 3
```

and the content of the model.

F.2 Introduction to SPICE

SPICE is an abbreviation for *Simulation Program for Integrated Circuit Emphasis*. PSPICE is the PC version of SPICE. Analog and digital electronic circuit designs are verified widely by both industries and the academia using PSPICE. It is used to predict the circuit behavior.

F.2.1 Passive Components: Resistors, Capacitors, and Inductors

```
Rname N+  N- Value [IC = TC1]
Lname N+  N- Value [IC = Initial Voltage Condition]
Cname N+  N- Value [IC = Initial Current Condition]
```

Examples:

```
R1 1 2 10K
L2 2 3 2M
C3 3 4 100P
```

F.2.2 Transformer

```
Lp Np+ Np- Lpvalue
Ls Ns+ Ns- Lsvalue
Kname Lp Ls Kvcalue
```

Example:

```
Lp 1 0 1mH
Ls 2 4 100uH
Kt Lp Ls 0.999
```

F.2.3 Temperature

.TEMP list of temperatures

Example:

```
.TEMP 27 100 150
```

F.2.4 Independent DC Sources

```
Vname N+ N- DC Value
%Vname N$^{+}$ N$^{-}$ DC Value
```

```
Iname N+ N- DC Value
%Iname N$^{+}$ N$^{-}$ Type Value
```

Examples:

```
Vin  1 0 DC 10
Is 1 0 DC 2
```

F.2.5 DC Sweep Analysis

```
.DC Vsource-name Vstart Vstop Vstep
```

Example:

```
.DC VD 0 0.75 1m
```

F.2.6 Independent Pulse Source for Transient Analysis

```
Vname N+ N- PULSE (VL VH td tr tf PW T)
```

Example:

```
VGS 1 0 PULSE(0 1E-6 0 1 1 10E-6 100e-6)
```

F.2.7 Transient Analysis

```
.TRAN time-step time-stop
```

Example:

```
.TRAN 0.1ms  100ms 0ms 0.2ms
```

F.2.8 Independent AC Sources for Frequency Response

```
Vname N+ N- AC Vm Phase
Iname N+ N- AC Im Phase
```

Example:

```
Vs 2  3 AC 2 30
Is 2  3 AC 0.5 30
```

F.2.9 Independent Sinusoidal AC Sources for Transient Analysis

```
Vname N+ N- SIN (Voffset Vm f T-delay Damping-Factor
Phase-delay)
Iname N+ N- SIN (Ioffset Im f T-delay Damping-Factor
Phase-delay)
```

Examples:

```
Vin  1 0 SIN (0 170 60 0 -120)
Is 1 0 SIN (0 2 120 0 45)
```

F.2.10 AC Frequency Analysis

```
.AC DEC points-per-decade fstart fstop
```

Example:

```
.AC DEC 100 20 20
```

F.2.11 Operating Point

```
.OP
```

F.2.12 Starting the SPICE Program

1. Open the PSpice A/D Lite window (**Start** > **Programs** > **Orcad9.2 Lite Edition** > **PSpice AD Lite**).
2. Create a new text file (**File** > **New** > **Text File**).
3. Type the example code.
4. Save the file as fn.cir (for example, Lab1.cir), file type: all files, and simulate by pressing the appropriate icon.
5. To include the Spice code of a commercial device model, visit the web site, e.g. http://www.irf.com, http:www.onsemi.com, or http://www.cree.com. For example, for IRF devices, click on (**Design** > **Support** > **Models** > **Spice Library**).

F.2.13 Example Program: Diode *I*–*V* Characteristics

```
VD  1 0 DC 0.75V
D1N4001 1 0 Power-Diode
.model Power-Diode D (Is=195pA n=1.5)
.DC VD 0V 0.75V 1mV
.TEMP 27C 50C 100C 150C
.probe
.end
```

F.3 Introduction to MATLAB®

MATLAB® is an abbreviation for MATrix LABoratory. It is a very powerful mathematical tool used to perform numerical computation using matrices and vectors to obtain two- and three-dimensional graphs. MATLAB® can also be used to perform complex mathematical analysis.

F.3.1 Getting Started

1. Open MATLAB® by clicking **Start** > **Programs** > **MATLAB** > **R2014a** > **MATLAB R2014a**. Follow a similar path for newer versions of MATLAB releases as well.
2. Open a new M-file by clicking **File** > **New** > **M-File**.
3. Type the code in the M-File.
4. Save the file as fn.m (e.g. Lab1.m).
5. Simulate the code by doing one of the following:
   - Click on **Debug** > **Run**.
   - Press F5
   - On the tool bar, click the icon **Run**.

   Use **HELP** by pressing F1.
   Use % at the beginning of a line for comments.

F.3.2 Generating an *x*-Axis Data

x=Initial-Value: Increment:Final-Value

Example:

```
x=1:0.001:5;
```

or
x=[list of all the values]

Example:

```
x = [1, 2, 3, 5, 7, 10];
```

or
x = linspace(start-value, stop-value, number-of-points)

Example:

```
x = linspace(0, 2*pi, 90);
```

or x = logspace(start-power, stop-power, number-of-points)

Example:

```
x = logspace(1, 5, 1000);
```

F.3.3 Semi-logarithmic Scale

```
semilogx(x-variable, y-variable);
semilogy(x-variable, y-variable);
grid on
```

F.3.4 Log–Log Scale

```
loglog(x, y);
grid on
```

F.3.5 Generate an *y*-Axis Data

y = f(x);

Example:

```
y = cos(x);
z = sin(x)
```

F.3.6 Multiplication and Division

A dot should be used in front of the operator for matrix multiplication and division.

```
c = a.*b;
c = a./b
```

F.3.7 Symbols and Units

Math symbols should be in italic. Math signs (like ( ), =, and +) and units should not be in italic. Leave one space between a symbol and a unit.

F.3.8 *x*-Axis and *y*-Axis Labels

xlabel('{\it x} (unit) ')
ylabel('{\it y} (unit) ')

Example:

```
xlabel('{\it v_{GS}} (V)')
ylabel('{\it i_{DS}} (A)')
```

F.3.9 *x*-Axis and *y*-Axis Limits

set(gca, 'xlim', [xmin, xmax])
set(gca, 'xtick', [xmin, step, xmax])
set(gca, 'xtick', [x1, x2, x3, x4, x5])

Example:

```
set(gca, 'xlim', [1, 10])
set(gca, 'xtick', [0:2:10])
set(gca, 'xtick', [-90 -60 -30 0 30 60 90])
```

F.3.10 Greek Symbols

Type: \alpha, \beta, \Omega, \omega, \pi, \phi, \psi, \gamma, \theta, and \circ to obtain: α, β, Ω, ω, π, ϕ, ψ, γ, θ, and $\circ$.

F.3.11 Plot Commands

plot (x, y, '.-', x, z, '- -') set(gca, 'xlim', [x1, x2]); set(gca, 'ylim', [y1, y2]); set(gca, 'xtick', [x1:scale-increment:x2]); text(x, y, '*symbol* = 25 V'); plot (x, y), axis equal

Examples:

```
set(gca, 'xlim', [4, 10]);
set(gca, 'ylim', [1, 8]);
set(gca, 'xtick', [4:1:10]);
text(x, y, '{\it V} = 25 V');
```

F.3.12 3D Plot Commands

[X1, Y1] = meshgrid(x1, x2);
mesh(X1, Y1, z1);

Example:

```
t = linspace(0, 9*pi);
xlabel('sin(t)')
ylabel('cos(t)')
zlabel('t')
plot(sin(t), cos(t), t)
```

F.3.13 Bode Plots

```
f = logspace(start-power, stop-power, number-of-points)

NumF = [a1 a2 a3];%~~~~%Define the numerator of polynomial in $s$-domain.
DenF = [a1 a2 a3];%~~~~%Define the denominator of polynomial in $s$-domain.
[MagF, PhaseF] = bode(NumF, DenF, (2*pi*f));

figure(1)
semilogx(f, 20*log10(MagF))
F = tf(NumF, DenF)%~~~~\%Converts the polynomial into transfer function.
[NumF, DenF] = tfdata(F)%~~~~\%Converter transfer function into polynomial.
```

F.3.14 Step Response

```
NumFS = D*NumF;
t = [0:0.000001:0.05];
[x, y] = step(NumFS, DenF, t);
figure(2)
plot(t, Initial-Value + y);
```

F.3.15 To Save Figure

Go to File, click Save as, go to EPS file option, type the file name, and click Save.

F.3.16 Example Program

```
clear all
clc
x = linspace(0, 2*pi, 90);
y = sin(x);
z = cos(x);
grid on
xlabel('{\it x}')
ylabel('{\it y }, {\it z }')
plot(x, y, '-.', x, z, '-~-')
```

F.3.17 Polynomial Curve Fitting

```
x = [0 0.5 1.0 1.5 2.0 2.5 3.0];
y = [10 12 16 24 30 37 51];
p = polyfit(x, y, 2) %Find the coefficients of a polynomial of degree 2
yc = polyval(p, x); %Polynomial is evaluates at all points.
plot(x, y, 'x', x, yc)
xlabel('x')
ylabel('y'), grid
legend('Actual data', 'Fitted polynomial')
```

F.3.18 Bessel Functions

J0 = besselj(0, x);

F.3.19 Modified Bessel Functions

I0 = besseli(0, x):
Example:

```
model = [1 2 3]:
rro = -1:0.00001:1;
kr = (1 + j)*(rodel)'*(rro);
JrJ0 = besseli(0,kr);
figure, plot(rro, abs(JrJ0))
figure, plot(rro,angle(JrJ0)*180/pi)
```

F.3.20 Example MATLAB Code

The following program is for duty cycle-to-output voltage transfer function T_p and duty cycle-to-inductor current transfer function T_{pi} for buck DC–DC converter. Similar procedure can be followed for other transfer functions and for other power converters. This program produces Figures F.2–F.7:

```
%%%%%%%%%%%%%%%%%%%%%%%%%%%%%%%%%%%%%%%%%%%%%%%%%%%%%%%%%%%%%%
clc; clear all; close all
s = tf('s');
set(cstprefs.tbxprefs,'FrequencyUnits','Hz');
```

```
%% Design Parameters
VI  = 28;       % Nominal input voltage
VO  = 14;       % Nominal output voltage
fs  = 100e3;    % Switching frequency
eta = 0.90;     % Efficiency of power stage
RL  = 10;       % Load resistor

%% Circuit Parasitics
rDS = 0.11;
RF  = 0.022;
rL  = 0.19;
Co  = 100e-12;
VF  = 0.7;
rC  = 0.11;

%% Output, Control and Input Parameters
MVDCnom  = VO/VI;
D        = MVDCnom/eta;
IL       = VO/(RL);
IO       = IL;
II       = D*IL;
MIDCnom  = IO/II;
IS       = D*IL;
VD       = D*VI;
PO       = VO*IO;

%% Circuit Power Stage Parameters
L = 301e-6;
C = 68e-6;

%% Parameters Needed for Small-Signal Modeling of Buck in CCM
r  = D*(rDS) + ((1-D)*RF) + rL;                       % Tot. parasitic resistance
Z1 = r + s*L;                                         % Series filter network
Z2 = RL*(rC + (1/(s*C)))/(RL + rC + (1/(s*C))); % Output filter network

%%%%%%%%%%%%%%%%%%%%%%%%%%%%%%%%%%%%%%%%%%%%%%%%%%%%%%%%%%%%%%%%%%%%%%%%%%%%
%%%%%%%%%%%%%%%%%%%%%%%    Tp Transfer Function   %%%%%%%%%%%%%%%%%%%%%%%%%
%%%%%%%%%%%%%%%%%%%%%%%%%%%%%%%%%%%%%%%%%%%%%%%%%%%%%%%%%%%%%%%%%%%%%%%%%%%%
%% Tp Transfer Function in term of impedances
Tp     = (VI*Z2)/(Z1+Z2);
Tp_zpk = zpk(Tp)

%% Tp Transfer Function in s-domain
wz   = 1/(C*rC);
fz   = wz/(2*pi);
Tpx  = (VI*RL*rC)/(L*(RL+rC));
zeta = (L + C*((r*(RL+rC))+ RL*rC))/(2*sqrt((L*C*(RL+rC))*(RL+r)));
w0   = sqrt((RL+r)/(L*C*(RL+rC)));
f0   = (1/(2*pi))*w0;
Tpo  = (VI*RL)/(RL+r);
Tp_in_sdomain     = Tpx*(s+wz)/(s^2+2*zeta*w0*s+w0^2);
zpk_Tp_in_sdomain =  zpk(Tp_in_sdomain)

%% Poles and zeros of Tp Transfer Function
Tp_PolesAndZeros  =  minreal(Tp, 0.01)
zpk_Tp            =  zpk(Tp_PolesAndZeros)
Poles_Tp          =  pole(Tp_PolesAndZeros)/(2*pi) % poles of Tp
Zeros_Tp          =  zero(Tp_PolesAndZeros)/(2*pi) % zeros of Tp

%% Plot Magnitude and Phase plots for Tp Transfer Function
f = logspace(-1, 6, 500);
Omega = 2 * pi * f;
```

```
[Tpmag, Tpphase, HOmega] = bode(Tp, Omega);
Tpmag   = squeeze(Tpmag);
Tpphase = squeeze(Tpphase);

figure(1);
subplot(2,1,1)
semilogx(f, 20*log10(Tpmag), 'r', 'Linewidth', 1.5); grid on;
ylabel('|{\itT_{p}}| (dBV)');
set(gca,'YLim',[-60 50])
set(gca,'YTick',[-60:20:50])
set(gca,'XLim',[10 1e6])
subplot(2,1,2)
semilogx(f, (Tpphase), 'r', 'Linewidth', 1.5); grid on;
ylabel('\phi_{\itT_{p}}(\circ)')
set(gca,'YLim',[-180 0])
set(gca,'YTick',[-180:45:0])
set(gca,'XLim',[10 1e6])
xlabel('{\it f} (Hz)');

%% Plot Step Response for Tp Transfer Function
t1     = 0:1e-6:15e-3;
uTp    = 0.1*step(Tp,t1);    % Step Change in Duty Cycle = 0.1
figure(2)
plot(t1*1000,VO+uTp,'r', 'Linewidth', 1.5); grid on;
set(gca,'XLim',[0 6])
set(gca,'XTick',[0:1:6])
set(gca,'YLim',[13 19])
set(gca,'YTick',[13:1:19])
xlabel('{\it t} (ms)');
ylabel('{\it v_O }(V)');

%% Plot Poles and Zeros for Tp Transfer Function as a func. of Duty Cycle
D_Var  = 0.3:0.05:0.99; % Duty Cycle range
for d = 1:numel(D_Var)

    % Impedances
    r     = D_Var(d)*(rDS) + ((1-D_Var(d))*RF) + rL;
    Z1    = r + s*L;
    Z2    = RL*(rC + (1/(s*C)))/(RL + rC + (1/(s*C)));

    % Tp Transfer Function
    Tp  = (VI*Z2)/(Z1+Z2);

    Tp         =  minreal(Tp, 0.001);
    zpk_Tp     =  zpk(Tp);
    Poles_Tp   =  pole(Tp) % poles of Tp
    Zeros_Tp   =  zero(Tp) % zeros of Tp

    % Plot Rootlocus of Tp Transfer Function
    figure(3)
    plot(real(Poles_Tp(1))/1000, imag(Poles_Tp(1))/1000,...
        'rx','linewidth',2,'markersize',10); hold on; grid on
    plot(real(Poles_Tp(2))/1000, imag(Poles_Tp(2))/1000,...
        'rx','linewidth',2,'markersize',10); hold on; grid on
    plot(real(Zeros_Tp(1))/1000, imag(Zeros_Tp(1))/1000,...
        'bo','linewidth',2,'markersize',7);  hold on; grid on
    xlabel('\sigma (krad/s)')
    ylabel('{\itj}\omega (krad/s)')

end
figure(3)
str1 = '{ {\it p_{\rm1}} }';
```

```
str2 = '{ {\it p_{\rm2}} }';
str3 = '{ {\it z_n} }';
text(-10, 6,str1)
text(-10,-6,str2)
text(-130,0.5,str3)
%%%%%%%%%%%%%%%%%%%%%%%%%%%%%%%%%%%%%%%%%%%%%%%%%%%%%%%%%%%%%%%%%%%%%%%%%%%%%
%%%%%%%%%%%%%%%%%%%%%%%   Tpi Transfer Function   %%%%%%%%%%%%%%%%%%%%%%%%%%
%%%%%%%%%%%%%%%%%%%%%%%%%%%%%%%%%%%%%%%%%%%%%%%%%%%%%%%%%%%%%%%%%%%%%%%%%%%%%
%% Tpi Transfer Function in term of impedances
Tpi     =  VI/(Z1+Z2);
Tpi_zpk = zpk(Tpi)

%% Tpi Transfer Function in s-domain
wzi   =  1/(C*(RL+rC));
fzi   = wzi/(2*pi);
Tpio = VI/(RL+r);
Tpix                = Tpio*w0^2/wzi;
Tpi_in_sdomain      = Tpix*(s+wzi)/(s^2+(2*zeta*w0*s)+w0^2);
Tpi_in_sdomain_zpk = zpk(Tpi_in_sdomain)

%% Poles and zeros of Tpi Transfer Function
Tpi_PolesAndZeros  =  minreal(Tpi, 0.01)
zpk_Tpi            =  zpk(Tpi_PolesAndZeros)
Poles_Tpi          =  pole(Tpi_PolesAndZeros)/(2*pi) % poles of Tpi
Zeros_Tpi          =  zero(Tpi_PolesAndZeros)/(2*pi) % zeros of Tpi

%% Magnitude and Phase plots for Tpi Transfer Function
f = logspace(-1, 6, 500);
Omega = 2 * pi * f;
[Tpimag, Tpiphase, HOmega] = bode(Tpi_in_sdomain, Omega);
Tpimag   = squeeze(Tpimag);
Tpiphase = squeeze(Tpiphase);

figure(4);
subplot(2,1,1)
semilogx(f, 20*log10(Tpimag), 'b', 'Linewidth', 1.35); hold on
ylabel('|{\itT_{pi}}| (dBA)'); grid on;
set(gca,'YLim',[-40 40])
set(gca,'YTick',[-40:20:40])
set(gca,'XLim',[10 1e6])
subplot(2,1,2)
semilogx(f, (Tpiphase), 'b', 'Linewidth', 1.35); hold on; grid on;
ylabel('\phi_{\itT_{pi}}(\circ)')
set(gca,'YLim',[-90 60])
set(gca,'XLim',[10 1e6])

%% Step Response for Tpi Transfer Function
t2      = 0:1e-6:15e-3;
uTpi    = 0.1*step(Tpi,t2);    % Step Change in Duty Cycle = 0.1
figure(5)
plot(t2*1000,IL+uTpi,'b', 'Linewidth', 1.5); hold on;
set(gca,'XLim',[0 6])
set(gca,'XTick',[0:1:6])
set(gca,'YLim',[1 3])
set(gca,'YTick',[1:0.4:3])
xlabel('{\it t} (ms)');
ylabel('{\it i_L }(A)'); grid on;

%% Plot Poles and Zeros for Tpi Transfer Function as a func. of Duty Cycle
D_Var  = 0.3:0.05:0.99;
for d = 1:numel(D_Var)
```

```
    % Impedances
    r      = D_Var(d)*(rDS) + ((1-D_Var(d))*RF) + rL;
    Z1     = r + s*L;
    Z2     = RL*(rC + (1/(s*C)))/(RL + rC + (1/(s*C)));

    % Tpi Transfer Function
    Tpi =  VI/(Z1+Z2);

    Tpi        =  minreal(Tpi, 0.001);
    zpk_Tpi    =  zpk(Tpi);
    Poles_Tpi  =  pole(Tpi) % poles of Tpi
    Zeros_Tpi  =  zero(Tpi) % zeros of Tpi

    % Plot Rootlocus of Tpi Transfer Function
    figure(6)
    grid on;
    plot(real(Poles_Tpi(1))/1000, imag(Poles_Tpi(1))/1000,...
    'color',[1 0 0.5],'marker','x','linewidth',2,'markersize',10); hold on;
    plot(real(Poles_Tpi(2))/1000, imag(Poles_Tpi(2))/1000,...
    'color',[1 0 0.5],'marker','x','linewidth',2,'markersize',10); hold on;
    plot(real(Zeros_Tpi(1))/1000, imag(Zeros_Tpi(1))/1000,...
    'color',[0 0.5 0],'marker','o','linewidth',2,'markersize',7);  hold on;
    xlabel('\sigma (krad/s)')
    ylabel('{\itj}\omega (krad/s)')

end
figure(6)
str1 = '{{\it z_i} }';
str2 = '{Trajectory of {\it p_{\rm 1}} with increase in {\itD}}';
str3 = '{Trajectory of {\it p_{\rm 2}} with increase in {\itD}}';
text(-1.45,0.5,str1)
text(-1.38,5.5,str2)
text(-1.38,-5.5,str3)
annotation('textarrow',[0.8 0.7],[0.83 0.83])
annotation('textarrow',[0.8 0.7],[0.20 0.20])
%%%%%%%%%%%%%%%%%%%%%%%%%%%%%%%%%%%%%%%%%%%%%%%%%%%%%%%%%%%%%%%%%%%%%%%%%%%%%%%
```

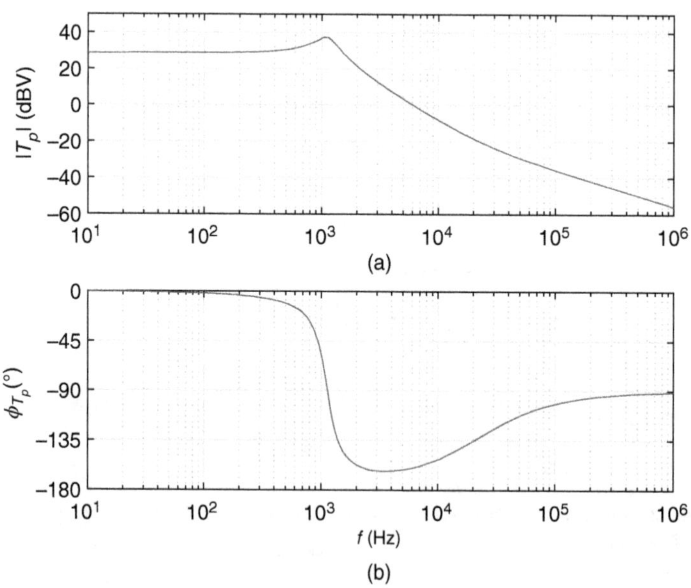

Figure F.2 Theoretically obtained plots of duty cycle-to-output voltage transfer function T_p. (a) Magnitude T_p. (b) Phase.

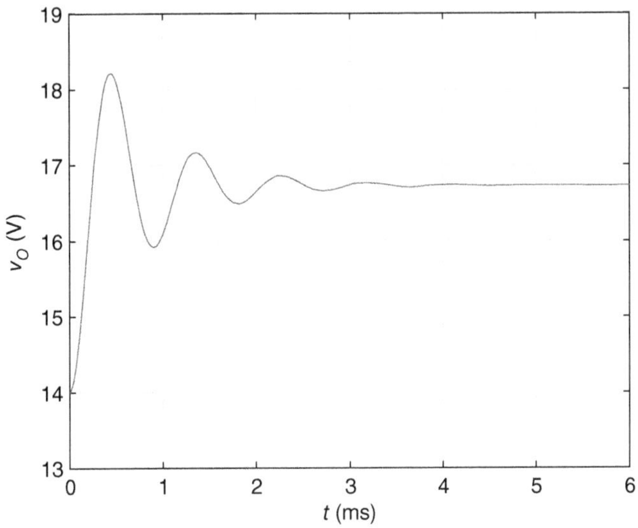

Figure F.3 Step response for duty cycle-to-output voltage transfer function T_p.

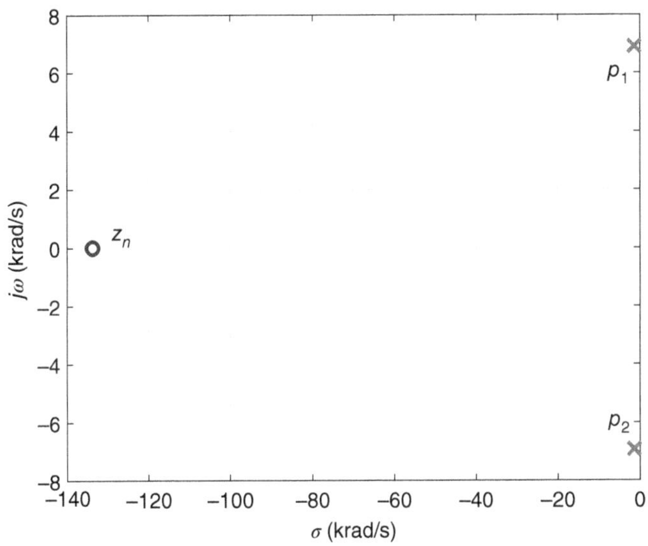

Figure F.4 Poles and zeros of T_p transfer function as functions of the duty cycle.

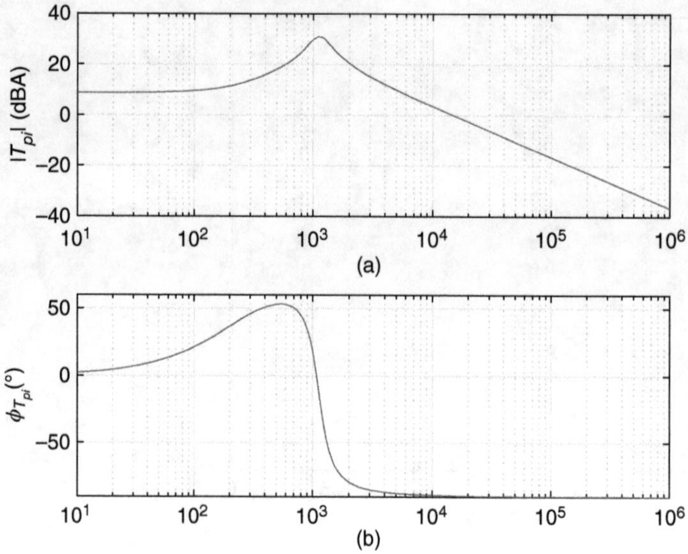

Figure F.5 Theoretically obtained plots of duty cycle-to-inductor current transfer function T_{pi}. (a) Magnitude. (b) Phase.

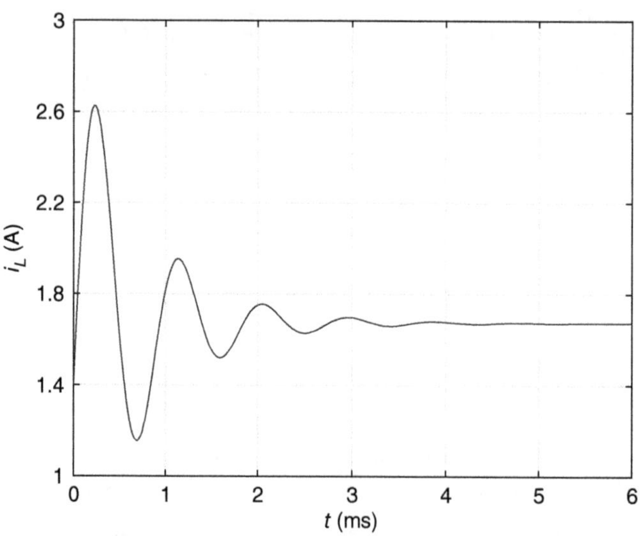

Figure F.6 Step response for duty cycle-to-inductor current transfer function T_{pi}.

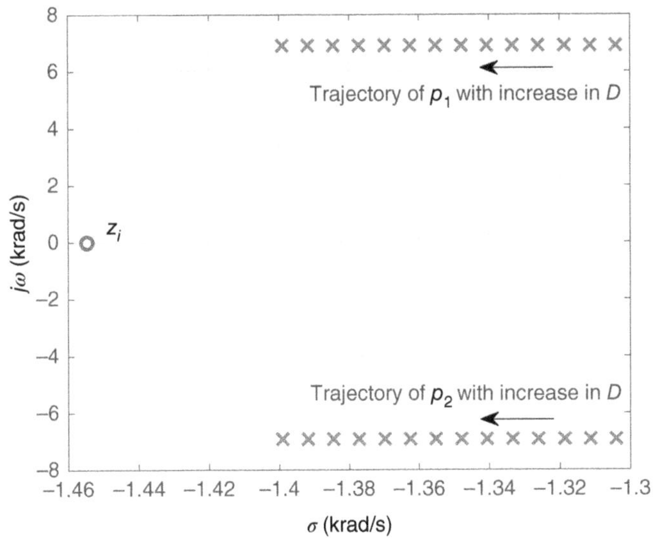

Figure F.7 Trajectory of the poles and zeros of T_{pi} transfer function as functions of the duty cycle.

F.4 Introduction to SABER Circuit Simulator

SABER is an intuitive, advanced, integrated environment for designing and analyzing power electronic circuits and systems. It enables schematic design, mixed-signal multi-domain circuit simulation and waveform analysis, and report generation.

F.4.1 Setting Up a Circuit on SABER

1. On the SABER circuit design window, construct the given circuit/schematic using the parts provided in the component window. The parts can be found in the component library at the bottom icon tray on your screen under "Select and place parts," and you can search for parts individually. Left click at the terminals to connect various parts in your schematic. Note that it is recommended to "save" your work regularly with a suitable file name. Make sure to save all your work in a designated folder, since SABER generates netlists that are required for circuit simulation.
2. Provide appropriate values to all the components on the schematic, and name the wires that are used to connect these components such as V_{out}, V_{in}, i_L, and so on. This can be performed by right clicking on the selected wire and then selecting "attributes."
3. Upon completion of your design, your schematic must look as the example shown in Figure F.8.

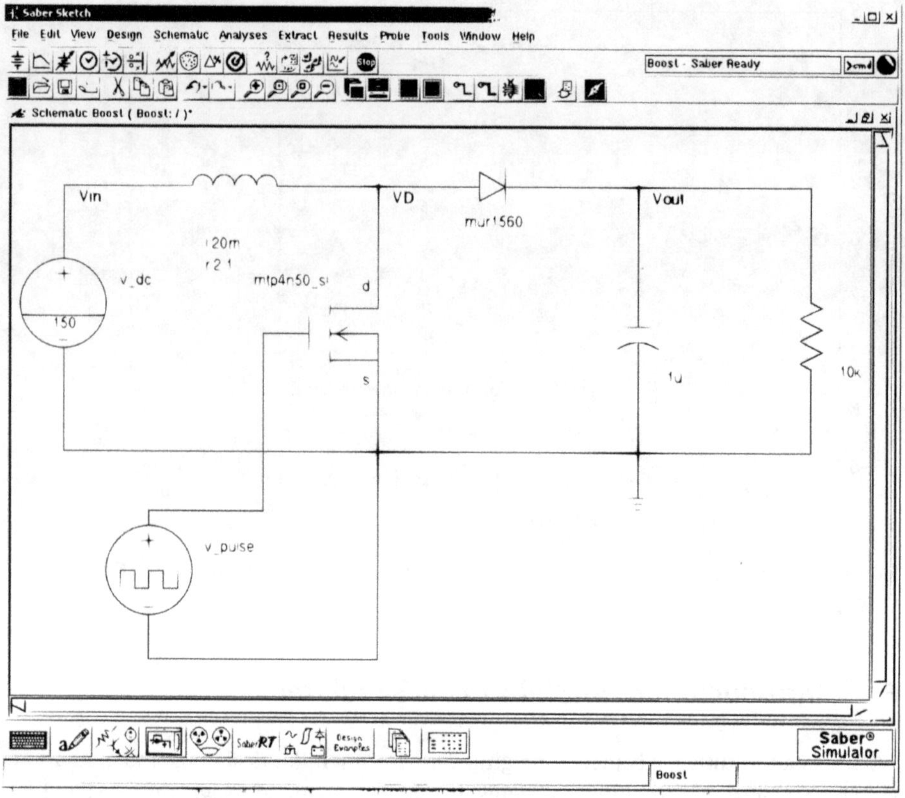

Figure F.8 Example circuit schematic in SABER.

F.4.2 Performing TRANSIENT Analysis on the Designed Circuit

1. Select the show/hide SABER Guide icon bar to view the simulation tool bar.
2. Select an Operating Point/Transient Analysis to run a transient analysis versus time.
   (a) Click the Basic Tab.
   (b) Set a simulation end time. Assign appropriate end times based on the time taken for the system to reach steady-state operation. Set a time step of ≈ 0.1 μs (0.01 μs is better for our duty cycle precision; however, if the end time is relatively high, it is recommended to increase the time step so that the simulation runs faster.)
   (c) Set Monitor Progress = 100. Plot after Analysis = Yes Open Only.
   (d) Click the Input/Output Tab. Select All Signals for the Signal List.
   (e) Select "Across and through variables" to save both voltage and current information.
   (f) Select "OK" to run the analysis. It will take a minute or two to complete the simulation. If you want to view the status of the simulation, click the Simulation Transcript and Command Line button in the top right.

(g) Once the analysis is complete, it should open Cosmos Scope to view the simulation results.

F.4.3 Plotting

1. The Cosmos Scope pops up with two more dialog boxes. In the Signal Manager window, you should now see, under Plot files, the file you have been working on which contains the results of the transient analysis.
2. All types of waveforms can be plotted, and various mathematical operations can be performed on them. You should be able to find the required waveforms under the drop-down option corresponding to the component under consideration. A simple example of how the Cosmos Scope would appear is shown in Figure F.9.

F.4.4 Printing

Your plots and schematic can be exported (saved) in the format you desire. Export option is available under the File tab. Your file will be saved in the directory you are working in.

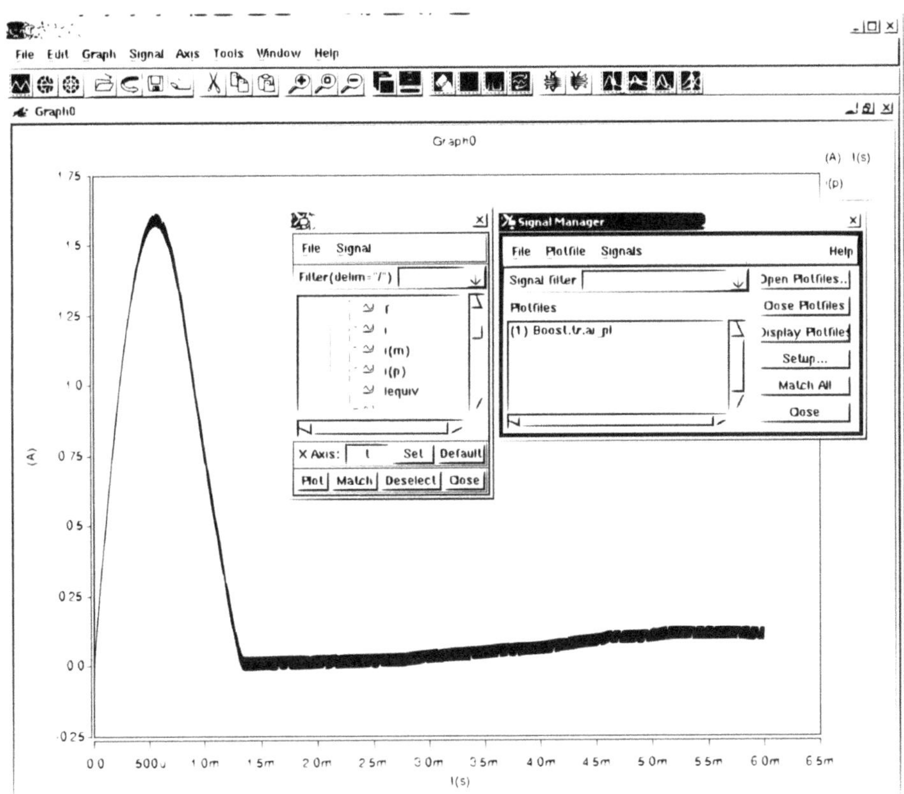

Figure F.9 Example of plot window in SABER.

Index

Average Current-Mode Control of DC-DC Power Converters, First Edition.
Marian K. Kazimierczuk, Dalvir K. Saini, and Agasthya Ayachit.

www.ingramcontent.com/pod-product-compliance
Lightning Source LLC
LaVergne TN
LVHW062207200926
848484LV00006B/1339